Organic Photochemistry: Principles and Applications

Jacques Kagan
Department of Chemistry,
University of Illinois at Chicago, USA

ACADEMIC PRESS

Harcourt Brace Jovanovich, Publishers

LONDON   SAN DIEGO   NEW YORK   BOSTON
SYDNEY   TOKYO   TORONTO

This book is printed on acid-free paper

ACADEMIC PRESS LIMITED
24–28 Oval Road
LONDON NW1 7DX

United States Edition published by
ACADEMIC PRESS INC.
San Diego, CA 92101

ISBN 0–12–394320–5

Typeset by Alden Multimedia Ltd
Printed in Great Britain by The University Printing House, Cambridge

Contents

Preface   iv

1. The fundamentals   1

2. Experimental techniques   26

3. Photochemistry of the carbon–carbon double bond   34

4. Photochemistry of the carbonyl group   55

5. Fragmentation reactions   65

6. Photochemistry with oxygen   82

7. Electron transfers in photochemistry   99

8. Synthetic applications   115

9. Photochemistry in biochemistry   141

10. Photochemistry and biology   166

11. Photochemistry and medicine   185

12. Applied photochemistry   201

13. Conclusion   222

Index   224

Preface

Invisible ultraviolet radiation, like visible light, is capable of inducing myriad chemical transformations. Organic chemists have described many of them, and the experimental results can often be extrapolated to predict whether untested conversions are likely to take place. Many photochemical transformations convert simple molecules into extremely complex products, with an ease not approached by the standard synthetic chemistry practiced in the laboratory.

The apparent complexity of photochemical transformations may have discouraged biologists, physicians and other scientists utilizing light sources, including lasers, from attempting to analyze photoinduced transformations in terms of simple, experimentally verifiable, steps. My primary goal has been to outline in the earlier chapters the principles, techniques, and some of the well-known reactions occurring in organic molecules, and later to illustrate more complex photochemical transformations occurring in organic chemistry (Chapter 8), biological systems (Chapters 9 and 10) and medicine (Chapter 11). A few practical applications of organic photochemistry are collected in Chapter 12. Some of the physical chemistry which parallels all these photochemical transformations has not been emphasized, since chemists or biologists preoccupied with making compounds or rationalizing photochemical transformations can often delay the acquisition of an intimate knowledge of the detailed photophysical phenomena, the nature of the excited states, their lifetime, etc. These studies often acquire a life of their own.

Organic photochemistry is a science which can be carried out with simple tools: artificial light sources are more convenient and give more reproducible results, but excellent synthetic and biological results have been obtained with sunlight and standard glassware (of course, more sophisticated and expensive equipment is necessary in order to get more sophisticated information, particularly of a mechanistic nature). A short section on experimental techniques is included in the book to encourage novices who wish to get started, and perhaps help them ask questions of more experienced photochemists.

Astronauts looking at the planet earth can never see it in its entirety, even when there are no obscuring clouds. Likewise, should all aspects of photochemistry and photobiology be made to fit neatly on a sphere, only a portion of it would be seen by an observer from his or her vantage point. What I saw and reported in this book is necessarily incomplete, and others might have chosen and described different features of the sphere. Just as maps distort the portions of the earth that they attempt to represent, my map of photochemical sciences presented in this book is flawed, since I have emphasized topics where structural transformations could be formulated chemically. Our knowledge of many important areas of photobiology has not sufficiently evolved from the stage of observation to that of chemical explanation. One exception is in the field of photosynthesis, which truly deserves a separate treatment and has been deliberately excluded from this book. Nature utilizes light in many other phenomena. The transmission of information through vision allows communication between many organisms, humans included. A number of organisms generate light biologically, while humans have developed this skill through chemical and physical processes. Even organisms which lack the ability to see with specific visual receptors can respond to light in other ways. Phototropism, either positive or negative, and photomorphogenesis are two examples. Subtle behavioral changes, for example in circadian rhythms, may also be induced by the duration of exposure to light. Numerous observations in these areas have been described, associated with many graphs and kinetic data, but few chemical transformations have been formulated. All these biological events will eventually be understood in terms of discrete chemical reactions.

In our daily experience, photochemistry is responsible for many noticeable changes. Colored fabrics and materials fade much more where they have been exposed to sunlight. People looking back at old color photographs of their parents or children cannot help noting a difference between the colors that they see and those remembered. The importance of this problem should not be underestimated, as much artwork which we hope to see preserved forever is doomed by the very fact that we wish to view it, and light used for observation induces photochemical reactions, for example through formation of the very reactive singlet oxygen. Of course, the bleaching power of sunlight was well known through the generations; even now, many shun electric driers because laundry dried outdoors is perceived as being distinctly whiter (an attitude cleverly recognized by the manufactor of the detergent Sunlight, which for good measure is guaranteed to be 100% phosphate-free).

It is perhaps through sunburn that most people really become aware of the physiological responses of biological cells to photons. The judgment of whether or not a suntan is desirable has varied with the times. Rational analysis certainly cannot explain the great popularity recently enjoyed by tanning booths. Popular wisdom, however, long ago recognized the virtues of sunlight exposure in the treatment of some widespread skin conditions, and correlated the administration of naturally available sources of photosensitizers and the success of the sunlight treatments. Modern PUVA therapy for vitiligo and psoriasis, and the photodynamic therapy of cancers, are simply refinements of the old approaches. Similarly, the benefits of sunlight exposure to increase vitamin D production were known long before the detailed photochemistry of the steroid precursors was elucidated.

Skin exposure to sunlight can bring several forms of dermatitis, such as berloque dermatitis, Club Med dermatitis, or bikini dermatitis, which result from photosensitized skin damage by components of perfumes or colognes, fruit extracts, or dyes. Many drugs also induce photosensitization.

Medical applications expand the field of organic photochemistry into challenging areas, but where the ability to conduct classical mechanistic experimentation is severely limited by ethical considerations. The photochemical treatment of infants suffering from neonatal hyperbilirubinemia is now widely used. Photodynamic therapy or PDT is an area of astonishingly rapid growth, as several forms of cancer appear to respond favourably to photosensitized treatments. This is certainly an area where understanding the basic phenomena will require increasingly close collaboration between physicians, biologists, and chemists.

The core of this book was covered in a short graduate course. I am grateful to the students who selected several of the examples and who commented on the earlier drafts of this text. The book emphasizes organic transformations. Organometallic photochemistry has been largely omitted, despite its importance and interest, in order to keep the length to a minimum.

The book does not pretend to be encyclopedic. I have attempted to keep it at a comfortably simple level, with enough examples to provide an introduction to the diversity of photochemical reactions, but without overwhelming by either depth or breadth of coverage. The order of topics covered, particularly in the last chapters, is totally arbitrary. I hope that readers in search of specific information will be tempted to browse through the book and will become stimulated by the material covered. Old and well-established reactions and processes are described without references, but the more unusual or recent transformations are fully referenced.

My own research, which has kept me increasingly interested in photochemical transformations, has been funded by the National Science Foundation, the National Institutes of Health and the Research Corporation, as well as by the University of Illinois at Chicago. A Fulbright Fellowship and a Visiting Professorship at the Museum d'Histoire Naturelle in Paris were instrumental in getting the writing of this book underway. I am very grateful for all the support received.

The support provided by P. A. Kagan, her suggestions and patience in editing the text as it evolved cannot be adequately acknowledged. It is unlikely that anyone will ever read this book as carefully as she did. I am also grateful to my postdoctoral, graduate, and undergraduate collaborators, and to I. A. Kagan for careful laboratory work (often unpublished), to Professors D. Crich and R. W. Tuveson for comments on the manuscript, and to the latter for a simulating scientific collaboration which greatly increased my interest in photobiology. The sadness brought by the news of Bob Tuveson's passing, one day before the proofs of this book were to be returned to the Editor, overwhelms my pleasure of having completed this undertaking.

Comments, suggestion or criticism from readers will be appreciated. They should be sent to the author at the University of Illinois at Chicago, Department of Chemistry, m/c 111, PO Box 4348, Chicago, IL 60680, USA.

1. The fundamentals

1.1. WHAT IS PHOTOCHEMISTRY?

There is an activation energy associated with all chemical reactions, and therefore no chemical transformations are possible without some input of external energy. This activation energy may be very small in some reactions, which therefore occur even at low temperatures. It may be so high in other cases that desired reactions cannot be performed in the dark no matter how high the temperature is raised. Instead, decomposition and production of other unwanted products are observed. Although frustrating in practice, this decomposition is easily explained by looking at the energetics of chemical bonds.

Absorption of energy into a molecule is quantized. Three different types of excitation can take place: rotational, vibrational, and electronic. A molecule at a very low temperature in the dark may perhaps possess enough energy for populating the lowest rotational and vibrational levels. As the temperature is increased, higher rotational and vibrational states are populated. Eventually there is enough energy available for a bond to become broken. However, this fragmentation may have occurred well before the activation energy barrier for the desired transformation was reached. Consequently, the desired transformation has been diverted toward an unwanted path.

As shown in Fig. 1.1, heating molecules represented at the energy level A, with the hope of forming B, is likely to lead instead to the isolation of C. The latter transformation has a lower activation energy barrier ($E_2 < E_1$), and C is also the most stable of the three states.

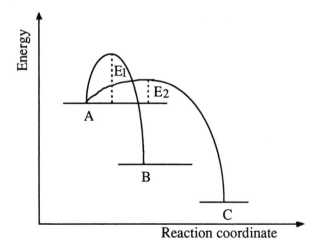

Fig. 1.1 Diagram showing the thermal transformation of A to B with the higher activation energy E_1, and the competing transformation of A to C with the lower activation energy E_2.

Unlike the rotational and vibrational energy levels, which are very closely spaced, the electronic levels in a molecule can have very large energy differences. Photochemistry is the art of taking advantage of these large energy differences which exist between electronic states to produce directly highly energetic molecules without having forced them to climb step by step through the various rotational and vibrational levels, thus avoiding thermally induced fragmentation reactions.

1.2. PHOTOCHEMISTRY: BASIC PRINCIPLES

A beam of light may be viewed as a collection of photons. *Photochemistry can take place only when a photon has been absorbed by a molecule*. Note that the photon must be *absorbed*, and that not all photons are necessarily absorbed by all molecules. It is perfectly possible for molecules to be exposed to a powerful beam of light possessing high-energy photons, and yet to be totally unaffected by it. This situation is encountered when the molecules do not contain proper *chromophores*, which are specific arrangements of atoms leading to absorption of photons at specific wavelengths within the emission spectrum of the light source.

1.3. ACTION SPECTRA

An *action spectrum* is obtained when the extent of production of a light-dependent phenomenon is recorded as a function of the wavelength of irradiating light. There are many different types of action spectra, ranging from recording the yield of a photochemical transformation as a function of wavelength to measuring the extent of growth of a seedling (or some other biological phenomenon) as a function of wavelength, or to measuring the fluorescence emission of a sample at a fixed wavelength as a function of excitation wavelength (this gives the excitation spectrum of the sample which, for a pure compound, is the same as its absorption spectrum).

The measurement of action spectra is very informative in photobiology, when it is desirable to establish whether it is a compound itself or its metabolite(s) which may be responsible for biological effect. When the compound itself is photochemically active, the action spectrum will match its absorption spectrum.

1.4. ABSORPTION OF LIGHT AND MOLECULAR ORBITALS

Quantum mechanics teaches that molecular orbitals can be created by properly combining all the atomic orbitals associated with the atoms making up the framework of a molecule. There are as many molecular orbitals as there are atomic orbitals. While such orbitals have an existence well defined by quantum mechanics, all do not necessarily contain electrons. It is customary to draw diagrams showing the relevant orbitals ranked in order of increasing energy. Note that when the combination of two atomic orbitals leads to a molecular orbital of lower energy, the latter is called the bonding orbital, and when the resulting orbitals has greater energy it is called the antibonding orbital (for each bonding orbital there is one antibonding orbital) (Fig. 1.2). Occasionally there is no energy change, resulting in a non-bonding orbital.

Two orbitals at adjacent atoms may combine in different manners. For example, the interaction between one p-orbital at each of two carbon atoms can lead to σ and π bonding orbitals and to the corresponding σ and π antibonding orbitals. The former are

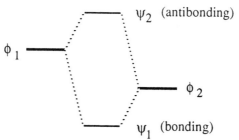

Fig. 1.2 Interaction between two atomic orbitals ϕ_1 and ϕ_2, producing the molecular orbitals ψ_1 (lower energy: bonding orbital) and ψ_2 (higher energy: antibonding orbital).

usually referred to as σ- and π-orbitals, and the latter as σ^* and π^* orbitals (pronounced 'sigma star' and 'pi star'). The orientation and representation of these orbitals are shown in Fig. 1.3.

Having listed all the molecular orbitals available as a framework for the electrons in the molecule, we must give a home to these electrons. We start with the orbital of lowest energy (usually representing a σ bond) and fill it with two electrons (the maximum allowed by Pauli's exclusion principle), then move on to the next higher molecular orbital, assigning two electrons to it, and so on until all the bonding electrons have been accounted for. Bonding molecular orbitals need not be full (half-filled bonding orbitals are found in molecules possessing free radicals; a stable biradical therefore has two half-filled bonding orbitals). The next important step is to assign a spin number, which is either $+\frac{1}{2}$ or $-\frac{1}{2}$, to the electrons in the molecular orbitals; according to Pauli's exclusion principle, two electrons which occupy the same orbital must possess different spin numbers.

It is now possible to get a picture of the orbitals and electrons in ethylene, which is essentially that shown in Fig. 1.3. The two electron spins are represented in Fig. 1.4 according to tradition, with arrows pointing in opposite directions. Going from the lowest to the highest-energy orbital, one finds first the σ-bonds (four carbon–hydrogen bonds, and the carbon–carbon bond), and next the π-bond. All are bonding orbitals, and they accommodate all the electrons present in the molecule. The π orbital is the highest-occupied molecular orbital (HOMO). The orbital having the next higher energy is the lowest-unoccupied molecular orbital (LUMO), which is the π^* orbital.

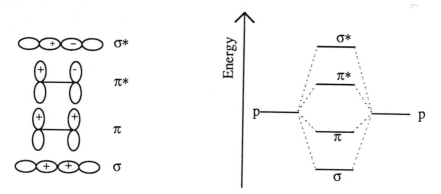

Fig. 1.3 Combination of two p-orbitals at carbon atoms in two different orientations, showing the two bonding orbitals (σ and π) and the two antibonding orbitals (σ^* and π^*) found in a carbon–carbon double bond (left), and the energy ranking of the orbitals (right). The spacing of the orbitals is not to scale.

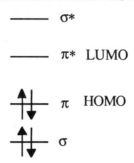

Fig. 1.4 Ranking of the molecular orbitals associated with the ground state of a carbon–carbon double bond, with a representation of the electrons in the bonding orbitals.

The orbitals of still higher energy are the antibonding orbitals corresponding to the four carbon–hydrogen σ-bonds and to the carbon–carbon σ-bond (note that orbitals which have the same energy are said to be degenerate orbitals).

The two molecular orbitals with the greatest importance in analyzing photochemical transformations are those designated as HOMO and LUMO. The electrons in the HOMO orbital are considered to be the most distant from, and therefore the least strongly attracted to, the nuclei. Consequently, they are the most likely to be expelled from their orbital when energy is provided, going then to the next higher orbital (on the energy scale), which is the LUMO. The transfer of one electron from the HOMO to the LUMO orbital necessarily requires energy, and one photon of exciting light must have precisely this energy (energy of HOMO minus energy of LUMO).

In summary, a molecule undergoing photochemistry must have electronic features conducive to the absorption of one photon which, when absorbed, promotes one electron from the HOMO to the LUMO. Photochemical reactions will be possible only if the incident light contains photons having exactly the energy corresponding to a specific HOMO-to-LUMO transition. If the available light does not contain such photons, photochemistry will not take place, even if many photons of greater energy are present.

In practice, most photochemical reactions involve the absorption of only one photon, and most photochemically active molecules contain π-systems. In many cases, the initial absorption of one photon cannot and does not lead to electronic excitation. For example, a single photon in the infrared range cannot do more than excite an organic molecule vibrationally and/or rotationally. Normally, these excitations do not lead to chemical changes. However, when infrared radiation emitted by a laser source irradiates a molecule, molecular changes reminiscent of those induced by ultraviolet or visible light sources are often observed. In these cases, a huge number of photons is pumped almost instantaneously into the molecule, raising it to very high rotational and vibrational levels before fragmentation processes compete significantly. These high energy levels could be similar to those reached when a single photon of ultraviolet light is absorbed by the molecule. In the case of alkenes, for example, *cis–trans* isomerization may be achieved with infrared laser radiation, as well as with ultraviolet light, but there may be differences in the stereochemistry of the cycloaddition reaction products.

1.5. SINGLET AND TRIPLET STATES OF MOLECULES

Two electrons in the same orbital must have opposite spins; they are said to be paired, and the molecule is said to be in its singlet state. When photochemical excitation occurs

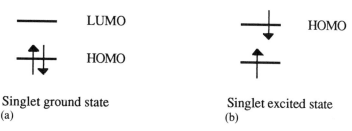

Singlet ground state
(a)

Singlet excited state
(b)

Fig. 1.5 Representation of (a) a singlet ground state (two electrons with opposite spins in the same orbital) and (b) a singlet excited state (two electrons with opposite spins in different orbitals).

and one of these electrons is promoted into the next higher orbital, spin conservation is usually observed, as required by the principles of quantum mechanics (note that exceptions do exist). In other words, the two electrons which had opposite spins in the ground state (normal) molecule still have opposite spins when the molecule is in its electronically excited state. In each case the molecule is in the singlet state.* More specifically, in the former it is in the *singlet ground state* and in the latter in the *singlet excited state* (Fig. 1.5).

Following electronic excitation, a very important step is called *intersystem crossing*. Once two electrons are in different orbitals, they no longer have to be paired because Pauli's exclusion principle no longer applies. Actually, the energy of the system where the two electrons are in different orbitals with the same spin is lower than when the electrons have opposite spin. Thus, the electrons which had opposite spin when they were in the same orbital still have opposite spin immediately following excitation when they are placed in two different orbitals, but spin inversion is now energetically feasible. This converts a molecule in a singlet electronically excited state into one which is in a triplet state. Following this intersystem crossing, the molecule is said to be in its *triplet excited state*. Since the relocated electron is now in the original LUMO orbital, this orbital occupied by one electron has become the HOMO of the excited molecule (Fig. 1.6). Because of Pauli's exclusion principle, this electron cannot possibly return into its original home orbital without undergoing another spin inversion (another intersystem crossing).

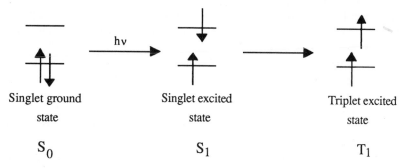

Singlet ground
state

S_0

Singlet excited
state

S_1

Triplet excited
state

T_1

Fig. 1.6 Representation of the electrons in the HOMO of the irradiated molecule going from the singlet ground state to the singlet excited state, and then to the triplet excited state.

*The term comes from the value of $n = 2(s_1 + s_2) + 1$, where s_1 and s_2 are the values of the spins of the electrons. With two electrons having opposite spins, $+\frac{1}{2}$ and $-\frac{1}{2}$, $n = 1$ (singlet state). With electrons having the same spin $+\frac{1}{2}$, $n = 3$ (triplet state).

1.6. FATE OF ELECTRONICALLY EXCITED MOLECULES

It is customary to outline the fate of a molecule which undergoes photon absorption on a *Jablonski diagram*, which is drawn (not to scale) on the background of the energy levels for the molecule (Fig. 1.7). According to quantum mechanics, energy absorbed by a molecule may contribute to vibrational, rotational, or electronic excitation. Since energy levels for vibrational and rotational excitation are very close, molecules in the ground state are really a collection of species with different rotational and vibrational levels. Electronic excitation can, and does, occur from any of these levels. Consequently, excited molecules (singlet or triplet) can also be created at different rotational and vibrational levels.

For convenience, let us follow one specific molecule (labelled A in Fig. 1.7) as it undergoes electronic excitation:

1. Because the mass of one electron is so much smaller than that of any nucleus, the initial step in the electronic excitation of a molecule takes place without changes in the position of the atoms; vertical excitation is said to have occurred. The absorption of one photon into A (a singlet ground state molecule) has now created a singlet excited state molecule B. The energy difference between B and A can be directly measured from the absorption spectrum of the starting material. Should the lifetime of B be long enough, relaxation with a change in the position of the atoms could take place: this is a simple photochemical reaction, possibly leading to a different molecule A′ (not shown in Fig. 1.6) after return to the ground state, in which the arrangement of the atoms is different from A. A *cis–trans* isomerization in an alkene (Chapter 3) is an example of conversion of A into A′:

$$R_1 \diagup\!\!\!\diagdown R_2 \quad \underset{h\nu}{\overset{h\nu}{\rightleftarrows}} \quad \left[R_1-CH=CH-R_2 \right]^* \quad \underset{h\nu}{\overset{}{\rightleftarrows}} \quad R_1 \diagup\!\!\!\diagup R_2$$

2. Molecular rearrangements or reactions are not the only possible transformations of a molecule in a singlet excited state. Energy could be lost instead, returning the molecule to the ground state (this is called *internal conversion*). This loss of energy could take place either by release of heat or by loss of a photon. The thermal route is easily pictured by considering the return from the high-energy electronically excited state through all the vibrational and rotational levels associated with the ground state. Overall, the heat released corresponds to the energy difference between the original level in the excited state and the final level in the ground state. The overall process from A through B and back to A (Fig. 1.7) is a conversion of light into heat.

The conversion of S_1 to S_0 is the reverse of the photon absorption step. However, the process could take place after relaxation to lower vibrational and rotational levels in the excited state, and the ground state molecule generated could be at more highly vibrationally and rotationally excited levels than when it was first excited. In other words, the energy gap between the excited molecule emitting the photon and the resulting ground state molecule could be smaller than in the original excitation. This means that the wavelength of the light emitted is longer than that of the exciting radiation. This luminescence process occurring from the singlet excited state is called *fluorescence*.

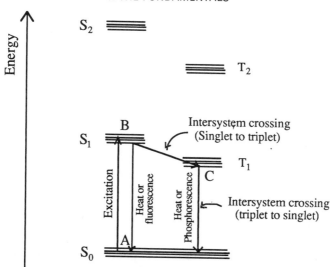

Fig. 1.7 A Jablonski diagram showing the excitation of a molecule A from its singlet ground state (S_0) to its first excited singlet state (S_1). This excited molecule (B) might return to the ground state with emission of heat or light (fluorescence), or undergo intersystem crossing to C, the molecule in its first triplet excited state, which would then return to the ground state (A again) with emission of either heat or light (phosphorescence).

3. A third option for the molecule in its singlet excited state is to change its electronic characteristics by undergoing *intersystem crossing* (a transformation thermodynamically favored, though forbidden in principle by quantum mechanics), forming the excited molecule C in its triplet state (the intersystem crossing occurs at the same energy level, but the excited triplet state initially created relaxes to lower vibrational/rotational levels). A molecule such as an alkene in its triplet state usually has a different geometry from that in either of the singlet states.

The electronically excited triplet state molecule again has three choices for returning to the ground state: (a) lose energy thermally, (b) lose energy in the form of a photon, or (c) undergo photochemical reaction. The luminescence process (b) is called *phosphorescence*. Obviously, the energy gap between the triplet excited state T_1 and the ground state S_0 is smaller than the gap between S_0 and S_1 (Fig. 1.7). Therefore, the light emitted by phosphorescence has a wavelength longer than both the initial incident light and the fluorescence emission (photon energies and wavelengths are inversely related, as shown in Section 1.8). The fundamental difference between fluorescence and phosphorescence rests in the multiplicity of the electronically excited states from which luminescence takes place: singlet state for fluorescence, triplet state for phosphorescence.

Summarizing the fate of electronically excited molecules, we find two possible types of behavior: (1) radiative decay (fluorescence or phosphorescence), and (2) non-radiative decay (S_1 to T_1, T_1 to S_0, or formation of photoproducts from either S_1 or T_1). In principle, higher electronically excited states could be obtained by further excitation of S_1 or T_1, but the probability of absorbing a second photon into a molecule is very low when normal light sources are used. This process is more feasible with properly tuned lasers. Note that a conversion of S_1 to T_1 could also be accompanied by emission of light, instead of heat.

1.7. MAKING INTERSYSTEM CROSSING MORE EFFICIENT: THE HEAVY-ATOM EFFECT

Because intersystem crossing is actually a violation of the spin selection rules, it is often not efficient. However, it can be enhanced by spin–orbit interaction between an excited molecule (in the singlet state) and atoms with high atomic number. This is called a *heavy-atom effect*. The most common occurrence is when the reaction solvent has halogen substituents and therefore induces a *heavy-atom solvent effect*. One should not be surprised, therefore, to observe very different photochemical results in changing the solvent from hexane to CCl_4, for example.

Heavy-atom effects may also be observed in an aqueous medium, by using micelles prepared with a brominated surfactant.[1]

1.8. TIME SCALE

The time scale for the initial events in photochemistry is the femtosecond ($1\,fs = 10^{-15}\,s$). Laser pulses as short as a few femtoseconds are now possible, so that the progress through the transition state of a reaction can be viewed directly.

In the example of the photofragmentation of the molecule ICN, it takes only a few femtoseconds to completely separate the fragments I and CN. Since their recoil velocity is typically $1\,km/s$, the fragments cover about $0.1\,nm$ (or close to the distance of a chemical bond) in $100\,fs$.[2]

1.9. RELATIONSHIP BETWEEN WAVELENGTH AND PHOTON ENERGY

The energy of a photon at a given wavelength is calculated from the expression $E = c/\lambda$, where c is the speed of light and λ the wavelength. The energy difference ΔE (in kilocalories per mole) for a molecule in its first excited state and in its ground state is $\Delta E = E_1 - E_0 = hc/\lambda$, where h is Planck's constant (9.534×10^{-14} kcal. s/mol), c is the speed of light (3×10^{17} nm/s), and λ is the wavelength in nanometers. Using these values, the expression becomes $\Delta E = 28\,600/\lambda$. Figure 1.8 shows energy values calculated at different wavelengths.

The ultraviolet region closest to the visible, from 315 to 400 nm, is the ultraviolet A (UVA) region (or near ultraviolet, NUV). At increasingly shorter wavelengths, one finds the UVB (280–315 nm) and UVC (100–280 nm) regions, also called the far ultraviolet (FUV). These boundaries were established by an international committee (Commission Internationale d'Eclairage, CIE)*, but they are often not used rigorously by practicing photobiologists, who frequently take 320 nm as the cut-off between the UVA and UVB regions.

An important reference wavelength is 253.7 nm, which corresponds to the main emission line in a low-pressure mercury lamp. The equivalent energy at this wavelength is 112.7 kcal/mol. Typical average energies for homolytic cleavage of selected chemical bonds in organic molecules are shown in Table 1.1.

Note that photons at wavelengths below about 250 nm ($\Delta E = 116$ kcal/mol) possess enough energy to break most of the carbon–carbon, carbon–hydrogen, carbon–

---

*The International Lighting Vocabulary may be found in CIE Publication No. 17.4 (1987).

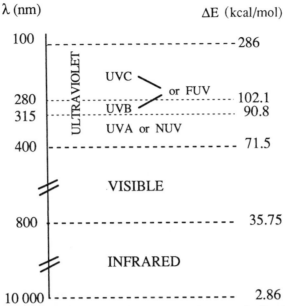

Fig. 1.8. Relationship between the wavelength of a photon and the energy (kcal/mol) of a mole of photons at that wavelength.

Table 1.1 Energies and corresponding wavelengths for homolytic fission of typical chemical bonds

Bond	Energy (kcal/mol)	λ (nm)
C=C	160	179
C–C	85	336
C–H	95–100	286–301
C–O	80–100	286–357
C–Cl	60–86	332–477
C–Br	45–70	408–636
O–O	35	817
O–H	85–115	249–336

halogen, carbon–oxygen, oxygen–oxygen, and oxygen–hydrogen bonds in organic compounds. When organic molecules are irradiated, however, bonds are seldom broken at random. Instead, the excited molecules undergo fairly selective bond breaking, rearrangements, or bimolecular reactions.

1.10. UNITS

The names and symbols of SI units important in photochemistry are listed in Table 1.2. One conversion which may be needed is from foot candle to lumen per meter, which requires multiplying the former by 10.76.

The chemical literature is replete with photochemical data in which the energy units are expressed in units other than kilocalories per mole. The other units, such as the

Table 1.2 Official units in photochemistry

Physical quantity	Name of unit	Symbol
Luminous intensity	candela	cd
Work energy	joule	J
Power, radiant lux	watt	W (J/s)
Fluence	kilojoule per square meter	kJ/m^2
Fluence rate	watt per square meter	W/m^2
Luminous flux	lumen	lm (cd · sr)
Luminance	candela per square meter	cd/m^2
Illuminance	lux	lx (lm/m^2)
Wavenumber	reciprocal meter	m^{-1}
Radiant intensity	watt per steradian	W/sr

Table 1.3 Conversion table for units of energy

kcal/mol	eV	joule	erg	cm^{-1}
1	4.3364×10^{-2}	4.184×10^{-3}	6.9473×10^{-14}	3.4976×10^2
2.8591×10^{-3}	1.2398×10^{-4}	1.9862×10^{-9}	1.9862×10^{-16}	1
23.060	1	1.6021×10^{-5}	1.6021×10^{-12}	8065.7
1.4394×10^{13}	6.2418×10^{11}	10^{-7}	1	5.0345×10^{15}
4.184×10^{-3}	6.2418×10^{18}	1	10^7	5.0345×10^{22}

joule, erg, electron-volt, or reciprocal centimeter, are more likely to be utilized by physical or theoretical chemists. A conversion table for all these units is given (Table 1.3).

1.11. BEER–LAMBERT'S LAW

The reader is reminded that when a beam of monochromatic light goes through a solution, the intensity I of the emerging beam is related to that of the incident light I_0 by the relationship $I = I_0 \times 10^{-\epsilon lc}$, where l is the thickness of the medium (expressed in centimeters) and c is the concentration (in moles per liter). The term ϵ (expressed in liters per mole per centimeter) is the molar absorption coefficient (or, formerly, molar extinction coefficient). The expression may be written as $A = \log_{10}(I/I_0) = \epsilon lc$, where A is the *absorbance* (or optical density) of the sample.

1.12. SOLVENT DEPENDENCE OF ABSORPTION SPECTRA; SOLVATOCHROMIC DYES

Small, but measurable, differences in the absorption maxima are often observed when the spectrum of a given compound is recorded in different solvents. The greatest differences are usually observed when comparing spectra in polar versus non-polar solvents, or protic versus non-protic solvents.

Dyes which show a significant change in their absorption spectra as a function of solvent composition are called solvatochromic dyes. One interesting application of such differences in absorption maxima as a function of solvent polarity is in the titration of water in organic solvents. For example, the maximum absorption of Reichardt's dye (**1**) recorded in a given organic solvent can be used to calculate the amount of water in the sample.[3]

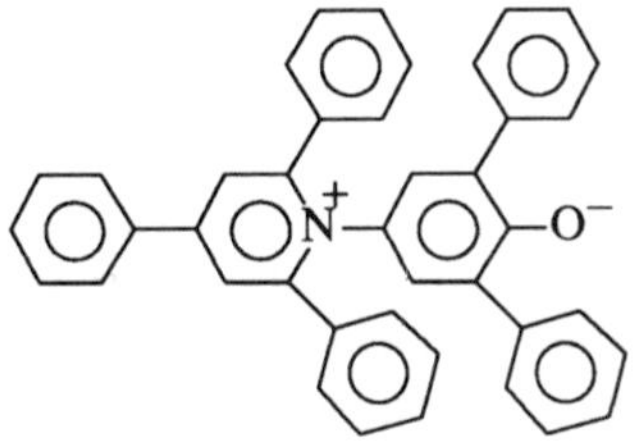

(**1**)

1.13. SOLVENT, TEMPERATURE, AND CONCENTRATION DEPENDENCE OF LUMINESCENCE SPECTRA

Luminescence spectra reflect the deactivation of electronically excited molecules. The luminescence intensity is reduced to the extent that an excited molecule can lose its energy in the form of heat through vibrational and rotational relaxation. Consequently, rigid molecules, such as polycyclic aromatic hydrocarbons, are more likely to show luminescence, and luminescence is usually enhanced in more rigid environments.

The temperature is often an important variable in luminescence studies, because a solution in a given solvent can change from being a very fluid liquid at high temperatures to becoming very viscous, glassy or crystalline at low temperatures.

Another factor affecting luminescence spectra is the concentration of the substrate in solution. One may imagine that, at extremely low concentrations, each excited molecule is far from other molecules of the same kind. As the concentration is increased, each excited molecule has a greater probability of being in close proximity to ground state molecules of the same kind. A complex between two identical molecules, one in an electronically excited state and one in the ground state, is called an *excimer* (excited dimer). The luminescence of an excimer occurs at longer wavelength with lower intensity compared to that of the compound in very dilute solution. The formation of a complex between an electronically excited molecule and a ground state molecule of a different kind (either of the solvent or another component in solution) is called an *exciplex* (excited complex).

The intermediacy of exciplexes and excimers has been postulated in many mechanistic schemes. In practice, it is important to keep in mind that the photophysical behav-

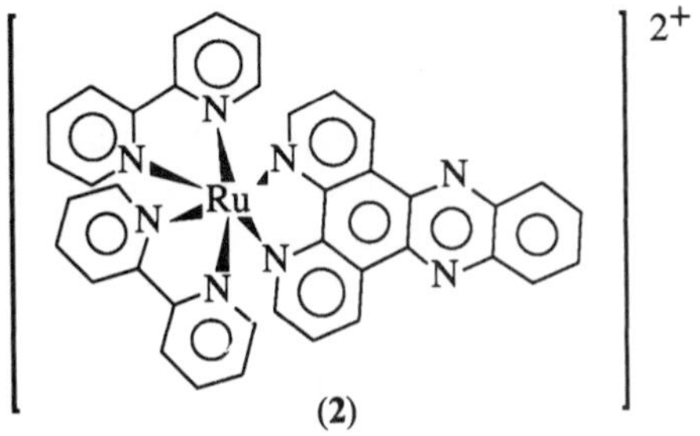

(**2**)

ior of an electronically excited molecule, and therefore the photochemical outcome of an irradiation, depends critically on the environment.

One striking demonstration is provided by the complex $Ru(bpy)_2(dppz)^{2+}$ (**2**). In aqueous solution at room temperature it shows no luminescence, but intense luminescence is observed in the presence of double helical DNA. In the presence of synthetic polynucleotides, the luminescence intensity depends on the polymer conformation.[4] This sensitive complex formation should be useful in assays for DNA.

In another example, the water content in an organic solvent can be calculated from the fluorescence emission maximum of a suitable molecule, such as that of the phthalimide (**3**).[5]

(3)

1.14. WAVELENGTH DEPENDENCE OF PHOTOCHEMICAL REACTIONS

The fact that molecules often have absorption spectra covering an appreciable range of wavelengths means that electronic excitation can be produced by any radiation within this range. As indicated earlier, equilibration in electronically excited molecules having different vibrational and rotational levels tends to take place faster than any other processes and, therefore, the same photochemistry is expected regardless of the excitation wavelengths (within the range of the absorption spectrum). Some exceptions are known, where product distribution depends on the excitation wavelengths. One example (for which a definite explanation is not known) has been included in Chapter 8. Perhaps there are cases where molecules in different vibrational and rotational excited states do undergo chemistry faster than equilibration.

1.15. QUANTUM YIELDS

Imagine a typical experiment in which a beam of light shines upon a vessel containing organic molecules. Not all the molecules instantaneously absorb light and become electronically excited. Among those that do, many will return from the singlet excited state to the ground state after emission of heat and/or light without leading to any chemical transformations. In addition, those excited molecules which do undergo intersystem crossing and reach their triplet state have a finite probability of returning to the original ground state molecule with emission of heat and/or light. Therefore, new products are obtained from only a small fraction of the molecules which were originally excited.

It is often important to know the fraction of the total number of photons specifically used for each step of interest (e.g. fluorescence, phosphorescence, or formation of a specific product). This fraction is the *quantum yield*. Quantum yields are usually not expressed as percentages. A fluorescence quantum yield of 0.1, for example, means that 10% of the photons absorbed by a pure sample have led to fluorescing molecules. This will be expressed as $\Phi_F = 0.1$.

The quantum yield for any process is the fraction of the photons used in the process, that is, the number of moles of compound undergoing the process divided by the number of moles of photons absorbed (1 mol of photons = 1 einstein). Therefore, the sum of the quantum yields for all the processes in which the photons have participated is equal to 1.

Occasionally, quantum yields of product formation exceed 1. From an energetic viewpoint, such reactions may be very valuable. Photons generated in the laboratory are usually very costly. If one photon can trigger the utilization of more than one molecule of starting material or the formation of more than one molecule of product, the energy cost per molecule of final product is obviously lower. This is particularly true in the case of *chain reactions*, where one photon initiates a reaction, which proceeds further through a chain process without need for additional photons (except to the extent that premature chain termination reactions take place).

A typical light-induced chain reaction is the photochemical reaction

$$Cl_2 + H_2 \rightarrow 2HCl$$

where the only light-dependent step is the homolytic cleavage of Cl_2 to give two $Cl\cdot$ radicals, which become involved in a chain reaction as shown:

$$Cl_2 \rightarrow 2Cl\cdot$$

$$Cl\cdot + H_2 \rightarrow HCl + H\cdot$$

$$H\cdot + Cl_2 \rightarrow HCl + Cl\cdot$$

New molecules of HCl are formed through the propagation steps of the chain reaction, involving either $H\cdot$ or $Cl\cdot$ reacting with either Cl_2 or H_2 molecules. The recombination reactions of $H\cdot$ with $H\cdot$, $Cl\cdot$ with $Cl\cdot$, or $H\cdot$ with $Cl\cdot$ are the termination steps.

1.16. QUANTUM YIELDS AND CHEMICAL YIELDS

The quantum yield for the formation of a desired reaction product is the number of excited molecules which gives rise to that product divided by the total number of molecules which have been electronically excited (this is the same as the number of quanta of light absorbed):

$$\Phi = \frac{\text{number of molecules produced}}{\text{number of quanta of light absorbed}}$$

When more than one process is considered, each quantum yield must be labeled separately, for example with subscripts. Note that a quantum yield may be defined for processes which are either photophysical (such as fluorescence, phosphorescence, or intersystem crossing) or photochemical (disappearance of starting material or formation of a selected product).

As mentioned earlier, few molecules are in their excited states at any given time, and there is not a direct relationship between the quantum yield and the actual chemical yield for the formation of a product. For example, the formation of a product may have a high quantum yield, but if one of the reaction products formed competitively has a high ϵ-value and absorbs nearly all the light, it effectively shields the remaining starting material. In this case the reaction rate will get smaller as the irradiation proceeds and the extent of conversion to the desired product will be low, despite the high quantum yield for its formation. Alternatively, a reaction may proceed with a very low quantum yield, and yet result in the quantitative conversion of a starting material into product(s) if the reaction time is long enough and if the initial reaction products do not undergo any other transformations.

A synthetic chemist is usually interested in getting high chemical yields. When cost is not an overriding consideration a low quantum yield is not a major handicap if the chemical yield is reasonably high. A photochemical transformation will be most cost effective when both the quantum yield and the chemical yield are high.

For quantum yield determination, the number of photons used in the experiment must be known. First, the number of photons per unit of time (usually per second) in the light beam crossing the reaction vessel must be measured. The total number of photons used in the experiment is then deduced from the duration of the exposure.

The energy of the light beam can be measured with a calibrated thermopile. Because it is extremely difficult to perform this type of absolute measurement routinely, indirect methods are usually preferred. They use as reference a reaction for which the quantum yield has previously been measured. By determining the extent of product formation for the known reaction under conditions absolutely identical to those used for the unknown conversion, it is possible to determine how many quanta of light have been used. The chemical yield of the product formation in the unknown reaction is measured, and divided by the number of quanta, to arrive at the yield of product for each quantum of light absorbed, the quantum yield.

1.17. ACTINOMETERS

The reference photochemical reaction mentioned above uses a starting material which is called an *actinometer*. In order to use an actinometer properly one must be certain that the quantum yield of the reference reaction and the reaction of interest are measured at the exact same wavelength(s). In this regard, the best actinometer is probably ferric oxalate. Upon irradiation, the ferric ion is converted into Fe^{2+}, along with oxalate oxidation. The extent of conversion is determined by adding 1,10–phenanthroline, and measuring the absorption of the Fe^{2+}–phenanthroline complex. The quantum yield has been carefully measured by Parker and Hatchard and found to vary little over a wide range of wavelengths.[6]

Any reaction for which the quantum yield has been reported could, in turn, be used in actinometry. A list of chemical actinometers and some detailed procedures have been published in *Pure and Applied Chemistry* (**61**, 188–210 (1989)). *One must be careful not to extrapolate away from the experimental conditions under which the original quantum yields were determined, as the values may be affected by the nature of the solvent, the temperature, the concentrations, and the wavelength of irradiation.*

A convenient actinometer, based on a completely reversible photoreaction, has been developed by H. G. Heller. The absorption spectra for the two compounds (**4**) and (**5**) have very little overlap. The starting material on the left (a fulgide available as Aberchrome 540) can be irradiated in the ultraviolet, to produce its photoproduct.[7]

(4)   (5)

Irradiation of the photoproduct with visible light regenerates the original fulgide. The quantum yield for each conversion has been determined in toluene as a function of the wavelength.

A related fulgide (Aberchrome 999P) allows quantum yields to be determined at wavelengths up to 633 nm. The sensitized photooxidation of the polycyclic aromatic hydrocarbon mesodiphenylhelianthrene by the laser dye $1, 1', 3, 3, 3', 3'$-hexamethyl-indotricarbocyanine iodide has been proposed for covering the range 670–795 nm, which is important in photomedicine.[8]

Quantum yield determinations using radiation which is not monochromatic, such as sunlight emission, are not very meaningful and are much more difficult to perform, because absorption coefficients are very dependent on the wavelength (as seen in the absorption spectrum of most molecules). A compound which possesses sharp absorption bands can differ considerably in absorption coefficients over a small wavelength span. If, in addition, the emission spectrum of the light source shows variations in intensity according to the wavelength, then the energetic balance for the overall chemical transformation is almost impossible to establish. Quantum yield determinations under these conditions are largely meaningless. Regardless, the measurement of quantum yields in polychromatic light has been attempted.

Note that, in practice, an actinometer is used to measure the number of photons actually available in a beam of light used to irradiate a vessel in which a photochemical reaction is taking place. It is essential to ensure that the experiment in which the actinometer is used and that in which the photochemical reaction is performed are absolutely comparable. This may be quite difficult when the light beam is wider than the vessel and the same fraction of the beam is not used in both experiments, or when the output of the light source is not stable.

The number of photons available from a light source may be quite different from that actually used in a specific vessel during a photolysis experiment. The emission of a light source in watts per square meter may be estimated directly with a radiometer. Some radiometers measure total energy, others have sensors measuring energy in specific wavelength ranges. Errors with these devices may be quite large.[9]

1.18. PHOTOSENSITIZED REACTIONS

As described earlier, an electronically excited molecule created upon absorption of a photon may react through a variety of pathways. Return to the ground state with emission of either light or heat is a simple possibility. Although ground state molecules will eventually always be formed, they may be the starting material, isomerized products, fragmentation products, or products of reactions with other molecules of starting material or with other reagents.

A crucial mode of energy loss for an electronically excited molecule is the transfer of its energy to a molecule in the ground state, which thereby becomes electronically excited. Since such a process is a bimolecular interaction, it is more likely to occur with a long-lived electronically excited energy donor. Thus, most sensitization reactions involve *triplet sensitizers*.

A typical reaction for a molecule A acting as a sensitizer from its triplet excited state and transferring energy to a singlet ground state molecule B (which is said to act as a *quencher*) must follow the quantum mechanical requirement for conservation of spin. It is written as

$$^3A + {}^1B \rightarrow {}^1A + {}^3B$$

Once created, the triplet state of B may undergo any reactions that would have occurred from the triplet state, had B accepted a photon directly. One important consequence is that, through sensitization, *chemical compounds can be made to undergo photochemical reactions in the presence of light which they do not absorb* (in apparent contradiction with the most fundamental principle of photochemistry listed in section 2). The sensitizer is regenerated after the energy transfer step is completed, and can thus be used catalytically. Another consequence is that any further reactions occurring from a molecule which has been electronically excited by the sensitizer involve the triplet excited state of this quencher. This may be particularly valuable in cases where the singlet excited state of a molecule is capable of giving other reaction products with a high quantum yield. In such a case, direct excitation leads to very few triplet excited molecules, and therefore very few molecules of reaction products may be obtained from that triplet excited state. Sensitization provides a powerful method for obtaining products from the triplet excited state without competition with products from the singlet state, since in this case the singlet excited state of the quencher molecule is not even generated.

Sensitizers are often assumed to be unreactive and to function catalytically, but this is seldom true; many sensitizers do decompose or undergo transformation as the reaction mixtures undergo irradiation. Carbonyl compounds such as acetone, acetophenone, and benzophenone often have a high-energy triplet state, and they are therefore frequently used as sensitizers. Mercury, which has the unique property of never undergoing photochemistry of its own, has also been used extensively.

An interesting aspect of photosensitization concerns the physical closeness of the interaction between a sensitizer and its substrate. For example, very close contact with an optically active sensitizer would be expected to produce asymmetric synthesis of the photoproduct. Actually, very low enantioselectivity was observed in several *cis–trans* isomerization reactions of cyclopropanes and alkenes,[10] but the optical yield was strongly dependent on the reaction temperature.[11] As discussed in Chapter 8, photosensitized deconjugation of an α, β-unsaturated ketone has yielded a product with over 70% enantiomeric excess.

It is useful to emphasize the fact that there is an alternative to using sensitization for obtaining products derived from a triplet state reaction in photochemical reactions: heavy-atom effects, as indicated in Section 1.7, may achieve the same result. However, the selectivity is not usually as good, as demonstrated in the example shown. Compound (6), derived from the singlet state, is the major product in normal irradiations, but compound (7), derived from the triplet state, occurs in a ratio as high as 2.5 : 1 over (6) when miscelles containing a brominated surfactant are photolyzed.[1]

Finally, laser flash spectroscopic techniques allow detailed studies of triplet excited states created through sensitization. The absorption spectra of triplet states may be quite sensitive to the nature of the solvent.[12]

(6)   (7)

1.19. MECHANISTIC STUDIES

It is often desirable to determine whether the synthesis of a photoproduct of interest is derived from a singlet or a triplet excited state (or from both). Two complementary approaches are convenient, both involving quenching reactions.

In the first approach, the photochemical reaction is performed in the presence of increasing concentrations of a quencher. The products are analyzed and any reduction in the yield of one product must be attributed to a decrease in the yield of formation of its electronically excited precursor. Since quenching usually affects only triplet state molecules, a product for which the yield decreases when formed in the presence of a quencher must therefore be derived from the triplet excited state of its precursor. Conversely, any photoproduct not affected by the presence of a quencher is assumed to be derived from the singlet excited state of the irradiated molecule.

The assessment of multiplicity derived from quenching experiments can be confirmed by sensitization experiments. When the starting material is not allowed to absorb light directly, but instead is electronically excited through a sensitizer, only the triplet excited state of the starting material is produced, and all the reaction products must be derived from it. The photoproducts which disappeared in the previous case, where the starting material was irradiated in the presence of an efficient quencher, will predominate in the sensitized irradiation.

In practice, there are examples where a given product originates exclusively from only one electronically excited state (which may be either singlet or triplet). In other cases, one or more products may be derived partially from each excited state. Piperylene (1,3-pentadiene) has often been used as a quencher in mechanistic studies.

The mechanism of energy transfer central to sensitization may be quite complex. In addition to the chemical quenching outlined above, one may encounter cases of physical quenching where the electronically excited sensitizer clearly returns to the ground state, but where there is formation of no new products, either from the sensitizer or from the quencher.

1.20. PHOTOSENSITIZED REACTIONS OF A TRIPLET GROUND STATE MOLECULE

One important exception to the rule that sensitization converts quenchers from their singlet ground state to their triplet excited state is the case of oxygen, which is a triplet in the ground state $(^3O_2)$. The sensitizer in its triplet state reacts with $^3O_2$ to produce a *singlet oxygen* molecule $(^1O_2)$, an electronically excited form of oxygen:

$$^3A + {^3O_2} \rightarrow {^1A} + {^1O_2}$$

Singlet oxygen is a very reactive and very important molecule, and its chemistry will be discussed in Chapter 6.

1.21. EXPERIMENTAL CONSIDERATIONS FOR PHOTOSENSITIZED REACTIONS

An ideal situation for performing a photosensitized reaction is when the molecule to be transformed (the quencher) has no absorption at the wavelength(s) used for exciting the sensitizer. If such a mixture is irradiated, only the sensitizer can become directly excited electronically, and any products derived from the quencher must necessarily result from energy transfer.

The situation is more complicated when the emission spectrum of the light source partially overlaps the absorption spectrum of both the sensitizer and the intended quencher. The irradiation of such a mixture must result in electronic excitation of both molecules, with attendant loss of selectivity. In such a case, the course of the reaction is controlled by increasing the probability of absorbing photons into the sensitizer rather than into the quencher. This is done by using a light source that is as monochromatic as possible, and by adjusting the concentrations so that the absorbance of the sensitizer in the emission range of the source greatly exceeds that of the quencher. Statistically, therefore, most of the photons will be absorbed by the sensitizer, and fairly clean photochemistry may ensue.

1.22. KINETICS

The change in concentration of a chemical species (either in the ground state or in an excited state) as a function of time can often be analyzed experimentally. Much can be learned by studying this change as a function of variables such as the nature and viscosity of the solvent, the concentration of various components in the reaction mixture (such as sensitizers or quenchers), the temperature, or the wavelength of irradiation. Likewise, it is often possible to follow the kinetics of luminescence processes (fluorescence or phosphorescence) as a function of similar variables. These experiments may be quite informative since they provide data on the behavior of electronically excited singlet and triplet states, respectively.

Kinetic experiments are most easily analyzed when the medium in which they are conducted is homogeneous. Unfortunately, there are many situations in which one cannot be sure that the medium is homogeneous. In other cases, it is definitely not homogeneous. Physical chemists usually have the luxury of optimizing the reaction conditions and of using the reagents in a homogeneous setting, but organic photochemists and photobiologists are often more limited when they are investigating the effect of a hydrophobic reagent in an aqueous medium or that of a hydrophilic reagent in an organic environment.

It is always essential to remember that the kinetic laws have been established in homogeneous media. No conclusions should be formulated from experiments in which deviation from a certain kinetic law is observed, if the medium is not homogeneous. For example, the failure of an aqueous solution of the enzyme superoxide dismutase to inhibit a photooxidation reaction does not necessarily mean that superoxide anion radical (O_2^-, see Chapter 6) is not formed, if the photooxidation is confined to an organic environment which does not mix with the aqueous medium containing the enzyme.

1.22.1 Lifetime of excited states

Electronically excited molecules spontaneously return to the ground state:

$$A^* \to A$$

When they are left undisturbed and they luminesce in dilute solution, the decay of their luminescence follows the simple first-order rate expression

$$-\frac{d}{dt}[M^*] = k_L^0 [M^*]$$

The lifetime of luminescence is related to the rate constant by the expression

$$\tau_L^0 = \frac{1}{k_L^0}$$

The subscript L is used to indicate that we are dealing with luminescence, and the superscript 0 is used to indicate that the lifetimes and rate constants refer to the cases where luminescence is the only mode of deactivation of the excited state considered. In practice, it is often possible to analyze separately the fluorescence decay and the phosphorescence decay, so that the rate constants and lifetimes of fluorescence and phosphorescence can be determined independently.

It must be noted that laser flash photolysis techniques allow the determination of the *absorption spectrum* of electronically excited molecules. Electronically excited triplet states are often analyzed in this manner, since their longer lifetime allows for a larger concentration to be present at the time of the analysis, when their absorption spectrum is determined. The measurement of the decay of the absorption spectrum of the triplet state as a function of time provides another method for determining the lifetime of electronically excited molecules.

When a molecule in a specific electronically excited state undergoes reactions and/or transformations in addition to luminescence (such as non-radiative emission, isomerization, quenching, or reaction with another molecule), the lifetime of the excited state is correspondingly shorter.

1.22.2. Fate of excited states

The quantum yield of luminescence (fluorescence or phosphorescence) is the number of molecules undergoing luminescence divided by the number of originally excited molecules.

Because electronically excited molecules have extremely short lifetimes, one must be concerned with their production and their fate as functions of time. In general, many reactions compete. For example, the first electronically excited singlet state of a molecule may undergo non-radiative internal conversion to the ground state, fluorescence, chemical reaction (such as isomerization), quenching, and intersystem crossing to the triplet excited molecule. One may define a quantum yield for each of these processes. It is often reasonable to assume that there may be a steady state under constant irradiation, where the rate of formation of the electronically excited state is equal to that of its disappearance. The quantum yield of formation of this excited state is then equal to the sum of the quantum yields of all the processes leading to its disappearance. Likewise, under steady-state conditions, the quantum yield of formation of a triplet excited state will be equal to the sum of the quantum yields for the processes leading to its disappearance, namely phosphorescence and non-radiative intersystem crossing to the singlet ground state, unimolecular and bimolecular chemical reactions, and quenching.

The quantum yield of luminescence is obviously affected when changes in conditions modify the partitioning of the electronically excited molecule toward the different processes associated with its deactivation. Studying how these quantum yields change with the experimental conditions can provide valuable information about the detailed mechanism of a photochemical reaction.

1.22.3. Quenching of excited states; luminescence quenching

The addition of a quencher to a sample being photolyzed should affect all the events subsequent to the formation of the excited state being quenched (usually a triplet state). Most notably, luminescence will be diminished and the yields of formation of selected reaction products will be decreased. A detailed kinetic analysis of all the quenching possibilities is beyond the scope of this book.

1.22.3.1. Luminescence quenching and the Stern–Volmer relation

The quantum yield of fluorescence is equal to the ratio of the rate constant for the radiative process converting the singlet excited state to the ground state divided by the sum of the rate constants for all the processes (fluorescence, internal conversion, intersystem crossing) associated with the singlet excited state:

$$\phi_F^0 = \frac{k_F}{k_F + k_{isc} + k_{ic}}$$

When a quencher Q is present, the contribution of the quenching effect must be added:

$$\phi_F = \frac{k_F}{k_F + k_{isc} + k_{ic} + k_Q[Q]}$$

The ratio is finally expressed as

$$\frac{\phi_F^0}{\phi_F} = 1 + \frac{k_Q[Q]}{k_F + k_{isc} + k_{ic}} = 1 + \tau_F k_Q[Q]$$

This is the Stern–Volmer expression for fluorescence quenching. Similar expressions may be derived for the quenching of other processes, such as phosphorescence or product formation.

1.22.3.2. Quenching of photooxidation reactions

Quenching experiments are often attempted in order to establish whether an oxidation reaction proceeded through singlet oxygen formation. A single experiment in which a presumed singlet quencher is added to a photolysis mixture is not capable of assessing whether any decrease in the yield of oxidation product(s) results from quenching of singlet oxygen. Instead, quenching of the triplet excited state of the sensitizer could have taken place, and singlet oxygen would not even have been formed.

The scheme shown here lists the pathways involving the sensitizer in its singlet and triplet excited states and their reactions with a substrate A, a quencher Q and with oxygen:[13]

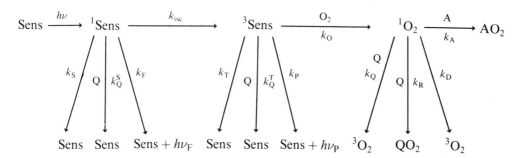

The quantum yield of formation of AO_2, the oxidized product from A, is expressed as a function of the reaction rates and concentrations as follows:

$$\phi_{AO_2} = \frac{k_{isc}}{k_S + k_F + k_{isc} + k_Q^S[Q]} \frac{k_0[O_2]}{k_0[O_2] + k_Q^T + k_T[Q] + k_P} \frac{k_A[A]}{k_A[A] + [k_R + k_Q]Q + k_D}$$

The three terms take into account the fate of the singlet sensitizer, the triplet sensitizer and singlet oxygen in the formation of the oxidized product. In this expression, the subscripts use the initials for singlet, quencher, fluorescence, triplet, phosphorescence, reaction (with oxygen), and deactivation.

In practice, the expression may be simplified considerably because quenching of a singlet excited state is negligible in view of the very short lifetime of this species: the term $k_Q^S[Q]$ may therefore be neglected. The first term in the quantum yield expression then turns out to be the expression for the quantum yield of intersystem crossing ϕ_{isc}. Furthermore, the contributions of a good triplet sensitizer to phosphorescence and to non-radiative return to ground state may be neglected.

When quenching affects solely the excited sensitizer, the terms involving the reactions of singlet oxygen may be neglected in comparison with those affecting the sensitizer. In other words, the relationship becomes

$$\phi_{AO_2} = \phi_{isc} \frac{1}{1 + k_Q^T[Q]/k_0[O]_2} \frac{1}{1 + k_D/k_A[A]}$$

The expression may be written as

$$\phi_{AO_2}^{-1} = \phi_{isc}^{-1} \left(1 + \frac{k_Q^T[Q]}{k_0[O_2]}\right)\left(1 + \frac{k_D}{k_A}[A]^{-1}\right)$$

When an experiment is run under conditions where [A] is nearly constant, the relationship between the reciprocal of the quantum yield of formation of the photooxidation product and $[A]^{-1}$ is linear. The intercept (corresponding to the value for $[A]^{-1} = 0$) depends on the quencher concentration as well as on the oxygen concentration (and therefore oxygen pressure).

When a series of experiments is performed under identical conditions, except for different quencher concentrations, one may compare the yields (rather than quantum yields) of oxidation product formed. A series of lines having no common intercept will characterize quenching of the triplet excited sensitizer (Fig. 1.9). The results should also be different at different oxygen concentrations (e.g. in air versus pure oxygen).

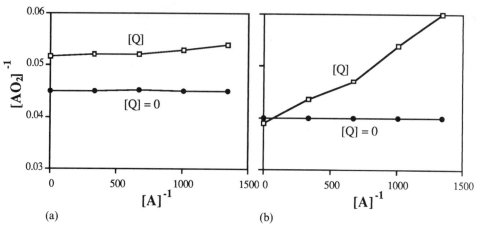

Fig. 1.9 Photooxidation of adamantylideneadamantane sensitized by diphenylbutadiyne. The quenchers were (**a**) p-benzoquinone and (**b**) β-carotene.

The other extreme situation is that in which the added quencher affects the triplet state of the sensitizer, without quenching the formation of singlet oxygen. With the same approximations, the master equation becomes

$$\phi_{AO_2}^{-1} = \phi_{isc}^{-1}\left(1 + \frac{k_D + (k_R + k_Q)[Q]}{k_A}[A]^{-1}\right)$$

In this case, the plot of the reciprocal of the quantum yields (or yields) of formation of AO_2 versus the reciprocal of the substrate concentration will show a series of lines sharing a commont intercept. This proves that the substrate A and the quencher Q compete for a common intermediate, and it is diagnostic for quenching of singlet oxygen (Fig. 1.9).

The results shown in Fig. 1.9 are derived from simple experiments in which p-benzoquinone and β-carotene were tested for quenching the oxidation reaction of an alkene (adamantylideneadamantane) into a dioxetane (cf. Chapter 6) using diphenylbutadiyne as a sensitizer.[14] In each set of experiments, the plot of $[AO_2]^{-1}$ versus $[A]^{-1}$ is shown at two quencher concentrations, one being [Q] = 0 (the two sets were run at different times under slightly different conditions, which explains why the results of [Q] = 0 are different). Clearly, one can deduce that p-benzoquinone quenches the excited state of the sensitizer, while β-carotene quenches the singlet oxygen generated.

1.23. ELECTRONICALLY EXCITED MOLECULES FROM DARK REACTIONS

While the electronic excitation of a molecule by absorption of one photon is fairly straightforward, several pathways have been discovered in which all transformations are conducted in the absence of light, and yet where at least one reaction product is created in an electronically excited state. These reactions must include at least one exothermic step from which energy is channelled into the formation of an electronically excited molecule. The typical thermal reactions that yield electronically excited molecules (Fig. 1.10) comprise aromatization reactions (e.g. isomerization of Dewar benzene into benzene, decomposition of an endoperoxide into an aromatic product and singlet oxygen) and conversion of cyclic molecules into unsaturated products (e.g. decomposition of 1,2-dioxetanes into two carbonyl derivatives, one of them in an excited state).

Electronically excited molecules can also be created through radical–radical recombination, a technique which is seldom used by the average synthetic organic chemist, and which is beyond the scope of this book.

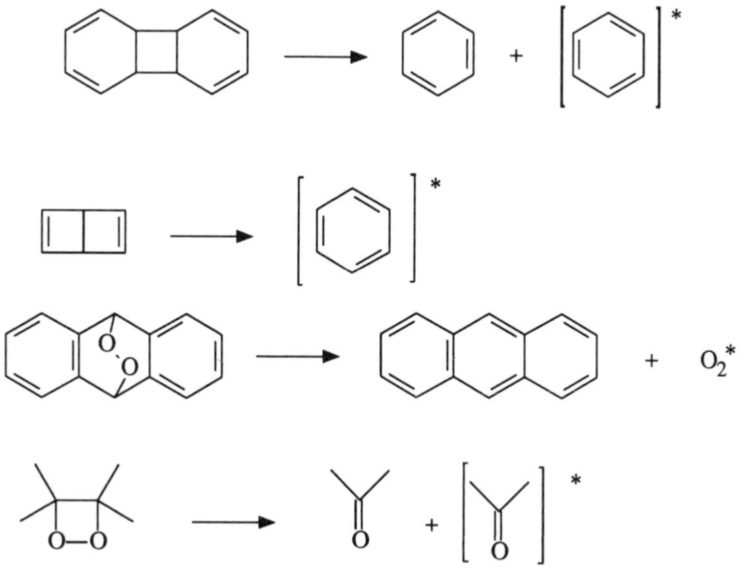

Fig. 1.10 Thermal generation of electronically excited molecules.

1.24. COMPLICATIONS

One fundamental problem in practical preparative photochemistry is that, almost always, the initial products of photochemical transformations are also exposed to the incident light. They may then undergo photochemistry on their own, and give products which may interfere with a specific photochemical transformation under study. In a mechanistic study, a partial solution to this problem is to stop the reaction as early as possible, when very little conversion has taken place, and to characterize the reaction products and/or reaction parameters utilizing sensitive probes, such as spectroscopic analysis of the transients formed upon a short flash with a laser source.

Another complication in photochemical studies is that not only can the initial photoproduct(s) undergo further transformation by thermal means, but they may completely escape detection if these transformations are very rapid. One technique widely used to analyze thermally unstable photoproducts consists of photolyzing the compound of interest at very low concentrations in an inert medium (therefore minimizing bimolecular reactions) and at very low temperatures (therefore minimizing thermal reactions). A rare gas matrix is ideal for this type of study. A good example is the photolysis of furan (**8**) and thiophene (**9**) in an argon matrix at 10 K, which led to the detection of the Dewar isomers of these two heterocyclic molecules. Even in such low-temperature studies, local heating may take place. This is due to the non-radiative deactivation of excited molecules, and is difficult to detect and measure. The lower the overall reaction temperature, the smaller the risk of inducing important secondary, thermal, reactions. A number of photochemical studies in liquid helium have been published, leading to the characterization of theoretically important molecules such as cyclobutadiene (10), Dewar benzene (11), and benzyne (12), as well as many other unstable compounds.

(8)   (9)   (10)   (11)   (12)

X = O (8), S (9)

Photochemical transformations of natural products have often been studied in the laboratory. In principle, these transformations may be models for reactions occurring in the environment, provided that the wavelength(s) of the radiation used correspond to a significant component of sunlight. Clearly, all transformations performed at 253.7 nm, the main line in the low-pressure mercury vapor lamp, have little environmental relevance since practically all sunlight radiations reaching the earth's surface have wavelengths greater than 280 nm. Likewise, photochemical transformations performed in the laboratory in an inert atmosphere have little relevance to natural processes, which usually take place in an aerobic environment.

Finally, photochemical reactions performed in dilute solution in organic solvents also have questionable relevance to natural processes where chemicals may be in highly ordered cellular surroundings in a partially aqueous environment.

1.25. RADICAL CATION AND RADICAL ANION FORMATION

A molecule in its electronically excited state can undergo many different reactions. Aside from returning to the ground state, with or without luminescence, or from undergoing simple photochemical transformations such as isomerizations, electronically excited molecules can participate in electron transfer reactions, and form either radical cations or radical anions. The well-known *type-I photodynamic* reactions are typical examples of electron transfer reactions from an electronically excited molecule M to oxygen:

$$^{3}M + {}^{3}O_2 \rightarrow M^{+} + O_2^{-}$$

Not only is it possible to observe the formation of the initial intermediates by flash photolysis, but it is also possible to create the same charged radical species in the absence of light, by using pulsed radiolytic techniques that create high-energy electrons in an aqueous solution. The hydrated electrons and hydroxyl radicals formed can react either with the molecules of interest or with other reagents to create different reactive intermediates. For example, hydroxyl radicals can produce superoxide anion radicals by reacting with formate ions in aqueous solution.

1.26. A COMMENT ON THE USE OF CIRCULARLY POLARIZED LIGHT

The principles described above assume that photochemical reactions, whether direct or sensitized, utilize normal, non-polarized, light. Circularly polarized light, on the other hand, can induce asymmetric transformations. These may be asymmetric syntheses, partial photoresolution, or asymmetric photodestruction. One notable example involved the synthesis of hexahelicene (see Chapter 8), in which irradiation of the stilbene precursor with either left or right circularly polarized light produced optically active helicenes with optical rotations of the same magnitude, but with opposite signs. The asymmetric synthesis competed with selective photodestruction.[15]

1.27. PHOTOCHEMISTRY FROM HIGHER EXCITED STATES

Most of the photochemical transformations described in this book are assumed to take place from the lowest possible excited states. There is increasing interest in photochemical reactions from higher excited states; one example is provided in Chapter 3, where the formation of carbene-like intermediates is ascribed to Rydberg states. Photochemical transformations of aromatic molecules have also been analyzed in terms of Rydberg-like states.[16]

REFERENCES

1. T. Wolff, B. Fröschle and G. v. Bünau, *J. Photochem. Photobiol. A*, **58**, 331–338 (1991).
2. A. H. Zewail and R. B. Bernstein, *Chem. Eng. News*, **66**, 24–43 (Nov 7 1988).
3. Langhals, H., *GIT Fachz. Lab.*, **35**, 766–768, 770–771 (1991).
4. A. E. Friedman, J. C. Chambron, J. P. Sauvage, N. J. Turro, and J. K. Barton, *J. Am. Chem. Soc.*, **112**, 4960–4962 (1991).
5. H. Langhals, *Anal. Lett.*, **23**, 2243–2258 (1991).
6. C. G. Hatchard and C. A. Parker, *Proc. R. Soc. London A*, **235**, 518–536 (1956).
7. H. G. Heller, School of Chemistry and Applied Chemistry, University of Wales, Cardiff.
8. H.-J. Adick, R. Schmidt and H. Brauer, *J. Photochem. Photobiol. A*, **54**, 27–30 (1990).
9. R. M. Sayre and L. H. Klingman, *Photochem. Photobiol.*, **55**, 141–143 (1992).
10. J.-I. Kim and G. B. Schuster, *J. Am. Chem. Soc.*, **112**, 9635–9637 (1990), and references cited therein.
11. Y. Inoue, T. Yokoyama, N. Yamasaki and A. Tai, *J. Am. Chem. Soc.*, **111**, 6480–6482 (1989).
12. R. V. Bensasson, O. Chalvet, E. J. Lands and J. C. Ronfard-Haret, *Photochem. Photobiol.*, **39**, 287–291 (1984).
13. C. S. Foote, Quenching of singlet oxygen. In H. H. Wasserman and R. W. Murray (eds) *Singlet Oxygen*, Academic Press, pp. 139–171. New York (1979).
14. J. Kagan, I. Prakash, S. N. Dhawan and J. A. Jaworski, *Photobiochem. Photobiophys.*, **8**, 25–33 (1984).
15. A. Moradpour, J. F. Nicoud, G. Balavoine, H. Kagan and G. Tsoucaris, *J. Am. Chem. Soc.*, **93**, 2353–2354 (1971).
16. G. Koehler, *Radiat. Phys. Chem.*, **37**, 667–671 (1991).

2. Experimental techniques

2.1. WARNING

Before attempting any photochemical work in the laboratory, it is important to be aware of the hazards associated with such work. Photons can be absorbed by biomolecules, inducing photochemical reactions leading to biological damage which may be irreversible. *During the course of any photochemical work it is particularly important, therefore, to protect the skin and eyes from exposure to ultraviolet light.* Eyeglasses need not be stylish or have fancy colors, but they must be transparent to visible light, while effectively filtering ultraviolet light.

2.2. LIGHT SOURCES

Sunlight is the most universal source for photochemical reactions in living organisms. Nature utilizes sunlight to carry out biochemical transformations, for example those associated with photosynthesis. All the earlier researchers also utilized sunlight to carry out synthetic photochemistry. Unfortunately, the use of sunlight suffers from major limitations since it is not continuously available and its spectral distribution and intensity vary, not only from location to location, but also according to the time of day and the day of the year at a single location. Thus, it is almost impossible to design truly quantitative and reproducible experiments with sunlight, particularly when long exposures are required.

Artificial light sources are readily available. Low-pressure mercury vapor lamps have very few emission lines. The most important is at 253.7 nm. One of the minor emissions is at 185 nm, a region where the excitation of electrons in several kinds of σ-bonds and in isolated double bonds becomes possible. However, most vessels are not transparent to such short wavelengths. Low-pressure mercury vapor lamps, which are also often sold as germicidal lamps, are convenient and rather inexpensive, and are widely used. Furthermore, they are efficient and emit little heat. Figure 2.1 shows the typical emission of a commercial lamp (on a logarithmic scale).

Probably, the device most widely used for photochemistry with low-pressure mercury lamps is the Rayonet apparatus from the Southern New England Ultraviolet Co. in Hamden, Connecticut. The presence of phosphorescent materials coating the walls of low-pressure lamps can affect their emission characteristics. Such modified lamps, also available for the Rayonet apparatus, can provide emission centered around either 300 or 350 nm (the latter, also known as BLB lamps, are available in several different sizes from manufacturers such as General Electric, Sylvania, or Westinghouse). These lamps have much broader emission ranges than the 253.7 nm lamps, a definite advantage in cases where the absorption maximum of the compound to be irradiated does

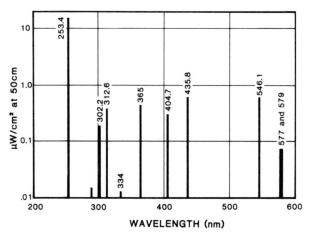

Fig. 2.1 Typical irradiance at 50 cm from a low-pressure mercury lamp (Oriel 6035 at 18 mA).

not coincide with one of the mercury emission lines. However, the use of these lamps (when not filtered) makes actinometric measurements and mechanistic studies more complicated, unless monochromatic radiation is isolated with filters. In contrast to the medium- and high-pressure mercury lamps discussed below, low-pressure mercury lamps in a Rayonet apparatus are much more energy-efficient and generate very little heat. The fan built in the reactor is usually sufficient to dissipate the excess heat.

As the pressure in a mercury lamp is increased, the emission spectrum changes. Most notably, the 253.7 nm emission disappears while the proportion of energy emitted at longer wavelengths in the ultraviolet, in the visible and in the infrared increases. These medium- and high-pressure mercury lamps are very inefficient light sources: they produce much heat, making water cooling indispensable. The emission spectrum of mercury lamps containing xenon is different, providing continuous emission from the ultraviolet to the infrared, a background onto which emission lines of mercury are superimposed. Both these lamps and xenon arcs, which provide only a broad emission spectrum, are useful for crude photochemistry, and for simulating sunlight. However, monochromatic light of any wavelength can be obtained from the broad-band emission by judicious use of monochromators or filters.

Mercury vapor arcs are traditionally designated according to their power requirements, and a specific power supply is required for each type. Medium-pressure lamps (400–500 W) are often used. It is important to note that little of this power is available in the form of photons at any given wavelength, since the electrical energy is converted into a broad range of electromagnetic radiation, from the ultraviolet to the infrared, most of which is useless for performing any specific photochemical transformation. Medium-pressure lamps are usually placed in a jacketed quartz well, through which cold water is running. The assembly is immersed in the solution to be irradiated, which is contained in a regular Pyrex vessel. Magnetic stirring and bubbling with inert gas are often applied. Emission spectra for medium- and high-pressure mercury, xenon, and xenon–mercury lamps are shown in Figs 2.2 and 2.3.

It is important to recognize that the output of any lamp varies with time. Although the extent of a preparative photochemical reaction is best determined analytically by sampling the reaction mixture as a function of irradiation time, one must be careful while doing mechanistic work not to rely on the previously measured actinometry of a lamp

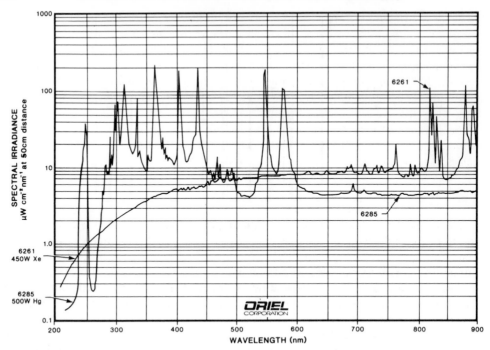

Fig. 2.2   Spectral irradiance of 450 W xenon and 500 W mercury lamps from Oriel.

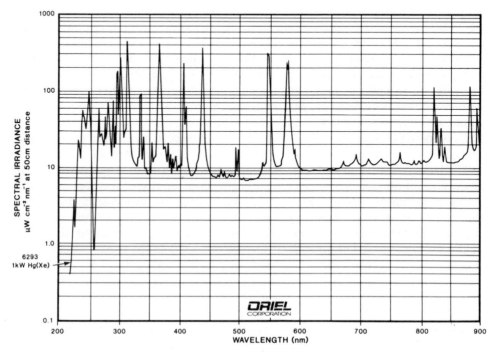

Fig. 2.3   Spectral irradiance of a 1000 W mercury–xenon lamp from Oriel.

when its optimum operating regime has not been reached. If a succession of short irradiations is required (as in kinetic experiments, for example), it is much preferable to leave the lamp on throughout the experiment, and to use a shutter or to withdraw the vessel rather than turning the lamp off and on for each sampling.

Several types of lamps are available for irradiating in the visible region; fluorescent lamps are the most common. It is important to note that high-intensity quartz halogen lamps may be quite attractive but they are not necessarily equivalent to fluorescent lamps because of their higher content in far ultraviolet radiation, which may induce changes in nucleic acid molecules and is therefore genotoxic and potentially carcinogenic.[1] The use of filters, such as Pyrex, should suffice to attenuate the problem.

2.3. VESSELS

Photochemical transformations may often be performed by exposing the molecules to the source of light without interposing any transparent surfaces. This is true of many reactions in nature, where the organisms are directly exposed to sunlight. To the human eye, most transparent and colorless materials look alike; for example, it is tempting to assume that greenhouse windows made of glass or polystyrene serve equally well. Actually, they do not; the outcome of many light-dependent reactions may depend on this small detail.

Each material has its own transmission characteristics, and only the wavelengths not absorbed by the material can induce photochemical reactions with the transmitted light. Take the human eye as an example. As discussed earlier, ultraviolet radiation is often the most useful in photochemistry, but the lens in the human eye is not transparent to it and *we are therefore unable to see objects illuminated with ultraviolet light*. People having undergone a cataract operation without receiving a lens implant, on the other hand, do see ultraviolet light, which now reaches their retina without being filtered off (potential military applications have apparently been considered).

The same is true of windows in the greenhouse. Polystyrene has a strong absorption in the ultraviolet (corresponding to the phenyl chromophore) and is unsuitable for many photochemical reactions. For a similar reason, the disposable Petri dishes almost universally used in microbiological and biochemical studies should be avoided for studies with ultraviolet radiation. Should these dishes be indispensable for other reasons, they should be exposed to ultraviolet light without a cover, or with a quartz or pyrex glass cover.

Any transparent material can be characterized by its transmission spectrum. Vessels made of quartz or fused silica, the most transparent, must be used for irradiation below 300 nm. The transmission of ordinary glass varies enormously with age and manufacturer. Pyrex is definitely transparent above 330 nm.

In summary, it is essential to select judiciously the vessels in which the irradiated samples will be placed. This choice is linked to the nature of the chromophores involved in the photoreaction which, in turn, dictates the choice of light sources.

2.4. STARTING AN EXPERIMENT

Any intended photochemical transformation must start with recording the absorption spectrum of the compound or compounds to be photolyzed. The selection of the appropriate light source and vessel depends critically on this knowledge. Most molecules have a rather broad absorption spectrum, often with several different absorption maxima. If at all possible, it is more efficient to select a light source with maximum emission matching one absorption maximum, and it is usually advantageous to select the maximum at the longest possible wavelength. Since the energy of photons decreases as the wavelength increases, the severity of the side reactions (such as further phototransformation of the initial photoproducts) may be decreased with longer wavelength light.

Strictly speaking, none of the light sources commonly available to the average scientist are monochromatic. Screening away the unwanted portion of the emission of a light source can be performed in three different ways.

2.4.1. Selecting the appropriate light source

Paying attention to whether a mercury arc operates under low, medium, or high pressure allows some control over the intensity of the 253.7 nm radiation available. Because of the high energy of this radiation (112.7 kcal/mol), it is usually beneficial to keep it out when it does not correspond to the λ_{max} of the starting material. Paying attention to whether a mercury arc also contains xenon is important, since the emission spectrum of the lamp is drastically affected (see Figs 2.2 and 2.3). Selectivity is more likely to be lost when using a mercury–xenon source, since the radiation is available over the whole spectrum. On the other hand, such sources are very convenient when a mixture of molecules having very different absorption spectra must be photolyzed.

As mentioned above, Rayonet low-pressure lamps are convenient and efficient for working in spectral areas centered at 253.7, 300, or 350 nm, or in the visible range.

2.4.2. Selecting a specific wavelength

2.4.2.1. Monochromators

If an appropriate laser is not available, monochromators are, in principle, perfect for isolating one narrow band centered at a selected wavelength. However, not all light sources are suitable for use with a monochromator, which requires a narrow light beam to be focused onto its entrance slit. Low-pressure mercury lamps used in a Rayonet apparatus, for example, are much too large to allow their focusing onto a monochromator. The smaller the light source, the smaller the lens system required for focusing; housing units equipped by a focusing lens are commercially available for medium- and high-pressure mercury and mercury–xenon lamps. One property unavoidably associated with these lamps is that the fraction of the total energy available at any one selected wavelength is usually fairly small. Since, in addition, a monochromator has slits which are quite small, the beam of light coming out of it is usually too weak for conducting preparative experiments. It may be perfect, however, for mechanistic work, particularly quantum yield measurements.

2.4.2.2. Filters

With a monochromator, the vessel must be selected to match the small size of the light beam. More flexibility is available when one is satisfied with radiation which is not perfectly monochromatic, as filters may then be used. Here again, the larger the vessel selected for the photochemistry to be performed, the more difficult it is to use elaborate filters.

Filters of different types are available. They are usually described according to their cut-off value. On one side of this wavelength threshold the light is transmitted, while on the other side it is not (the manufacturer may instead specify that at least 90% or some other percentage value of the light is filtered on one side of the designated threshold). The definition must be examined carefully, to ensure that it states specifically whether it is above or below the cut-off value that light is transmitted. A long-pass filter transmits light above the cut-off value, while a short-pass filter transmits below it.

Short-pass filters have already been encountered. For example, Pyrex glass has a cut-off value somewhere between 300 and 350 nm, and it filters radiation of shorter wavelength than this value. Long-pass filters are less common, perhaps, although water, which is often used as a filter against infrared radiation, is an example.

A large selection of filters may be purchased commercially, for example from Oriel (PO Box 872, Stratford, CT 06497), which supplies a catalog containing detailed specifications. To be effective, these filters must be placed in a well-defined light beam, ruling out their use in Rayonet equipment. It is wise not to exceed the experimental parameters of a filter, particularly the maximum temperature and the maximum power.

Although a combination of one low-pass filter with one high-pass filter allows the isolation of the wavelength range between the two cut-off values, this is often unsatisfactory for precisely selecting one single wavelength at which to conduct a photochemical transformation. However, when a broader range is acceptable, such an arrangement may be quite convenient.

One may find that selecting a wavelength with interference filters is often competitive with monochromators. Two kinds exist, depending upon whether it is a specified wavelength which is not allowed to be transmitted, or it is one wavelength which is transmitted, the rest of the spectrum being filtered off. The latter kind is more common, and filters may be purchased for the selection of almost any wavelength in the ultraviolet and in the visible regions. Typically, the filters have a bandwidth of 10 nm.

Many methods have been described for obtaining fairly monochromatic radiation from mercury arcs. Some are not very convenient, such as those given for isolating 253.7 or 265 nm radiation, since they require chlorine gas. Others are made more easily, for example by combining a glass filter and solutions of common chemicals such as nickel sulfate, potassium chromate, and potassium biphthalate in order to isolate 313 nm radiation. Since the bandwidth of such filters is fairly large, they are best used with light sources which provide emission lines well separated from their neighbors. Obviously, mercury lamps would be more useful than mercury–xenon lamps for such applications; see ref. 2 for two excellent sources of practical information.

Finally, there are occasional situations where the intensity of the selected radiation is too high and must be attenuated. Since the output of a single light source cannot be easily modulated in intensity, one must prevent part of the beam from reaching the sample. This can be done simply by using either metallic screens of different mesh sizes or commercial neutral density filters. Alternatively, the light beam may be split with mirrors, and only part of it used for an irradiation.

2.5. MEDIUM

The medium in which the molecules are irradiated can make a major contribution to the course of the photochemistry. The substrate is most often irradiated in solution, using a chemically inert solvent which is transparent at the wavelengths used. Note that a solvent which is usually inert in chemical reactions is not necessarily so in photo-chemical reactions. For example, diethyl ether and tetrahydrofuran are excellent sources of hydrogen radicals in photochemical reactions.

Solvents, either pure or as mixtures, which behave like glasses upon cooling are selected for reactions carried out at extremely low temperatures, in order to maintain a homogeneous environment. In a recent study on the spectroscopy of the aryl cation and aryl radical formed upon photolysis of a diazonium salt at 77 K, the following solvents were found to be satisfactory: 1-butanol, diethyl ether, ethyl acetate, ethanol, glycerol, ethylene glycol, lithium chloride–water, lithium chloride–water–acetone, phenyl acetate, 1- or 2-propanol, 35% sulfuric acid, toluene, water, and xylene.[3] Considerations of sub-strate solubility govern the choice of solvent systems in specific cases.

In situations where the molecular motions must be restricted, it may be advantageous to irradiate crystals of a compound. Alternatively, irradiation of a compound in a frozen solvent may achieve a similar result.

The environment of molecules may also be controlled during irradiation by placing them in normal or reverse micelles, thereby maintaining an organic microenvironment in an aqueous solvent, or an aqueous environment in an organic solvent. They may also be incorporated in a polymeric matrix, or they may be adsorbed onto a solid support, such as silica gel. A powerful and still developing technique involves the incorporation of molecules into zeolites which have known topology that makes them either capable or incapable of accepting a specific solvent or other molecules of interest.[4]

The techniques required for qualitative and quantitative work may differ consider-ably. In quantitative studies designed to describe the photophysical parameters of a molecule, a sample in dilute solution is irradiated and its spectral characteristics measured, but the following problems need to be considered:

1. It is difficult to control the homogeneity of the beam, particularly over large cross-sections. Consequently, different molecules in different positions of the solution may receive different intensities as well as wavelengths. Thorough mixing of the solution provides a partial answer to this problem.

2. Regardless of the size and shape of the vessel containing the sample, not all the molecules will receive the same amount of light. A gradient will necessarily exist in the direction of the beam (the molecules closer to the entrance window will receive more light than those further away). A gradient will also exist perpendicularly to the direction of the light beam if the beam is smaller than the window of the vessel, or if it is not homogeneous itself. Thus, if a small sample is withdrawn from the reaction vessel (for high-performance liquid chromatography (HPLC) analysis, for example) one must question how representative it is. A sample in the photolyzed zone will yield conclusions quite different from one taken away from it. Here again, the solution being photolyzed must be well mixed, consistent with the duration of the exposures.

3. Another problem requiring considerable attention is caused by the reaction prod-ucts formed. These have their own photophysical and photochemical properties, and if not spatially separated from their precursor molecules they may therefore undergo simultaneous light-induced reactions. These may even predominate when the absorb-ance of the product(s) becomes significantly greater than that of the starting material.

4. Finally, the presence of oxygen often needs to be controlled in photochemical reactions. Oxygen may participate in a number of pathways, leading initially either to deactivation of electronically excited states, with or without formation of singlet oxygen,

or with electron transfer to produce the superoxide ion. The formation of other products usually ensues.

2.6. SENSITIZED REACTIONS

There is an almost infinite choice of sensitizers and quenchers available for photochemical studies. The following (with approximate triplet energy in kcal/mol) have been frequently used: acetophenone (73.6), carbazole (70); triphenylamine (70); benzophenone (69); fluorene (68); triphenylene (67); thioxanthone (65.5); phenanthrene (62); piperylene or 1,3-pentadiene (58.5); biacetyl (56); fluorenone (53); pyrene (49). Remember that the selection of the concentration of sensitizers and quenchers is important: a sensitizer must absorb all (or most) of the incident light in order to prevent competition with unsensitized reactions, while a quencher should absorb none (or as little as possible) of the incident light. Thus, not only are the triplet energy data important, but the absorption spectra must be analyzed carefully in relation to the incident light available.

REFERENCES

1. S. De Flora, A. Camoirano, A. Izzotti and C. Bennicelli, *Carcinogenesis (London)*, **11**, 2171–2177 (1990).
2. J. G. Calvert and J. N. Pitts, Jr, *Photochemistry*. Wiley, New York (1967). S. L. Murov, *Handbook of Photochemistry*. Marcel Dekker, New York (1973).
3. H. B. Ambroz, T. J. Kemp and G. K. Przybytniak, *J. Photochem. Photobiol. A: Chem.*, **60**, 91–99 (1991).
4. V. Ramamurthy, D. R. Corbin, N. J. Turro, Z. Zhang and M. A. Garcia-Garibay, *J. Org. Chem.*, **56**, 255–261 (1991), and references cited therein.

3. Photochemistry of the carbon–carbon double bond

3.1. INTRODUCTION

The carbon–carbon double bond is one of the most familiar chromophores capable of undergoing photochemistry. Its electronic representations, in the ground state and its first electronically excited states, are reproduced in Fig. 3.1.

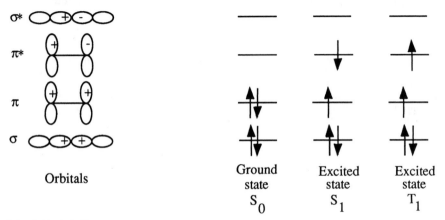

Fig. 3.1 Orbitals (left) and electronic configuration for a carbon–carbon double bond in the ground state, and in the singlet and triplet electronically excited states.

3.2. GEOMETRIC ISOMERIZATION

The well-known geometric rigidity associated with an alkene in its ground state, responsible for the existence of (E) (*trans*) and (Z) (*cis*) isomers, may be viewed as a consequence of the bonding overlap found in the p-orbitals which contain the two π-electrons. The geometry is different in the electronically excited states, where the highest occupied molecular orbital is antibonding and where these two electrons are in half-filled orbitals; rotation about the carbon–carbon bond held by its s-bonding orbital becomes possible. The net effect is to allow geometric isomerization, with (E)–(Z) (*cis*–*trans*) interconversion.

Isomeric (E)- and (Z)-alkenes usually have very similar absorption spectra, and the photoproduct formed, which is unavoidably exposed to the irradiating light, usually becomes electronically excited itself. In the absence of secondary reactions, a photoequilibrium is therefore obtained.

A (Z)-alkene usually can be isomerized by non-photochemical means. It is the more stable (E) isomer which predominates upon protonation of the double bond, because the carbocation intermediate may be deprotonated after rotation about the $C(sp_2)$–$C(sp_3)$ bond has taken place.

The photochemical isomerization process is especially valuable: not only is the less stable (Z) isomer formed, but it tends to predominate when photoequilibrium is reached. A reason for this predominance is that both the wavelength for maximum absorption (λ_{max}) and the probability for the molecules to be electronically excited (related to the coefficient ϵ in Beers' law) are different in the two diastereomers. The (E) isomer usually has a larger ϵ-value and absorbs at longer wavelength. Therefore, it may be electronically excited and converted into the (Z) isomer more readily than the (Z) isomer is converted into the (E) isomer.

The photochemical (E)–(Z) isomerization of alkenes can also be performed with a sensitizer. Regardless of whether an alkene is photoisomerized directly or by a sensitized process, its practical value is very high. Of course, a quantitative conversion of one diastereomer into the other cannot be expected, since a photoequilibrium is established; a separation step must necessarily follow. Crystallization or chromatographic separation (high-performance liquid chromatography (HPLC), column chromatography, or thin-layer chromatography (TLC)) are the methods of choice, as distillation or gas–liquid chromatography (GLC) often leads to thermal isomerization in the direction of the (E) isomer.

The photochemical (E)–(Z) isomerization reaction may create extremely reactive intermediates, as in the transformation of a cyclohexene into its *trans* isomer, which reacts immediately with alcohols to give ethers, or else dimerizes (try to build *trans*-cyclohexene with a set of molecular models!):[1]

Important examples of photochemical (E)–(Z) isomerization reactions producing reactive intermediates are also found in biology, for example in the biochemistry of vision, where the first step is the conversion from (Z) to (E) of the double bond at position 11 in the retinal portion of rhodopsin. This conversion of rhodopsin into bathorhodopsin occurs within 200 fs.[2]

In recent years, it has become increasingly obvious that the nature of the environment of photochemical isomerization reactions strongly affects their course. For example, the photoequilibrium of a stilbene favoring slightly the (Z) isomer may be switched to produce the (E) isomer exclusively, in the presence of a zeolite in which the lattice accepts the (E) but not the (Z) isomer. Thus the (E) isomer is withdrawn from the photolyzed mixture into the zeolite as irradiation proceeds. Because it is held in the lattice of the zeolite, where the (Z) isomer does not fit, the (E) isomer is no longer photoisomerized, even though electronic excitation undoubtedly occurs. Eventually, pure (E)-stilbene can be eluted from the zeolite with an appropriate solvent. It is reasonable to expect that a similar, but much more useful, process will be discovered where it is the (Z) isomer that will be retained in a suitable matrix.

Cis–trans $((E)$–$(Z))$ isomerization has important biological consequences. Your ability to see this material provides one example. Visual pigments contain a polyunsaturated moiety which undergoes *cis–trans* isomerization in the presence of light and is enzymatically regenerated. Another example is in the treatment of newborn

babies, particularly premature ones, who are often jaundiced because they have difficulty excreting bilirubin derived from the metabolism of heme. Bilirubin is toxic, and
high concentrations of these molecules can induce brain damage or even death. It is therefore important to treat neonatal hyperbilirubinemia. It turns out that exposing the infants
to fluorescent light dramatically reduces the concentration of bilirubin in their system.

There are four bilirubin diastereomers, $(4Z,15Z)$-bilirubin (**1**) being the native isomer.
It is interesting to note that there are many conformations available for bilirubin; one of
them, shown here as structure (**2**), is reminiscent of a porphyrin. The naturally occurring
bilirubin undergoes (E)–(Z) isomerization at each double bond upon irradiation and a
photoequilibrium is established. It is believed that a reason for the efficacy of the photochemical treatment in babies is the fact that at least one of the photoisomers is more
readily excreted; the photoequilibrium is therefore displaced to the point where the concentration of bilirubin drops below the critical threshold. In time, the infants develop
the metabolic capability to cope with the problem and no longer need irradiation.[3]

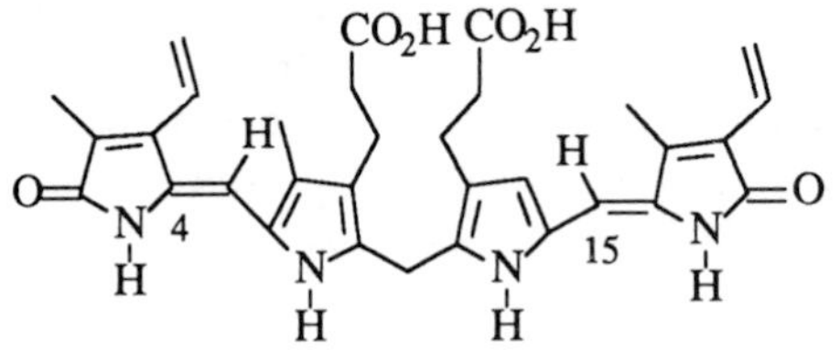

(1) 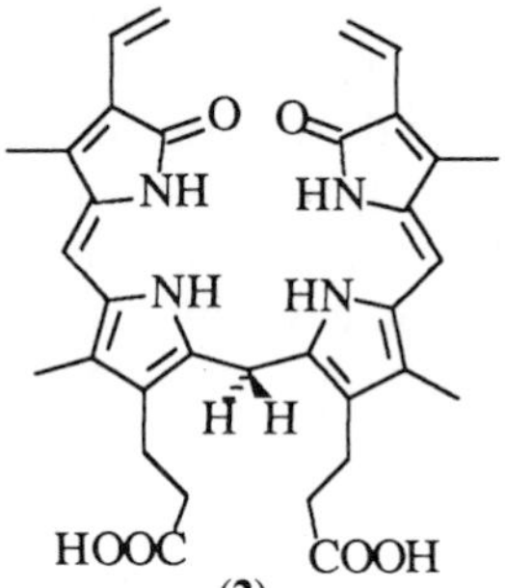   (2)

3.3. CYCLOADDITION REACTIONS

The products of cycloaddition reactions, such as the photochemical equivalent of the
thermal Diels–Alder reactions which are so well known in organic synthesis, result
from the interaction of the highest occupied molecular orbital (HOMO) of one reagent
with the lowest unoccupied molecular orbital (LUMO) of the other, as stated in the
Woodward–Hoffman rules and their alternative formulations. Cycloadditions are common in photochemistry; rings are formed by interaction of a carbon–carbon double
bond either with another carbon–carbon double bond or a triple bond, or with other
types of double bonds, such as carbon–oxygen, carbon–nitrogen, or nitrogen–nitrogen.
Many of these cycloaddition reactions are stereospecific, and the stereochemistry can be
readily predicted from simple electronic considerations. The first step involves drawing
the electronic configuration of each of the reactants, and identifying the HOMO and
LUMO orbitals which are to interact in forming new bonds.

In the example below, we assume that (Z)-2-butene is to undergo cycloaddition with
(E)-3-hexene; the orbital representations are shown in Fig. 3.2. Two observations must
be made: (1) the orbital representations would have looked identical had we decided to
excite the former rather than the latter alkene; (2) there is no longer a π-LUMO in the
excited alkene, as the higher orbital available now contains one electron.

The next step is to represent the actual HOMO + LUMO interaction between the

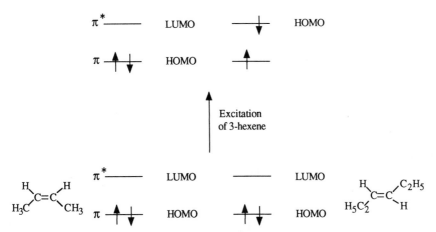

Fig. 3.2 Electronic representation of the π-orbitals in a mixture of (Z)-2-butene and (E)-3-hexene in the ground state, and of those present upon excitation of the (E)-3-hexene component.

two alkenes (the postulate that a ground state molecule interacts with an excited state molecule follows from the negligible probability of finding two electronically excited molecules capable of interacting within their very short life span). Of course, the same product with the same stereochemistry would have been obtained had we drawn the interaction of the HOMO of (Z)-2-butene with the LUMO of (E)-3-hexene. A review follows in Section 3.3.1.

Another important point concerns the cyclobutane formed in the cycloaddition. Two opposite σ-bonds have been created, their electrons having originated with the π-electrons of the interacting alkenes. Obviously, the cyclobutane could have been synthesized by generating the opposite set of bonds, pointing to the choice of two molecules of 2-pentane as starting materials (Fig. 3.3).

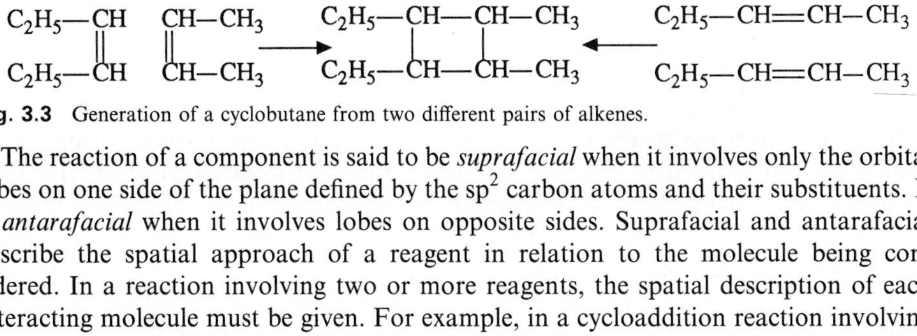

$$C_2H_5-CH \quad CH-CH_3 \qquad C_2H_5-CH-CH-CH_3 \qquad C_2H_5-CH{=}CH-CH_3$$
$$C_2H_5-CH \quad CH-CH_3 \longrightarrow C_2H_5-CH-CH-CH_3 \longleftarrow C_2H_5-CH{=}CH-CH_3$$

Fig. 3.3 Generation of a cyclobutane from two different pairs of alkenes.

The reaction of a component is said to be *suprafacial* when it involves only the orbital lobes on one side of the plane defined by the sp^2 carbon atoms and their substituents. It is *antarafacial* when it involves lobes on opposite sides. Suprafacial and antarafacial describe the spatial approach of a reagent in relation to the molecule being considered. In a reaction involving two or more reagents, the spatial description of each interacting molecule must be given. For example, in a cycloaddition reaction involving two components, the stereochemistry in the final product depends on the complete description of the mode of reaction in each of the components.

It is easy to see how the suprafacial–suprafacial interaction described above involves a transition state that is readily achieved. Figure 3.4 depicts the molecules with the interacting orbitals drawn in the plane of the paper. The cyclobutane produced is therefore itself in the plane of the paper. We could have produced the final product via an antarafacial–antarafacial interaction; however, the orbital overlap in the transition state would be greatly diminished.

For a given alkene undergoing either a thermal or a photochemical reaction, the HOMO and LUMO are different. In a thermal cycloaddition reaction, the suprafacial–suprafacial

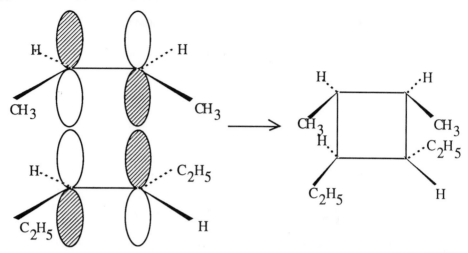

Fig. 3.4 Suprafacial–suprafacial interaction of the HOMO of excited (E)-3-hexene with the LUMO of ground state (Z)-2-butene, showing the bonding overlap of the p-orbitals involved in the transition state (the same product could have been obtained in the photochemical cyclization of (Z)-2-pentene and (E)-2-pentene).

approach leads to a bonding interaction in only *one* of the bonds to be formed. The thermal process therefore requires a suprafacial–antarafacial interaction of the two alkenes and is consequently expected to produce a diastereomeric cyclobutane. However, it is difficult to achieve a good orbital overlap in this transition state, and the concerted thermal cyclo-addition of two alkenes to give a cyclobutane is understandably a rare occurrence. Instead, cyclobutanes, being strained molecules, usually decompose thermally into alkenes more easily than they are produced from alkenes. Photochemistry is clearly a method of choice for the synthesis of cyclobutanes from alkenes.

The general principles of [2 + 2] cycloaddition reactions explained above apply to intra-molecular reactions as well as intermolecular reactions. As shown in Fig. 3.5 for intra-molecular cycloaddition reactions, two orientations for the interacting double bonds may lead to two different products. If the two double bonds are separated by only one carbon atom, then the smallest ring in the product is not four membered, but three membered, and is fused to the four-membered ring. An isomeric product, [1.1.1]bi-cyclopentane, is also allowed, corresponding to the 'crossed' mode of cycloaddition.

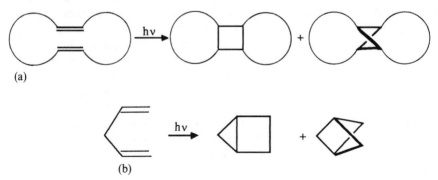

Fig. 3.5 Intramolecular [2 + 2] cyclizations. (a) General scheme; (b) specific case of the formation of [2.1.0]bicyclopentane and [1.1.1]bicyclopentane from 1,4-pentadiene.

The cycloaddition reactions shown depend only on having two double bonds suitably located. They may be connected through atoms other than carbon, as shown below:[4]

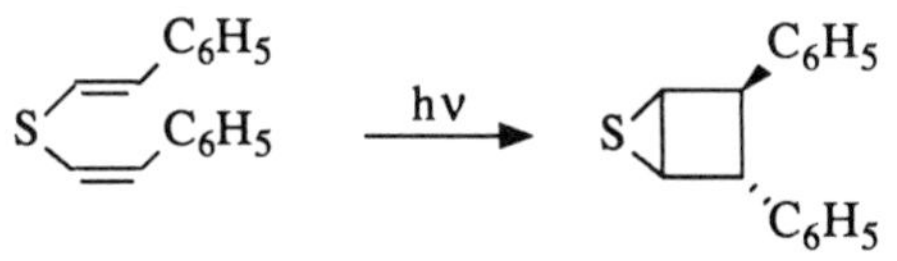

At the limit, when the interacting alkenes are not separated by intervening atoms, the formal cycloaddition reaction is usually described as an *electrocyclic reaction*. For example, two products are formed by irradiation of 1,3-butadiene (**3**), one monocyclic (**4**) and the other bicyclic (**5**).

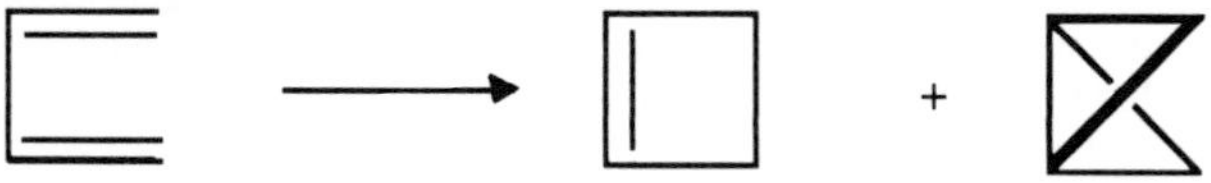

3.3.1. Dealing with orbitals: a review

At this point, it may be useful to digress, and to remind the reader that there are different methods for determining when a reaction involving unsaturated components is *concerted* or *symmetry allowed*, implying that it is expected to have a rather low activation energy. The original approach, immortalized in the honor of its originators as the Woodward–Hoffman rules, considers symmetry elements in the orbitals and in the reaction itself; other approaches are more convenient. Two important approaches are now considered:

1. When the molecules involved are reasonably small, the various molecular orbitals fairly easily remembered and it is more convenient to consider the HOMO and LUMO of the interacting components. The orbital representations for molecules having a small number of interacting p-orbitals can be memorized. Since one needs to know only the sign, rather than the magnitude of the contribution, of each p-orbital in each molecular orbital, the table shown in Fig. 3.6 provides all the practical information needed.

The use of the information presented in Fig. 3.6 is straightforward. For example, there are six interacting p-orbitals in the conjugated system of a 1,3,5-hexatriene. The HOMO in the ground state is ψ_3, but it becomes ψ_4 after electronic excitation. The orbital representation of ψ_4 is given in Fig. 3.6 by the sequence (+ − − + +−), which translates into the molecular orbital shown in Fig. 3.7.

2. A different, and generally more powerful approach, is based on the concept that *allowed reactions involve 'aromatic' transition states*. Just as Hückel's rule shows that a conjugated, cyclic, molecule containing $4n + 2$ electrons is aromatic, a *transition state with parallel p-orbitals involving either 0 or an even number of sign inversions and containing $4n + 2$ electrons will be considered aromatic in a **thermal** reaction*. The thermal reaction involving such a transition state will be favored. For a **photochemical** reaction, *however, the transition state involving $4n$ electrons will be favored.*

Another type of interaction between the p-orbitals can be drawn, in which either one or an odd number of sign inversions is introduced in creating the transition state. This is

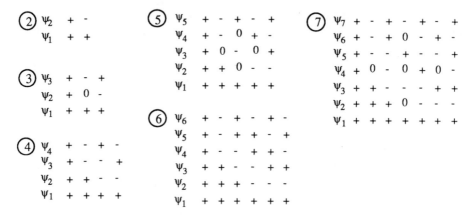

Fig. 3.6 Schematic representation for the contribution of each atomic orbital in each molecular orbital for linear polyenes having the indicated number of carbon atoms. The signs are placed where each carbon atom is contributing a p-orbital to the polyene. In each group of molecular orbitals, the energy increases from bottom to top.

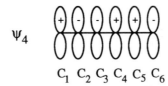

Fig. 3.7 The molecular orbital representation of the HOMO of a 1,3,5-hexatriene after electronic excitation.

a *Möbius transition state*, which is aromatic if $4n$ electrons participate in a thermal reaction, or if $4n + 2$ electrons participate in a photochemical reaction. A pictorial description of Hückel and Möbius transition states in the case of a $[4 + 2]$ cycloaddition is given in Fig. 3.8. Here, we are just looking at the set of atomic orbitals involved, not at specific molecular orbitals.

In summary, the stereochemistry of concerted reactions is predicted by the following sequence of steps:

a. draw the components in the reactions, with each $+$ lobe of a p-orbital next to the $+$ lobe of the adjacent p-orbitals;
b. count the number of electrons involved in the transition state;
c. determine whether the reaction under consideration is thermal or photochemical;
d. conclude whether the transition state must be Hückel ($4n + 2$ electrons, thermal; $4n$ electrons, photochemical) or Möbius ($4n$ electrons, thermal; $4n + 2$, photochemical);
e. formally move or twist the bonds in the reactants, in order to achieve the desired transition state;
f. deduce the stereochemistry of the product(s).

The conversion of the elements of transition states, as pictured in Fig. 3.8, into products in which two new σ-bonds have been created comprises steps in which two specific p-orbitals, one each from the alkene or the diene, are rotated by 90° in order to become aligned. A new σ-bond may have been created by rotating the two p-orbitals in the same direction (clockwise or counterclockwise): this process is described as *conrotatory*. Alternatively, when the rotation of both p-orbitals is performed in the opposite manner (one clockwise and the other counterclockwise), the process is *disrotatory*. The conversion

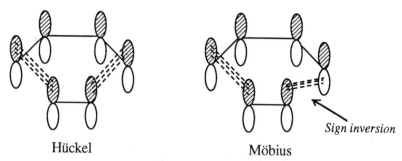

Sign inversion

Hückel   Möbius

Fig. 3.8 Hückel and Möbius transition states for a [4 + 2] cycloaddition reaction. Only the p-orbitals at the carbon atoms involved have been shown. It is important to draw the substituents in order to visualize the stereochemistry of the final products.

of the Möbius transition state shown in Fig. 3.8 into a cyclohexene would include one disrotatory and one conrotatory process, as shown in Fig. 3.9. Two points must be noted:

a. since each conrotatory and disrotatory bond formation can take place in two different manners, clockwise or counterclockwise, there are four different modes for closing the ring, as shown;
b. the terminology applies to bond formation as well as to bond-breaking processes, for example the concerted photochemical isomerization of a cyclobutene into a 1,3-butadiene to be described in Section 3.3 will be called a disrotatory process.

The steps involved in analyzing the stereochemistry of a photochemical Diels–Alder cycloaddition reaction are detailed in Fig. 3.10. The reaction involves a total of six electrons in the transition state $4\pi + 2\pi$, and since this is a photochemical reaction the transition state must be of a Möbius type. After the p-orbitals in the two reactants are drawn, the partners are brought into approximate contact as if the reaction were taking

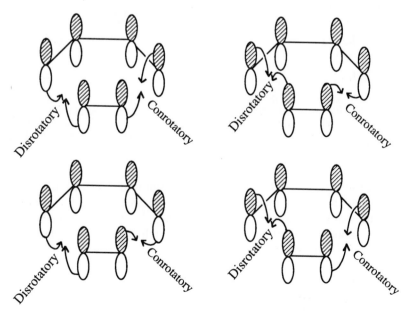

Fig. 3.9 Alternative Möbius transition states for the [4 + 2] cycloaddition reaction depicted in Fig. 3.8.

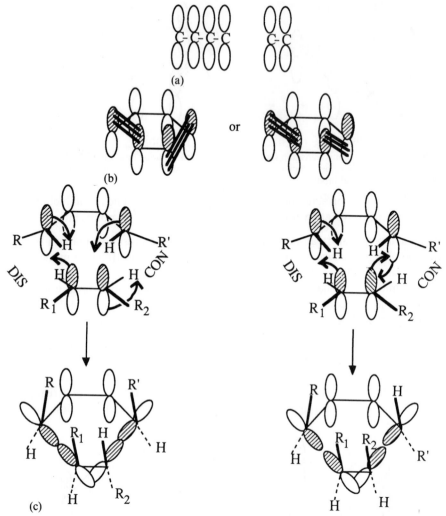

Fig. 3.10 The stereochemistry of the photochemical Diels–Alder reaction, visualized via the required Möbius aromatic transition states. (a) The reagents and their p-orbitals involved in the cyclization. (b) The Möbius transition state, showing the bonds to be created. (c) Bond formation, assuming (E, E)-butadiene and (Z)-alkene. Now the substituents must be drawn (two out of the four possible combinations of conrotatory and disrotatory motions are illustrated).

place, and the p-orbital at one terminus of one reagent is aligned with that of the other reagent (matching the signs of the lobes). Then, one produces the Möbius transition state by making sure that the interacting p-orbitals at the other termini of the bond-forming process are on opposite sides of the ideal plane in which the atoms are drawn. From the drawing obtained, one deduces how the new σ-bonds are created, and the remaining π-bond is therefore identified.

Note that the orbitals shown in Fig. 3.10 are just the set of *atomic* p-orbitals; they *do not* represent specific molecular orbitals. Only the p-orbitals at the reacting centers are actually involved in the analysis. The other signs could have been omitted from the drawings. Regardless of the direction chosen for each conrotatory and disrotatory motion, one observes that R and R' are *cis* in the final product when R_1 and R_2 are *trans*, or vice versa. A conclusion identical to that shown in Figs 3.8–3.10 would have been obtained by properly interacting the HOMO of one reactant with the LUMO of the other, as the reader is encouraged to verify.

Because the cycloaddition reaction was dissected above in terms of separate events occurring at each bond to be formed, Fig. 3.10 showed whether the motion of the p-orbitals involved was disrotatory or conrotatory. However, if you look at the overall process, considering the reacting molecules, you will find it equivalent to a suprafacial–antarafacial interaction. As anticipated, it is just the opposite of the thermal Diels–Alder reaction, which is usually suprafacial–suprafacial (although it is also allowed with antarafacial–antarafacial interactions).

Note that the outcome of cycloaddition reactions (stereochemistry and yields) may be quite sensitive to the experimental conditions. For example, the structure and stereo-chemistry of the major cyclobutane products in the dimerization of cinnamic acid ($C_6H_5CH=CHCOOH$) differ according to whether the crystalline form, an organic solution, or a solution in water (liquid or frozen) is irradiated. This is also the case for stilbenes (ArCH=CHAr'), which dimerize much more rapidly in hydroxylic solvents (water or methanol) than in other solvents, where cyclization to phenanthrenes pre-dominates. The formation of aggregates in hydroxylic media is a reasonable expla-nation for the rate increase.[5]

Cycloaddition reactions depend on the two carbon–carbon double bonds being close enough and in the proper orientation for one to react with the short-lived excited state of the other. The presence of the C=C moieties within the same molecule facilitates their reactions, and it is not surprising that the first example of a [2 + 2] cycloaddition reaction described in the literature was an intramolecular process.[6] Carvone (6), a com-mon terpene found in many plants, was isomerized upon exposure to sunlight into the complex structure shown (7). The conformation of the carvone molecule suitable for cycloaddition is shown on the right (6'). An identical cycloaddition reaction was observed in the photochemical isomerization of the acyclic terpene citral (8) to form photocitral B (9) via the conformation (8').[7]

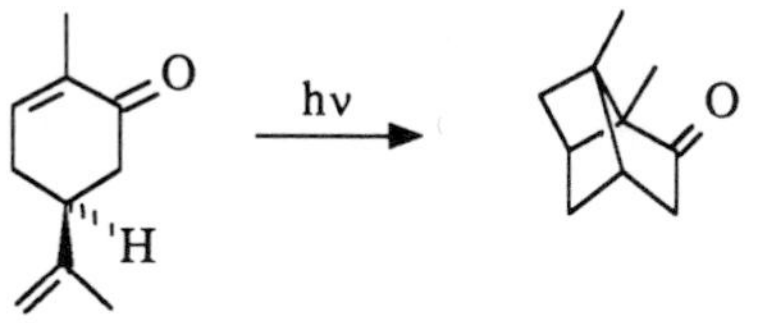

(6)   (7) 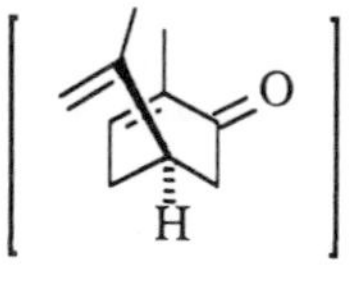   (6')

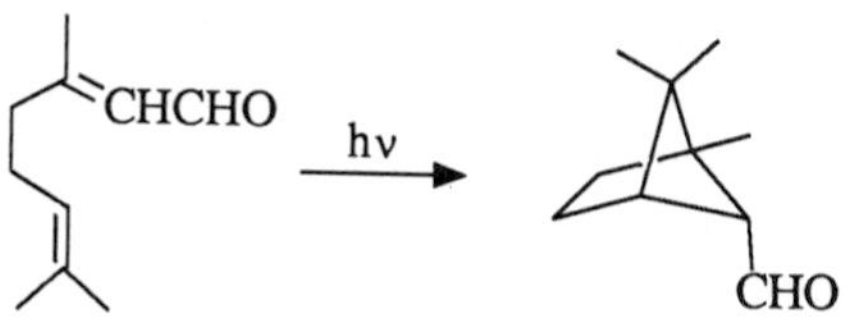

(8)   (9) 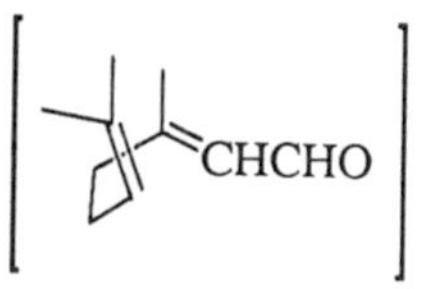   (8')

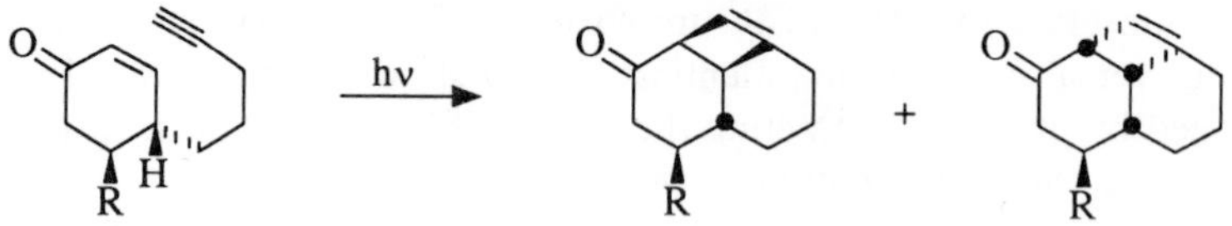

Fig. 3.11 Example of a [2 + 2] cycloaddition with an alkyne.

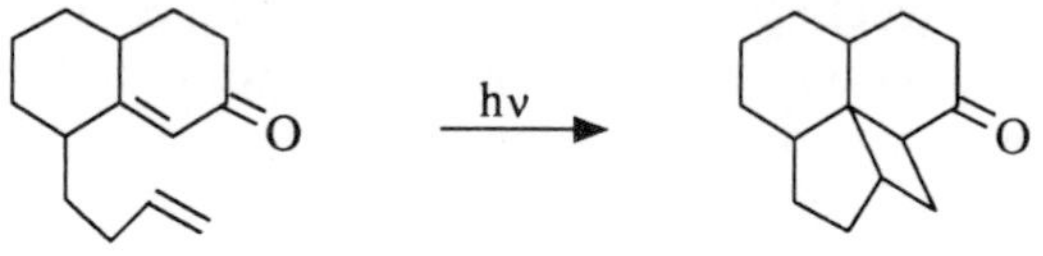

Similar cycloaddition reactions may involve acetylenic bonds. The example given in Fig. 3.11 has been chosen to demonstrate that when a four-membered ring is formed, its stereochemistry must be analyzed carefully in relation to the rest of the molecule. The photolysis produces a mixture, because there are two conformationally distinct transition states, positioning the chain carrying the triple bond either above or below the double bond.[8]

It is important to note how a fairly simple starting material submitted to just one synthetic operation, exposure to ultraviolet light, is converted into complex structures possessing several chiral centers. Another striking demonstration is where compound (**10**), a fenestrane (a structure in which four adjacent rings have one common carbon atom), is created from the bicyclic precursor (**11**).[9]

A simple synthesis of [2.2]metacyclophanes involving two successive cycloaddition steps is shown here. The last step is the conversion of the [2.2]metacyclophane (**12**) into the [4.4]metacyclophane (**13**) by reduction with sodium in liquid ammonia.[10] The numbers 2 and 4 in the brackets above refer to the number of carbon atoms in the smallest links between the *meta* positions in the cyclophanes.

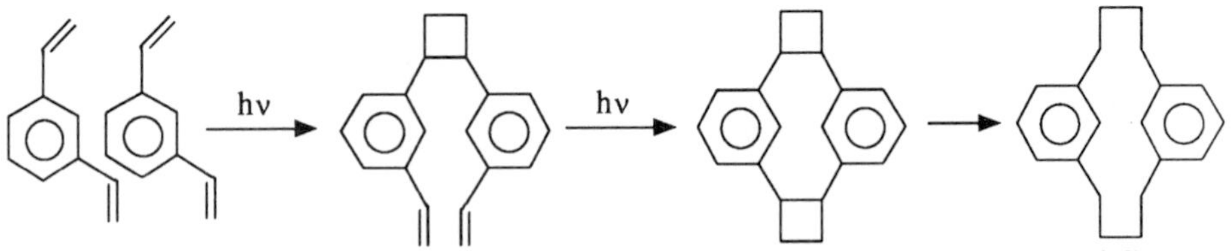

The mere presence of two double bonds within reasonable distance is not sufficient to ensure photoaddition, due to restriction imposed by the σ-frame on the interaction of the π-units.[11] For example, cubane (**15**) is not produced from its obvious precursor (**14**), even though the more substituted analog (**16**) does undergo the desired cyclization.[12]

It must be noted that cycloaddition reactions can be photosensitized, and they can also be catalyzed by copper(I) derivatives.[13] The reader is directed to a recent survey of intramolecular [2 + 2] cycloadditions reactions in organic synthesis for additional examples.[14]

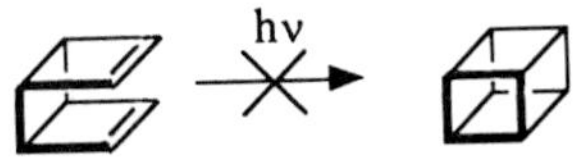

(14)   (15)

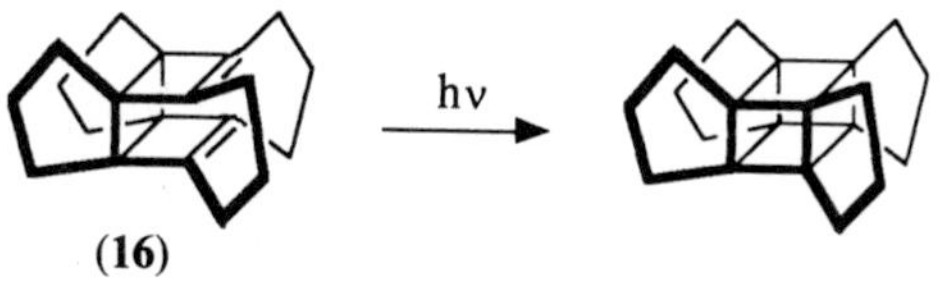

(16)

3.4. CYCLOREVERSION REACTIONS

For the purpose of synthetic transformations, the ability to create rings from alkene moieties is usually much more valuable than the reverse process. The Woodward–Hoffman and related rules for concerted reactions apply both to the forward and reverse transformations. Of course, the fundamental requirement that light be absorbed by a molecule limits the photoreactivity of cyclobutanes, since they do not absorb at wavelengths available from the usual lamps (253 nm or above) unless they are substituted with proper chromophores. The opening of cyclobutanes is often obtained in sensitized irradiations or through electron transfer reactions (Chapter 7).

The photochemical conversion of cyclobutenes into 1,3-butadienes is often non-stereospecific, but not always, as in the following examples which show good stereoselectivity toward the products of disrotatory ring opening predicted by orbital symmetry control:[15]

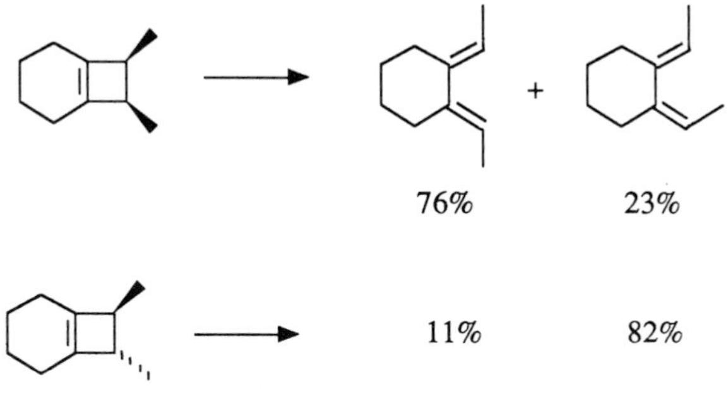

76%   23%

11%   82%

3.5. ISOMERIZATION REACTIONS: DOUBLE BOND SHIFTS

Standard methods also suffice to analyze orbital interactions involving one π-system and one σ-system. To illustrate yet another method of handling orbitals, we consider that the σ-bond to be broken contributes one electron to each fragment (homolytic cleavage). Because the reaction is photochemical, the HOMO involved is ψ_3 in the three-carbon system, as represented in Fig. 3.6. In Fig. 3.12, the p-orbitals involved in the rearrangement are shown first. We then draw the HOMO of the three-electron,

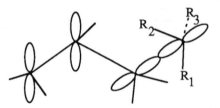

Relevant p-orbitals in the starting material

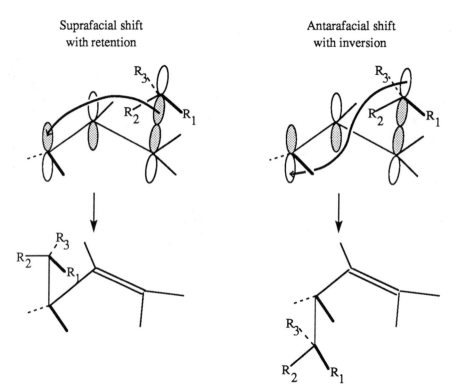

Fig. 3.12 The photochemical 1,3-migration of a double bond associated with the sigmatropic shift of an allylic substituent.

three-carbon system (ψ_3) interacting with the p-orbital of the alkyl substituent. It is clear that a bonding interaction will be maintained when the alkyl group migrates on the top side (suprafacial migration) while retaining its configuration, or when it migrates to the bottom side (antarafacial migration) with inversion.

Once again, identical results would have been obtained by using procedures based on the aromaticity of the transition state, or on the HOMO + LUMO interaction.

3.6. ISOMERIZATION REACTIONS: CYCLOPROPANE FORMATION

The isomerization of an alkene into a cyclopropane, which requires the migration of a substituent, could involve two π-electrons and two σ-electrons in a Möbius transition

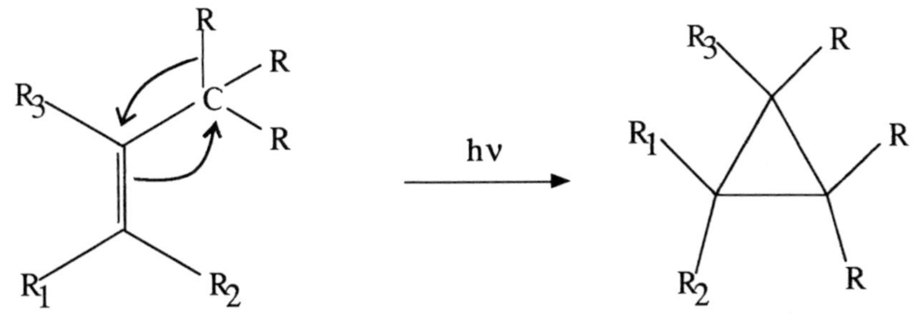

Fig. 3.13 Scheme illustrating the photoisomerization of an alkene into a cyclopropane.

state (Fig. 3.13). The resulting product absorbs light at significantly shorter wavelengths than the starting material and is therefore frequently protected from further direct light-induced reactions.

When any of the substituents are capable of absorbing light (as in the case of aromatic substituents), or when a sensitizer is used, a cyclopropane can also open. This may lead to the reverse of the above reaction. However, if the ring closes again, the stereochemistry of the cyclopropane can be changed. For example, cis- and trans-1,2-diphenylcyclopropane can be interconverted photochemically.

3.7. ISOMERIZATION OF ALKENES INTO CARBENES

Alkenes have been found to isomerize photochemically into isomeric alkenes and cyclopropanes. This process has been interpreted in terms of the rearrangement of an excited state in which one of the π-electrons has been promoted to a higher molecular orbital having approximately the same shape and size as the 3s atomic orbital of carbon. This excited state, which is called a $\pi,R(3s)$ Rydberg excited state, has a radical cation character, and can undergo rearrangement to a carbene. The subsequent reactions of this carbene produce an alkene (isomeric to the starting material) or a cyclopropane, each by insertion into a carbon–hydrogen bond, or an ether if the solvent is hydroxylic:

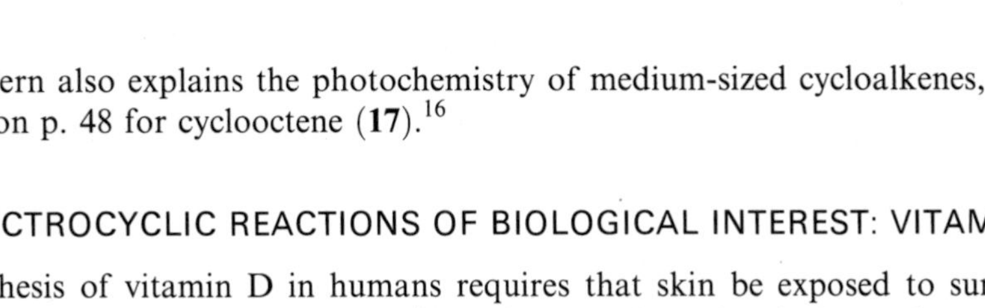

This pattern also explains the photochemistry of medium-sized cycloalkenes, as summarized on p. 48 for cyclooctene (**17**).[16]

3.8. ELECTROCYCLIC REACTIONS OF BIOLOGICAL INTEREST: VITAMIN D

The synthesis of vitamin D in humans requires that skin be exposed to sunlight. A vitamin D_3 precursor present in the skin undergoes photochemical transformations induced by the ultraviolet component (290–315 nm) of sunlight, and one of the photo-

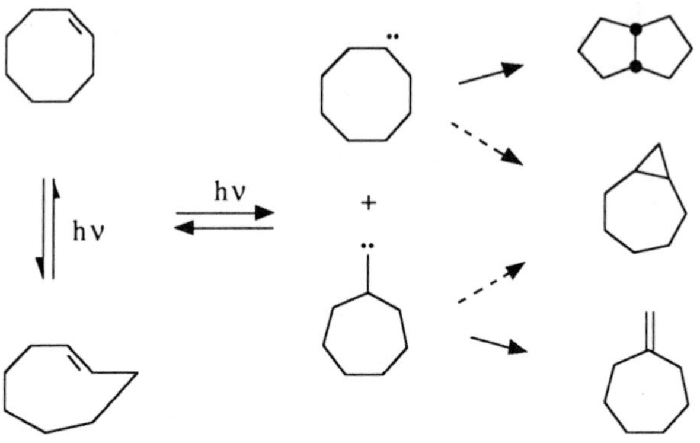

(17)

products is then carried into the bloodstream with the help of a binding protein. This photochemistry, which has been studied extensively, is summarized in Fig. 3.14.

It is in the epidermis that previtamin D_3 is formed, because very little ultraviolet radiation penetrates beyond it. For simplicity, an enzymatic step in the formation of the real vitamin has been omitted in Fig. 3.14. It involves the hydroxylation of the side-chain, which yields the biologically active product.

The initial ring opening of 7-dehydrocholesterol is allowed in a disrotatory manner. Its reverse action, also photochemical, can also produce the isomeric lumisterol if the electrocyclic reaction takes place in the opposite direction. The production of tachysterol is simply the result of a photochemical (E)–(Z) isomerization. Finally, the step converting

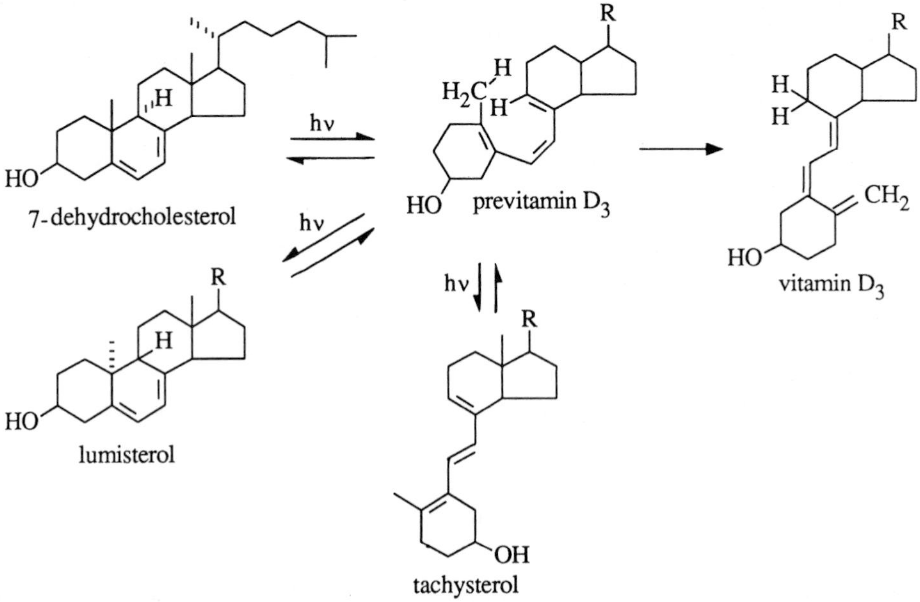

Fig. 3.14   The photochemical conversion of previtamin D_3 into vitamin D_3, lumisterol and tachysterol.

previtamin D_3 into vitamin D_3 takes place thermally at skin temperature. It is a 1,7-sigmatropic shift, which must occur antarafacially. This step could have taken place (suprafacially) photochemically, but does not *in vivo*.

Careful studies on the photochemistry of previtamin D_3 revealed that the quantum yields of formation of lumisterol, 7-dehydrocholesterol and tachysterol in ether were very dependent on the wavelength used for irradiation. This was explained by invoking the participation of different excited states.[17] In ethanol, the optimum wavelength for production of previtamin D_3 is 296 nm, but it is preferable not to reach the steady state if one wants to minimize the formation of undesirable secondary products. The cost of photons generated by the laser light sources presently available is still too high to make the photoisomerization at 296 nm commercially very attractive.[18]

The opening of the B-ring of the steroid described above requires electronic excitation, but not necessarily the presence of light (in some species vitamin D can be formed in complete darkness).[19] Treatment of isobutyraldehyde with a peroxidase in the presence of oxygen produces a transient dioxetanol, which is cleaved to produce electronically excited acetone. This in turn acts as sensitizer and leads to the opening of the B-ring.[20]

3.9. ANOTHER REACTION OF DIENES: THE DI-π-METHANE REARRANGEMENT

A predictable reaction can occur in dienes which have the two double bonds separated by one atom, most commonly a carbon atom. The process may be described as an allowed $2\pi + 2\sigma$ interaction, as shown in Fig. 3.15. The cyclopropane formed by this process is predicted by the pattern described in Section 3.6, except that a vinyl group consistently migrates in preference to any hydrogen or alkyl substituents.

One simple example, which illustrates the dramatic structural changes resulting from this simple $2\pi + 2\sigma$ interaction is shown in Fig. 3.16. Contrary to the initial temptation to think that the isomerization involves two 1,2-shifts of the phenyl groups, a di-π-methane rearrangement best explains this example of a common ring transformation, as shown with the numbers relating the atoms in the final product to those of the

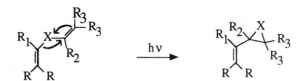

Fig. 3.15 The di-π-methane rearrangement, illustrated with the assumption that it is the alkene containing the R and R_1 substituents which migrates from X to the adjacent carbon atom. An isomeric product would be obtained if the other alkene migrated instead.

Fig. 3.16 Photochemical di-π-methane isomerization of a cyclohexadienone. Note that the bond between atoms 3 and 4 was broken, and a new bond between atoms 4 and 6 has been formed.

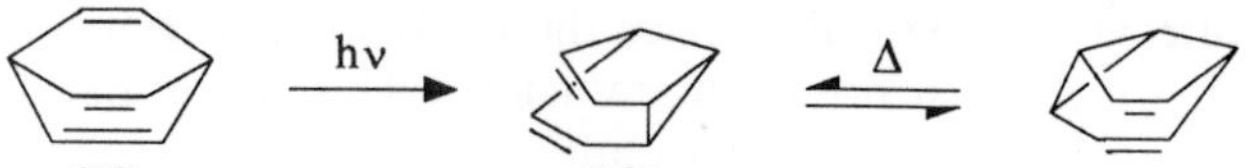

(18)   (19)

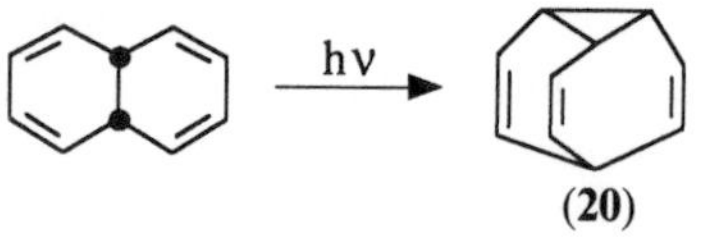

(20)

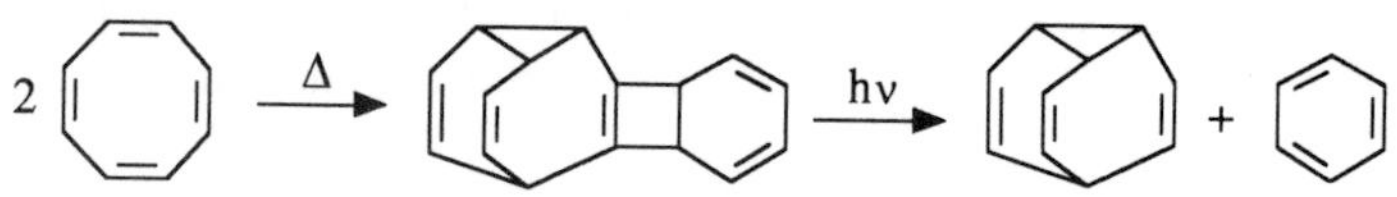

(21)   (22)

starting material. The conversion of barellene (18) into semibullvalene (19) is another illustration of the di-π-methane rearrangement. Here, the photoproduct is capable of undergoing a thermal rearrangement which produces an identical molecule (degenerate rearrangement). Bullvalene (20) itself is an even more fascinating molecule, because its higher degree of symmetry produces more extensive degenerate rearrangements. A pretty synthesis of bullvalene involves a di-π-methane rearrangement of a dihydronaphthalene.[21] Another synthesis uses a photochemical step, the formal reverse of a [2 + 2] cyclo-addition reaction, on the product (22) obtained from the thermal dimerization of cyclooctatetraene (21).[22]

3.10. COMPETITION BETWEEN SKELETAL REARRANGEMENT AND GROUP MIGRATION

In solution, the migration of a methyl group in methylindenes was shown to involve skeletal rearrangement. Instead of a di-π-methane rearrangement, stepwise cyclization and electrocyclic reactions have been proposed to account for the fact that the methyl group remained attached to the same carbon atom throughout (Fig. 3.17). In the gas phase, however, another mechanism did compete, producing indene itself as well as products of actual methyl group migration.[23]

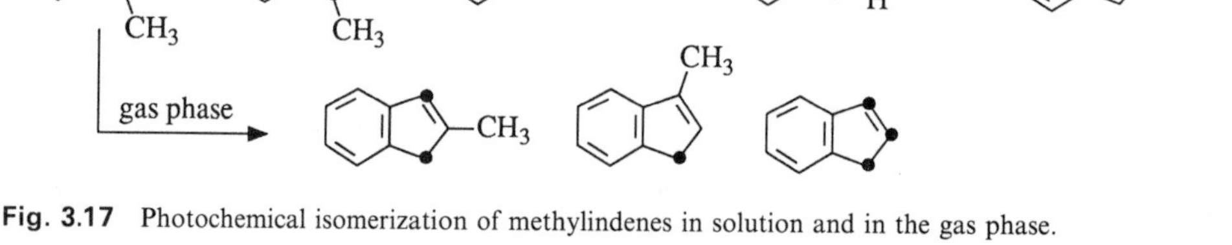

Fig. 3.17 Photochemical isomerization of methylindenes in solution and in the gas phase.

3.11. PHOTOCHEMISTRY OF AROMATIC COMPOUNDS

If a benzene ring is viewed as a collection of three double bonds, it is not surprising to find that electronic excitation may cause two of them to interact. If the interaction is 'straight', Dewar benzene (**23**) is formed; if 'crossed', the isomeric benzvalene (**25**) is produced. Further $2\pi + 2\pi$ cyclization of Dewar benzene gives Ladenberg benzene, or prismane (**24**). A fourth product, fulvene (**26**), is also formed, although its mechanistic origin is less obvious.

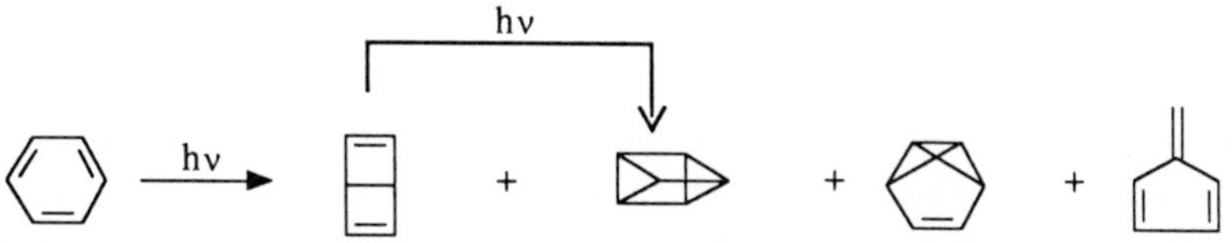

(23)   (24)   (25)   (26)

One explanation for the formation of fulvene starts with benzvalene, reverting to a carbene, which might have been a precursor in a planned chemical synthesis of benzvalene. Stabilization of the carbene by insertion into the adjacent carbon–hydrogen bond completes the conversion (Fig. 3.18).

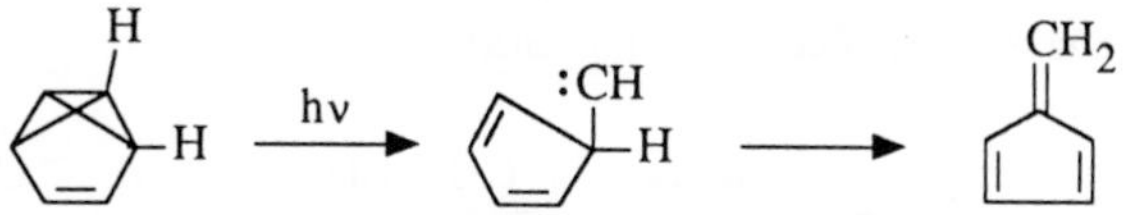

Fig. 3.18 Scheme accounting for the photochemical synthesis of fulvene.

The photochemical interconversion of benzene and benzvalene provides the explanation for a truly amazing isomerization of aromatic compounds illustrated in Fig. 3.19. The case suggests that a substituent has migrated from a *meta* position. However, labeling of the carbon atoms in the molecule showed that each methyl group in the product was still held by its original aromatic carbon atom in mesitylene.

The simplest explanation rests with the photochemical isomerization of the phenyl ring into benzvalene, which involves the formation of two new σ-bonds. Reversing the process by breaking the *other* two bonds in the bicyclobutane formed leads to a rearranged benzene ring (the isomerization is known to occur from the singlet excited state).

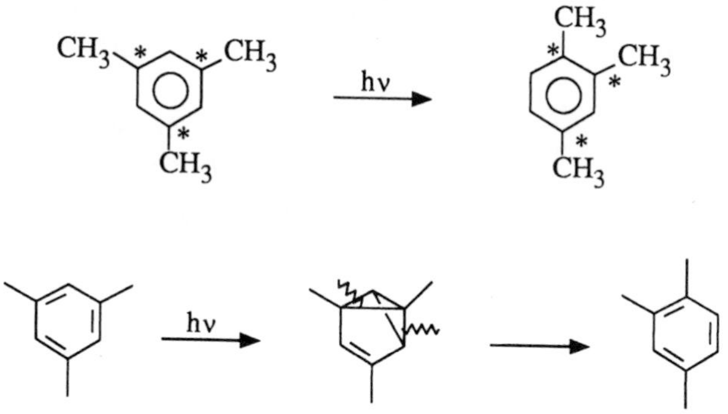

Fig. 3.19 Photochemical isomerization of mesitylene.

The efficiency of the isomerization depends on the nature of the substituents. In the following example the yields were very high:[24]

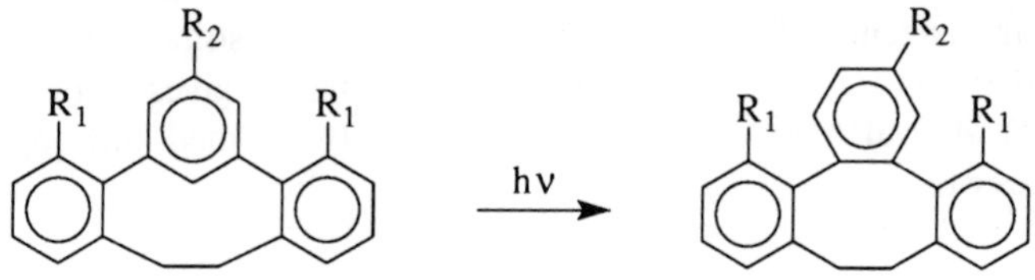

Alkenes may undergo photochemical cycloaddition with benzene rings, yielding three different products, in which either *ortho*, *para*, or *meta* carbon atoms are used (Fig. 3.20). While the first two products may occur from $[2+2]$ and $[4+2]$ cycloadditions respectively, the third may originate with an isomerization product from benzene, followed by a cyclization involving 2s and 2π electrons (Fig. 3.21), which breaks the bond designated by 'a'. The process may be equally well described as a $[4+2]$ cycloaddition.

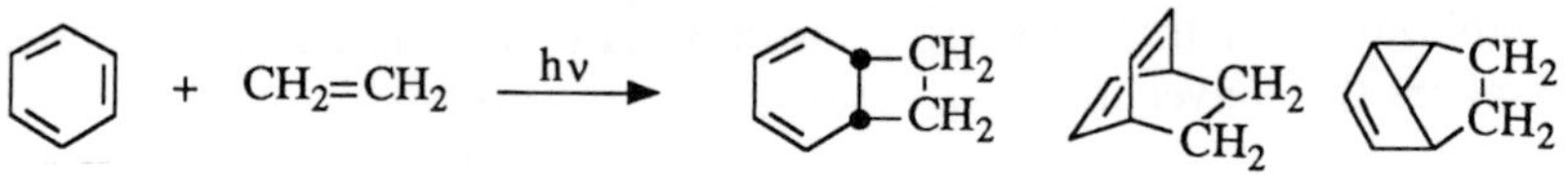

Fig. 3.20 Cycloaddition reaction of ethylene with benzene.

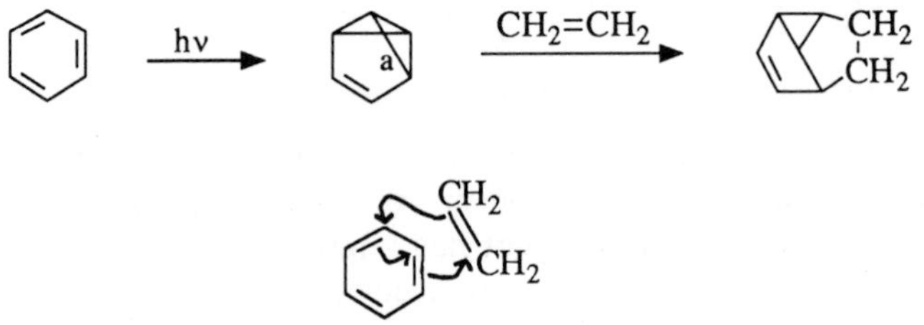

Fig. 3.21 Cycloaddition reaction of ethylene with benzvalene.

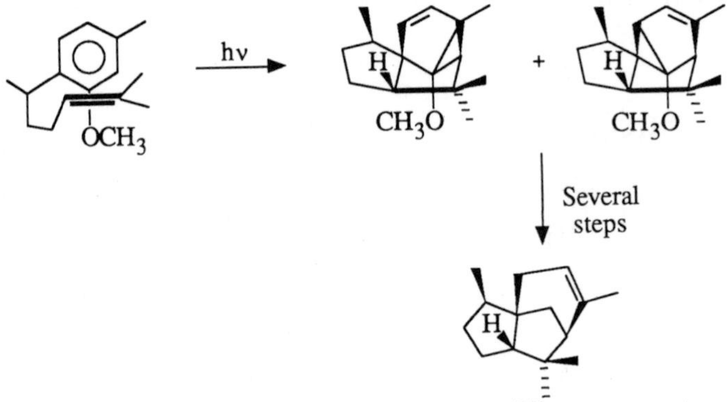

(27)

Because the *meta* cycloaddition process may fix up to six stereogenic centers in a single step, it is an attractive method for creating complex polycyclic molecules, as was first proved in a synthesis of the sesquiterpene cedrene (**27**).

Converting the photoproducts into cedrene requires opening the cyclopropane ring, followed by minor adjustments.[25] Polycyclic systems of even greater complexity may be constructed by incorporating the alkene portion into a cyclic or polycyclic moiety.[26]

3.12. DIMERIZATION OF AROMATIC COMPOUNDS

The photodimerization of anthracene (**28**), which is thermally reversible, is a very old reaction which has been considered as a possible scheme for utilizing solar energy. This reaction is surprisingly clean in the absence of oxygen, occurring even in preference to cyclodimerization of an alkene substituent.[27]

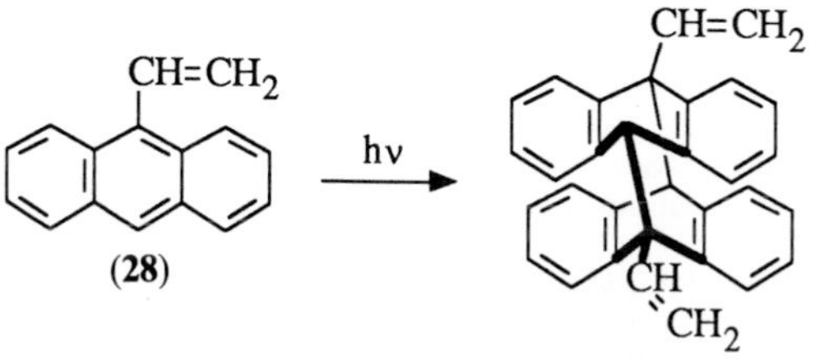

3.13. THE ISOMERIZATION OF HETEROCYCLIC COMPOUNDS

The photoisomerization of benzene derivatives has a parallel in heterocyclic molecules, particularly in furans and thiophenes. A representative example is the conversion of a 2-substituted thiophene into its 3-substituted isomer (Fig. 3.22). Three mechanisms have been considered, and these are also shown in Fig. 3.22. The actual mechanism

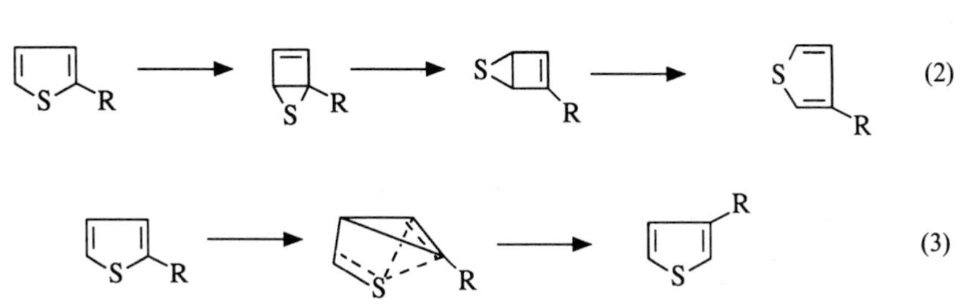

Fig. 3.22 The photochemical isomerization of a 2-substituted thiophene, and the three possible reaction mechanisms.

may well depend on the nature of the substituent, but isomerization into the [2.1.0] bicyclic system shown in mechanism (2) has been observed at very low temperatures. Theoretical analyses have been performed.[28] The reverse reaction, going from the 3- to the 2-substituted thiophene, is extremely rare, but not unknown.[29]

REFERENCES

1. J. Saltiel and G. R. Marchand, *J. Am. Chem. Soc.*, **113**, 2702–2708 (1991), and references cited therein.
2. R. W. Schoenlein, L. A. Peteanu, R. A. Mathies and C. V. Shank, *Science*, **254**, 412–415 (1991).
3. D. A. Lightner and A. F. McDonagh, *Acc. Chem. Res.*, **17**, 417–424 (1984). J. F. Ennever, *Photochem. Photobiol.*, **47**, 871–876 (1988).
4. E. Block and E. J. Corey, *J. Org. Chem.*, **34**, 896–900 (1969).
5. Y. Ito, T. Kajita, K. Kunimoto and T. Matsuura, *J. Org. Chem.*, **54**, 587–591 (1989). M. S. Syamala and V. Ramamurthy, *J. Org. Chem.*, **51**, 3712–3715 (1986).
6. G. Ciamician and P. Silber, *Chem. Ber.*, **41**, 1928–1935 (1908).
7. S. Wolff, F. Barany and W. C. Agosta, *J. Am. Chem. Soc.*, **102**, 2378–2386 (1980).
8. E. R. Koft and A. B. Smith, *J. Am. Chem. Soc.*, **106**, 2115 (1984).
9. V. Georgian, *Tetrahedron Lett.*, **42**, 4315 (1972).
10. J. Nishimura, Y. Horikoshi, Y. Wada, H. Takahashi and M. Sato, *J. Am. Chem. Soc.*, **113**, 3485–3489 (1991).
11. R. Gleiter and W. Schäfer, *Acc. Chem. Res.*, **23**, 369–375 (1990).
12. R. Gleiter and M. Karcher, *Angew. Chem., Int. Ed. Eng.*, **27**, 840–841 (1988).
13. R. Hertel, J. Mattay and J. Runsink, *J. Am. Chem. Soc.*, **113**, 657–665 (1991), and references cited therein.
14. D. Becker and N. Haddad, *Org. Photochem.*, **10**, 1–162 (1969).
15. W. J. Leigh and K. Zheng, *J. Am. Chem. Soc.*, **113**, 4019–4020 (1991).
16. G. Haufe, M. W. Tubergen and P. J. Kropp, *J. Org. Chem.*, **56**, 4292–4295 (1991), and references cited therein.
17. W. G. Dauben, B. Disayanaka, D. J. H. Funhoff, B. E. Kohler, D. E. Schilke and B. Zhou, *J. Am. Chem. Soc.*, **113**, 8367–8374 (1991).
18. M. Braun, W. Fuss, K. L. Kompa and J. Wolfrum, *J. Photochem. Photobiol. A*, **61**, 15–26 (1991).
19. E. H. White, J. D. Miano, C. J. Watkins and E. J. Breaux, *Angew. Chem. Int. Ed. Engl.*, **13**, 229–243 (1974).
20. Brunetti *et al.*, quoted in G. Cilento and W. Adam, *Photochem. Photobiol.*, **48**, 361–368 (1988).
21. W. v. E. Doering and J. W. Rosenthal, *J. Am. Chem. Soc.*, **88**, 2078–2079 (1966).
22. G. Schröder and J. F. M. Oth, *Angew. Chem. Int. Ed. Engl.*, **2**, 481–482 (1963).
23. R. J. Duguid and H. Morrison, *J. Am. Chem. Soc.*, **113**, 3519–3525 (1991).
24. B. S. Udayakumar and G. B. Schuster, *J. Org. Chem.*, **57**, 348–352 (1992).
25. P. A. Wender and J. J. Howbert, *J. Am. Chem. Soc.*, **103**, 688–690 (1981).
26. P. A. Wender, L. Siggel and J. M. Nuss, *Org. Photochem.*, **10**, 357–473 (1989).
27. H. K. Sinha, A. J. Lough and K. Yates, *J. Org. Chem.*, **56**, 3727–3728 (1991).
28. K. Jug and H.-P. Schluff, *J. Org. Chem.*, **56**, 129–134 (1991), and references cited therein.
29. N. Jayasuriya, J. Kagan, J. E. Owens, E. P. Kornak and D. M. Perrine, *J. Org. Chem.*, **54**, 4203–4205 (1989).

4. Photochemistry of the carbonyl group

4.1. ELECTRONIC EXCITATION

Bond breaking is a dominant feature in the photochemistry of carbonyl-containing molecules, in contrast to the carbon–carbon double bond which often acts as a glue in photochemical reactions and leads to products of cycloaddition. A carbonyl group may be electronically excited in two different manners. By analogy to a carbon–carbon double bond, electronic excitation can promote an electron from the highest occupied molecular orbital (HOMO) of the π-system to the original lowest unoccupied molecular orbital (LUMO). This is referred to as a π, π^* excitation. Alternatively, one of the non-bonding electrons can be promoted into the vacant π-orbital. This is now an n,π^* excitation (because there are no antibonding orbitals corresponding to a non-bonding orbital, n,n* excitation does not exist). The nature of the solvent affects these two excitations differently. In n,π^* excitations, solvents which can protonate the carbonyl group, and therefore make the excitation of the n-electrons more difficult, lead to a well-known shift in the λ_{max} in the absorption spectrum of the molecule, compared to spectra recorded in non-protic solvents. The π, π^* excitation is much less sensitive to the nature of the solvent.

4.2. NORRISH TYPE I REACTIONS

When a carbonyl group is electronically excited, non-productive events can occur: either luminescence or non-radiative energy transfer which returns the starting molecule to the ground state (but which is an important tool for producing the triplet excited state of other molecules). Productive reactions take many forms. The simplest is referred to as α-*fission*, or the *Norrish type I reaction*. In this reaction, a bond adjacent to the carbonyl is broken and a pair of radicals is formed:

$$R_1-\overset{\overset{\textstyle O}{\|}}{C}-R_2 \xrightarrow{h\nu} R_1-\overset{\overset{\textstyle O}{\|}}{C}\cdot \ + \ \cdot R_2$$

Radical reactions follow, and succeeding events depend very much on the conditions used in the irradiation. In the gas phase, for example, the two fragments are much more likely to separate and to avoid recombination than in a condensed medium where the fragments are constrained to remain in close proximity. Thus, the photolysis of acetone in the gas phase produces mainly methane, ketene, carbon monoxide, and ethane (Fig. 4.1). The formation of ethane can be explained either by reaction of a methyl radical with an acetyl radical or acetone itself, or by recombination of two methyl radicals.

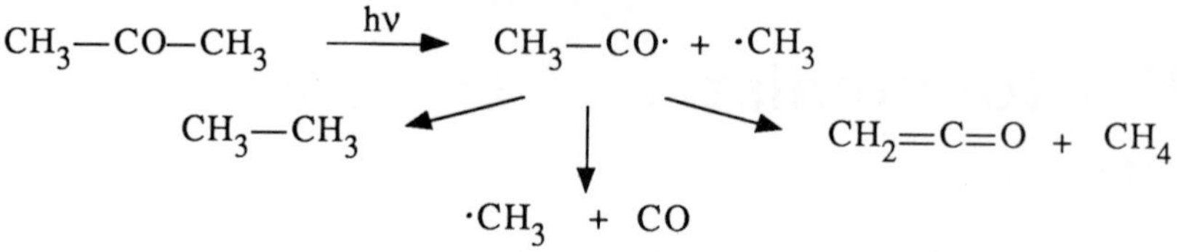

CH3—CO—CH3
hν
CH3—CO· + ·CH3
CH3—CH3
CH2=C=O + CH4
·CH3 + CO

Fig. 4.1 Main photochemical reactions of acetone in the gas phase.

As mentioned above, recombination of the initially formed radicals is a reaction expected in the condensed phase. The photolysis of liquid acetone, for example, produces mostly dimers and polymers. Similarly, the diffusion of the initial radicals is obviously restricted in cyclic ketones, and additional fragmentation does not always occur. This reaction is not necessarily trivial, as illustrated in the photoepimerization of the methyl ether of the steroid estrone (**1**). The stereochemistry at a quaternary center has been changed in going from the starting material to the product (**2**), a lumiestrone derivative; this is a transformation which is otherwise very difficult to achieve. Note that α-fissions almost always favor breaking the bond between the carbonyl group and the more substituted adjacent carbon atom.

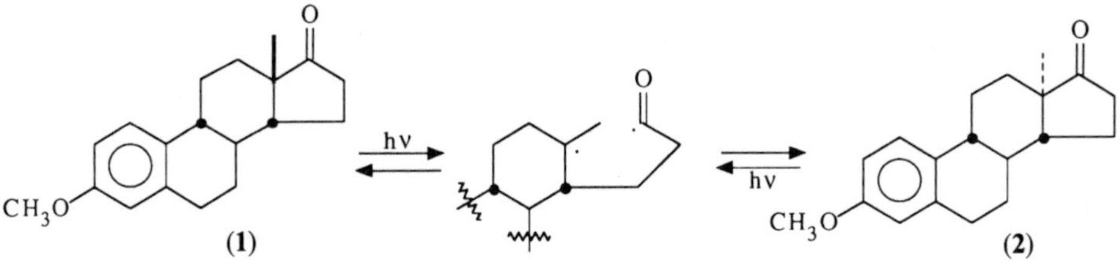

O
CH3O
hν
hν
(1)
(2)

The inclusion of the carbonyl compound in the lattice of a solid cyclodextrin reduces the ease of radical recombination dramatically. This was demonstrated with the benzylic ketone (**3**), which undergoes decarbonylation in solution (93% yield). In cyclodextrin, the benzylic radicals formed are so unreactive that electron spin resonance (ESR) spectroscopy shows them to be stable for days after the light has been turned off (when oxygen is absent).[1] Products identical to those obtained in solution were observed when the photolyzed inclusion complex was dissolved in water.

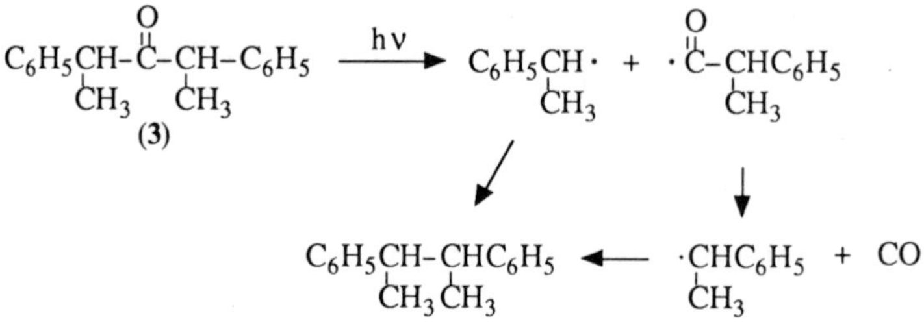

C6H5CH-C-CH-C6H5
CH3 CH3
(3)
hν
C6H5CH·
CH3
·C-CHC6H5
CH3
C6H5CH-CHC6H5
CH3CH3
·CHC6H5 + CO
CH3

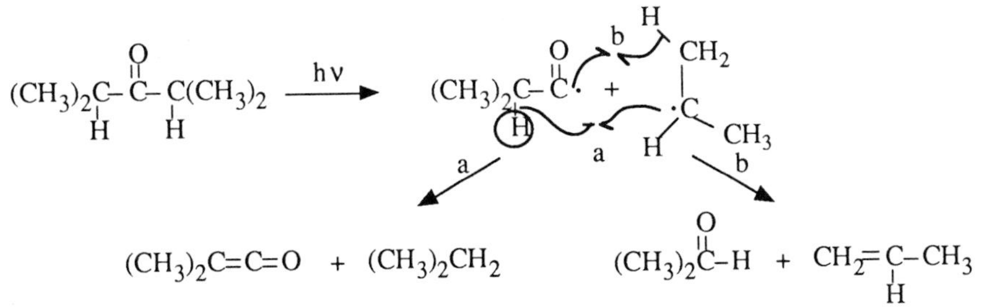

Fig. 4.2 Norrish I reactions of diisopropyl ketone.

As we follow the course of the photochemistry of a typical carbonyl group, the radical pair initially generated may seek to produce stable products. A typical method is for a radical fragment to abstract a hydrogen from the other fragment, producing two pairs of neutral products, either a ketene and a saturated fragment, or an aldehyde and an alkene (Fig. 4.2). The ketene, if formed, can of course undergo secondary reactions. For example, a carboxylic acid and an ester, respectively, are produced in nucleophilic solvents such as water and an alcohol. However, the ketene can be readily observed by its distinctive infrared spectrum (a sharp peak near $2100 \, cm^{-1}$) when the photolysis is performed at low temperature in a non-nucleophilic environment.

The photochemistry of a molecule containing both a carbonyl group and a suitably located hydroxyl group may lead to a deep-seated isomerization, as found in the case of parthenin (**4**), a sesquiterpene lactone component of some ragweeds.[2] Although the photoproduct has not been found in the same plants, one of its isomers was isolated in a different plant, raising the question of whether its biosynthesis involves a photochemical step.

The ability to trap a hydroxyl group intramolecularly by a ketene generated photochemically may be quite useful in synthesis. A synthesis for the oviposition attractant pheromone (**5**) of the mosquito *Culex pipiens quinquefasciatus* provides one

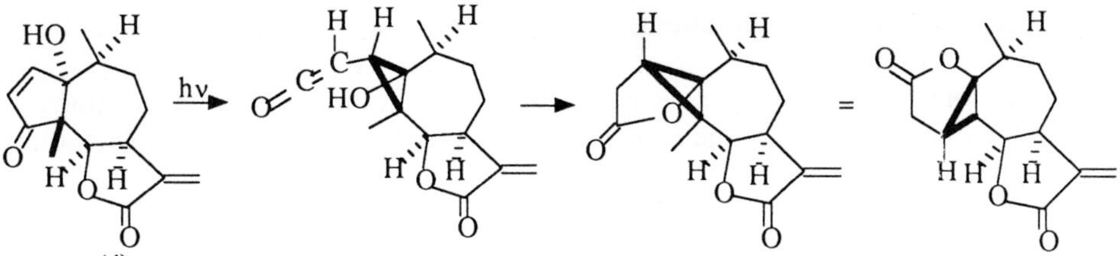

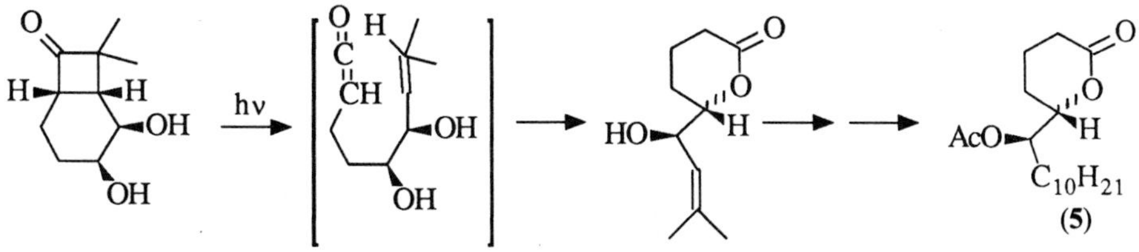

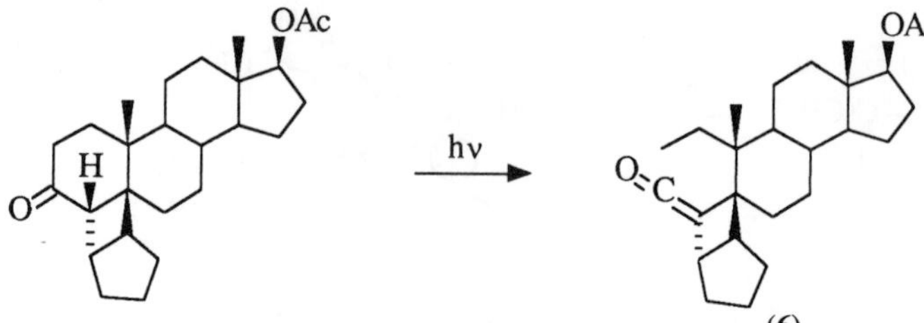

(6)

example.[3] Here, the key reaction may be viewed either as a Norrish type I, followed by fission of a carbon–carbon bond at the β-position, or as the reverse of a $[2+2]$ cycloaddition reaction, producing a ketene and an alkene. Intramolecular trapping of the ketene provides a lactone with a six-membered ring. The final transformations involved ozonolysis, a Wittig reaction of the aldehyde generated, and catalytical hydrogenation of the alkene formed.

As mentioned earlier, Norrish type I reactions tend to cleave preferentially the bond toward the more substituted α-carbon atom. One exception was reported to cleave toward the less substituted α-carbon, and to produce the unusually stable ketene (6).[4] The isomer in which the cyclobutane is *cis* fused also undergoes Norrish type I fission in the same direction, but produces an unsaturated aldehyde instead.

4.3. NORRISH TYPE II REACTIONS

Another common type of photoreaction for carbonyl compounds is conveniently represented with a six-membered ring transition state (Fig. 4.3). It involves the abstraction of a hydrogen atom by the carbonyl oxygen atom, associated with the fission of one of the carbon–carbon bonds attached to a carbon atom adjacent to the carbonyl group. Of course, the two fragmentation modes can compete in molecules possessing a hydrogen on a β-carbon atom.

The reactions outlined above have exact equivalents in the electron impact fragmentation of ketones occurring in the mass spectrometer, where they are called McLafferty rearrangements.

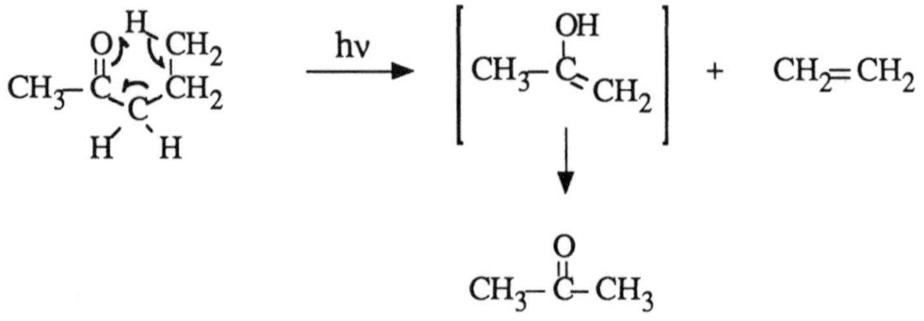

Fig. 4.3 Norrish type II fragmentation of 2-pentanone.

4.4. VARIATIONS AND EMBELLISHMENTS

The photochemistry of carbonyl compounds is very rich, and the preferred paths usually depend on the structure of the starting materials. Additional examples are presented in the remainder of this chapter.

4.4.1. Decarbonylation of cyclic ketones

As explained above, a Norrish type I fragmentation reaction breaks a bond adjacent to the carbonyl group. At higher temperatures in solution, or at lower pressures in the gas phase, irreversible loss of CO may occur. Recombination of the diradical produces the decarbonylated product. An example is the formation of cyclopentane (8) from cyclohexanone (7).

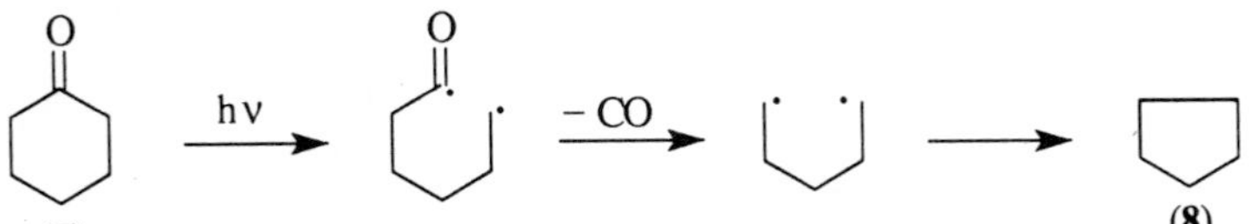

4.4.2. Cyclobutanone to tetrahydrofuran conversions

The pattern described in Section 4.4.1 is not followed with small ring ketones, particularly cyclobutanones, in which a ring *enlargement* takes place, as illustrated below:

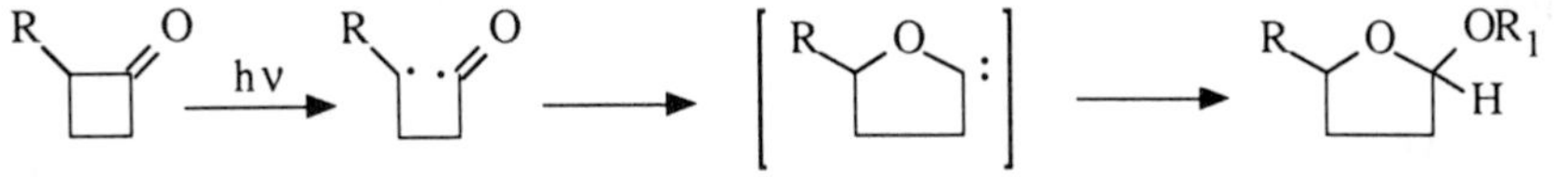

Interaction between the oxygen and the alkyl radical produces a cyclic carbene, which is stabilized by reaction with a protic solvent (R_1OH).

4.4.3. Cyclobutanol formation

If Norrish type II fragmentation is assumed to occur stepwise, a diradical intermediate may be theorized after the carbonyl has abstracted a hydrogen. When it is not fragmented to the normal alkene and enol products, recombination can yield a cyclobutanol:

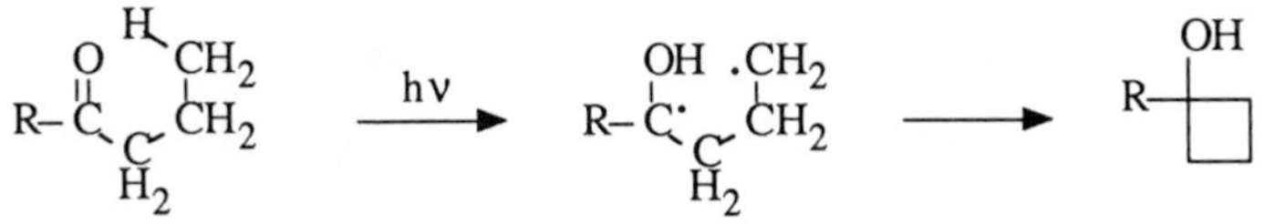

Alternatively, the cyclobutanol may be seen as the product of [2 + 2] cycloaddition of the primary products of Norrish type II fragmentation, an alkene and an enol.

4.4.4. Cycloaddition between a carbonyl and an alkene (Paterno–Büchi reaction)

The $[2 + 2]$ cycloaddition reactions described in the previous chapter apply even when one of the π-systems is a carbonyl group. The reaction product is an oxetane, and here again the reversible cyclization affords a powerful method for chemical transformations. Figure 4.4 outlines the photochemical synthesis of oxetanes in the most general way, neglecting the stereochemistry of the products. In the worst case, where all the diastereomers corresponding to the two modes of cycloaddition are formed, one must deal with a mixture of eight products! Fortunately, the reaction is usually rather selective.

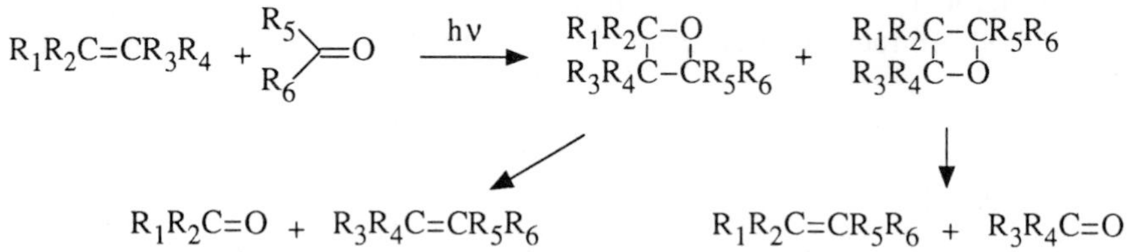

Fig. 4.4 Photochemical formation and decomposition of oxetanes.

An interesting example of $[2 + 2]$ cycloaddition comes from the field of nucleic acids. Ultraviolet light, particularly far-ultraviolet light, is mutagenic, and many different photochemical reactions of DNA have been uncovered. These may involve a single base, or two bases either in the same or in different strands. One example is the $(6 \rightarrow 4)$ photoadduct of thymidylyl$(3' \rightarrow 5')$ thymidine (TpT) (**9**), which decomposes to give the final product (**10**), with a covalent bond between the two adjacent pyrimidine bases.[5] This reaction has a parallel in cycloaddition reactions between an imine group (the nitrogen analog of a carbonyl group) and a double bond.

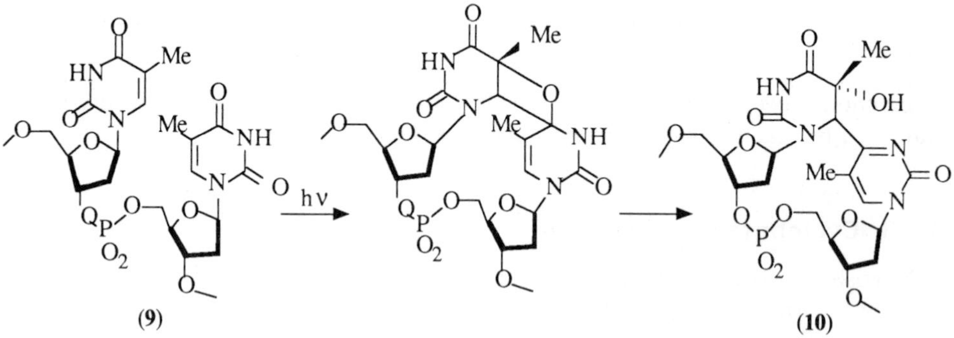

4.4.5. Photoreduction to alcohols and pinacols

The photoreduction of a carbonyl into an alcohol occurs upon photolysis in the presence of electron donors, such as amines, and hydrogen donors. This reaction is usually of little synthetic utility in view of the large number of chemical methods available to achieve it.

A solvent able to provide hydrogen radicals is required for the photopinacol reaction (Fig. 4.5), which starts with the excited carbonyl abstracting a hydrogen from the sol-

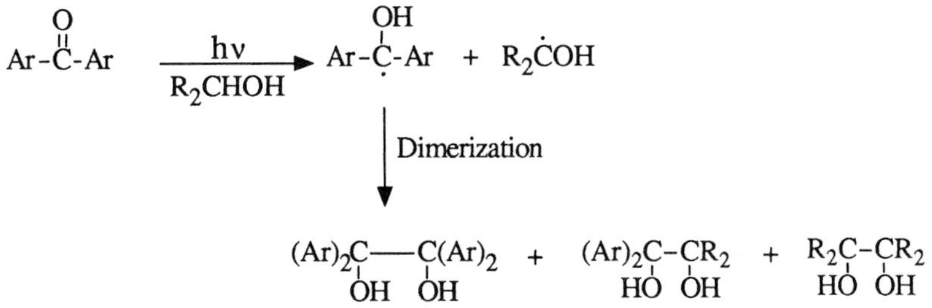

$$Ar\text{-}\overset{\displaystyle O}{\overset{\|}{C}}\text{-}Ar \quad \xrightarrow[R_2CHOH]{h\nu} \quad Ar\text{-}\overset{\displaystyle OH}{\overset{|}{\underset{\cdot}{C}}}\text{-}Ar \;+\; R_2\dot{C}OH$$

Dimerization ↓

$$(Ar)_2\overset{}{\underset{OH}{C}}\text{---}\overset{}{\underset{OH}{C}}(Ar)_2 \;+\; (Ar)_2\overset{}{\underset{HO}{C}}\text{-}\overset{}{\underset{OH}{C}}R_2 \;+\; R_2\overset{}{\underset{HO}{C}}\text{-}\overset{}{\underset{OH}{C}}R_2$$

Fig. 4.5 Photopinacol reduction, showing the three possible radical recombination products.

vent to give a tertiary radical (usually more stable than an alkoxyl radical). Dimerization follows, as in the metal-catalyzed reduction of ketones. Typically, secondary alcohols such as 2-propanol and alkylbenzenes such as toluene are excellent solvents for the photopinacol reaction. A classical example is the photoreduction of benzophenone into benzopinacol.

In general, the formation of the radical derived from the hydrogen donor creates a potential complication. This radical could (1) dimerize, (2) combine with the radical derived from the original ketone, or (3) act itself as the donor of a hydrogen radical. Complications are avoided when the alcohol is the reduced form of the starting ketone, for example $(C_6H_5)_2CHOH$ in the reduction of $C_6H_5COC_6H_5$. In this case, only a single pinacol product can be obtained.

Finally, since the photopinacol reaction is a redox reaction, it is not surprising to find that the reverse reaction can also be achieved: a pinacol can be cleaved into two carbonyl products upon irradiation in the presence of a ketone (such as acetone or p-benzoquinone).

4.4.6. Attachment of a carbonyl group in an allylic position

In this case, the reverse of Norrish type I fission may yield either the starting material or the allylic isomer (through a symmetry-allowed 1,3-shift). An example where the latter takes place is found in a synthesis of corrin (**11**).[6] The second step is a simple tautomerization.

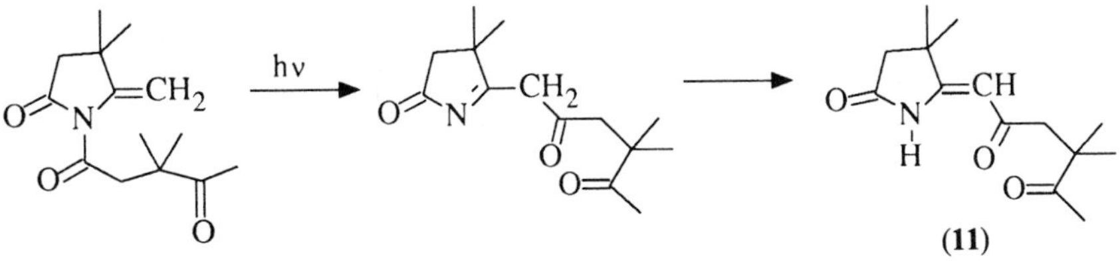

(11)

4.4.7. Norrish type I reaction, breaking a carbon–oxygen bond

The acid-catalyzed rearrangement of a phenyl ester in the presence of an acid is called the Fries rearrangement. Although it usually requires a very strong acid, a similar rearrangement takes place photochemically under strictly neutral conditions (Fig. 4.6). Attack by the acyl radical at the *para* position is another possibility, but the product from *ortho* attack predominates. The potential for Fries rearrangements under mild

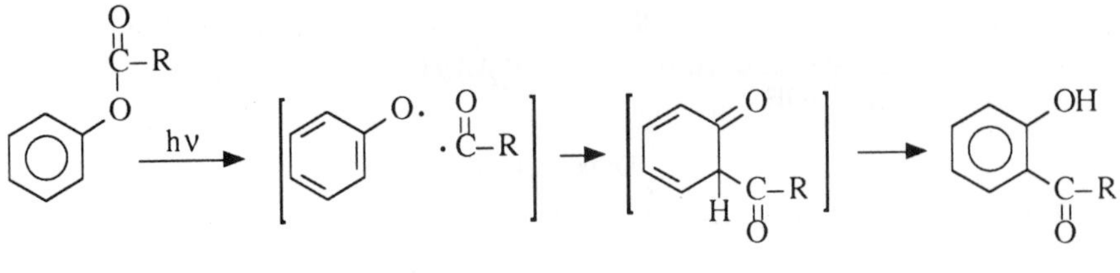

Fig. 4.6   The photo-Fries rearrangement, showing the product of migration to the *ortho* position.

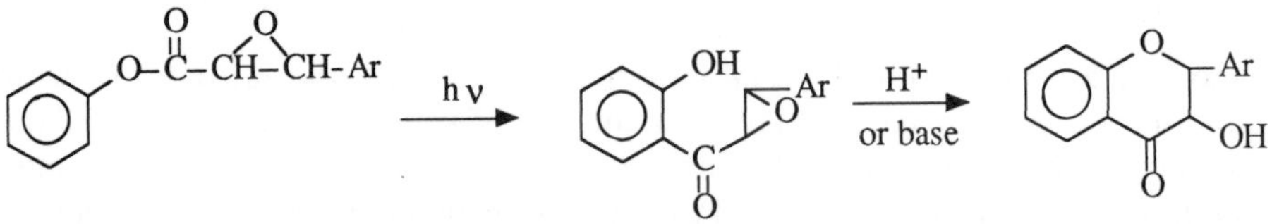

Fig. 4.7   The photo-Fries rearrangement of an epoxycinnamic ester to an *o*-hydroxy epoxycinnamate, and its conversion into a 3-hydroxyflavanone.

conditions is illustrated by the synthesis shown in Fig. 4.7 of labile epoxychalcones, which normally cyclize in the presence of a trace of acid or base.

4.4.8. Remote functionalization

The presence of a carbonyl group in a molecule can lead to hydrogen abstraction upon irradiation. Instead of intermolecular reactions, as in the sequence leading to the photo-pinacol reaction, an intramolecular reaction can take place. This is an extension of the rearrangement described in Section 4.4.3. For example, instead of a four-membered ring, transannular reaction in a 10-membered ring can form six-membered rings, as shown in Fig. 4.8. The atom selectively attacked in this case is five carbon atoms away from the carbonyl group. More remote functionalization can also occur, as demonstrated by a benzophenone chromophore attached to flexible alkyl chains[7] and to steroids (Fig. 4.9).[8] The position that is attacked depends predominantly upon the length of the chain. If $n = 0$, the short connecting link to the benzophenone does not

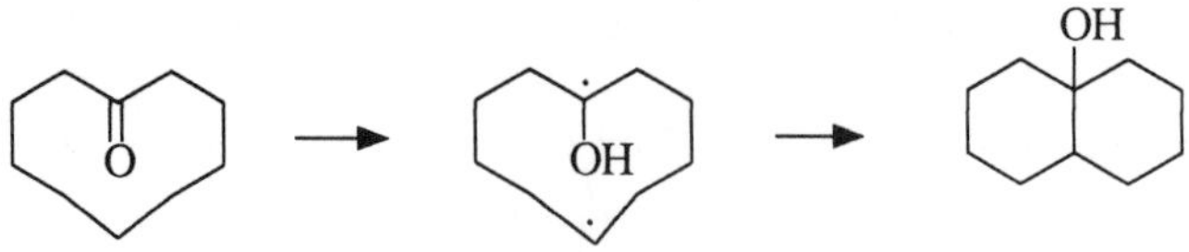

Fig. 4.8   Transannular hydrogen abstraction in a 10-membered ring ketone.

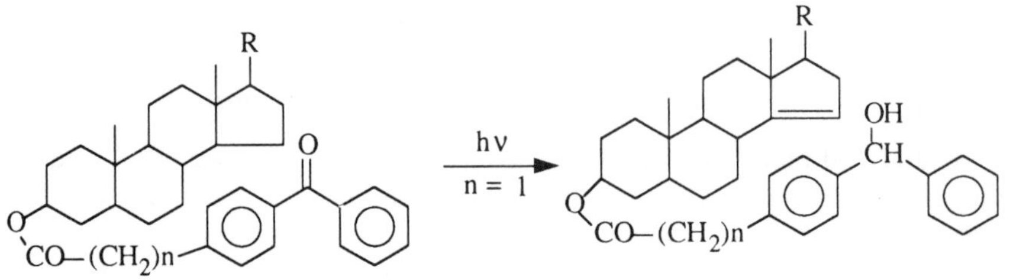

Fig. 4.9   Remote functionalization of a steroid.

allow enough flexibility to fit the carbonyl group close to any of the α-side hydrogens. If $n = 1$, however, the structure shown is the only detectable steroidal product. As n becomes larger, mixtures of alkenes are produced.

4.4.9. Note on hydrogen abstraction reactions

As described above, hydrogen atoms are often abstracted by photochemically generated radical species, particularly oxygen-centered radicals. When an alcohol R_1R_2CHOH is the hydrogen donor, one usually assumes that the carbon-bound hydrogen atom is abstracted rather than the oxygen-bound one. However, it may well be the latter which predominates. Thus, when a mixture of di-t-butyl peroxide and 2,2-dimethylpropanol (t-butylmethanol) was irradiated in the cavity of an ESR spectrometer, hydrogen abstraction from the hydroxyl group was found to predominate by a factor of 4 at 25°C, and by a factor of 15 at −40°C.[9] Hydrogen bonding in solution protects the hydroxyl hydrogens in these abstraction reactions.

4.4.10. Cyclophane synthesis

When Norrish type I fission of a ketone generates a benzylic radical, recombination may occur at any position in the ring where resonance theory predicts high electron density. This is similar to the process involved in the photo-Fries rearrangement described in Section 4.4.7. An interesting example is provided by the photochemistry of large-ring 2-phenylcycloalkanones, which are converted into cyclophanes in good yield (Fig. 4.10).[10]

One consequence of this isomerization reaction is that the overall shape of the molecule is changed. The 'ship in a bottle' strategy was demonstrated impressively by using a zeolite with a pore opening large enough to allow passage of the 15-carbon cyclic ketone starting material, but too small for the isomerized product. The internal cavities of the zeolite, however, were large enough to house the product. The starting material introduced onto the zeolite could be reversibly extracted. In contrast, photolysis reduced the amount of material extracted, which contained no isomerized product. When the zeolite was dissolved in acid, however, the isomerized product which had been formed was extracted in good yield.[11] The outcome of photochemical reactions on zeolites could be further changed by introducing additives which effectively modify the size of cavities. Even as simple an operation as treatment with benzene vapor or with water modified yields and isomer distribution.[12,13]

The same strategy, here called reptation, was used with a cycloalkanone too large to enter the zeolite framework. The diradical formed on the surface, however, could enter the zeolite, where it cyclized and where the product was trapped. Again, dissolving the zeolite led to the isolation of the cyclophane in excellent yield.[11] Conversely, the

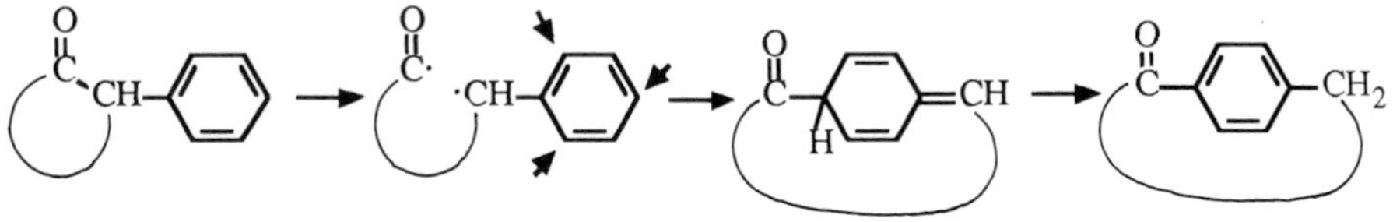

Fig. 4.10 Norrish type I opening of a cycloalkanone and ring closure at the *para* position, one of the alternative sites possessing high electron density (shown with arrows).

photochemical outcome could be diverted away from cyclophane formation by adsorbing the 2-phenylcycloalkanone in cyclodextrin cavities.[14]

4.4.11. Protection of polymers via intramolecular hydrogen abstraction

Polymers tend to degrade upon exposure to ultraviolet radiation. Impurities generate radicals which allow processes to generate peroxidic species. Their decomposition generates new radicals, and the polymer's properties are further degraded. One approach toward photochemical protection is the incorporation of small amounts of 2-hydroxy-benzophenone (11), which undergoes intramolecular hydrogen transfer in the presence of light. Tautomerization of the resulting ketone regenerates the starting material.

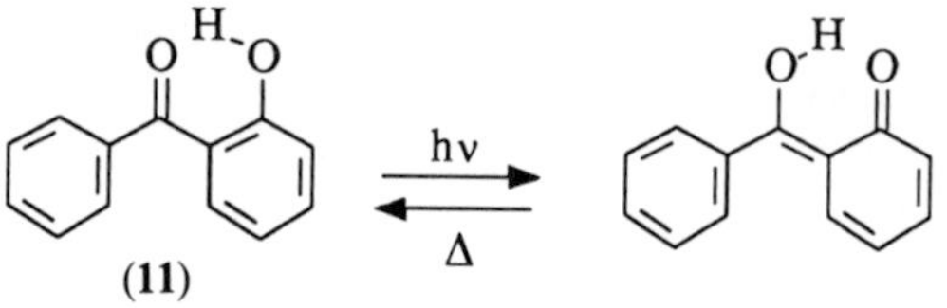

(11)

REFERENCES

1. V. P. Rao, M. B. Zimmt and N. J. Turro, *J. Photochem. Photobiol. A*, **60**, 355–360 (1991).
2. J. Kagan, S. P. Singh, K. Warden and D. A. Harrison, *Tetrahedron Lett.*, 1849–1851 (1971).
3. S. S. Rahman, B. J. Wakefield, S. M. Roberts and M. D. Dowle, *J. Chem. Soc. Chem. Commun.*, 303–304 (1989).
4. N. A. Kaprinidis, J. Woning, D. I. Schuster and N. D. Ghatlia, *J. Org. Chem.*, **57**, 755–757 (1992).
5. R. E. Rycyna and J. A. Alderfer, *Nucleic Acids Res.*, **13**, 5949–5963 (1985).
6. I. Felner, A. Fischli, A. Wick, M. Pesaro, D. Borman, E. L. Winnacker and A. Eschenmoser, *Angew. Chem. Int. Ed.*, **6**, 864–868 (1967).
7. R. Breslow, J. Rothbard, F. Herman and M. L. Rodriguez, *J. Am. Chem. Soc.*, **100**, 1213–1218 (1978).
8. R. Breslow, S. Baldwin, T. Flechtner, P. Kalicky, S. Liu and W. Washburn, *J. Am. Chem. Soc.*, **95**, 3251–3262 (1973).
9. D. Griller and K. U. Ingold, *J. Am. Chem. Soc.*, **96**, 630–632 (1974).
10. N. Han, X. Lei and N. J. Turro, *J. Org. Chem.*, **56**, 2927–2930 (1991).
11. X.-G. Lei, C. E. Doubleday, Jr, M. B. Zimmt and N. J. Turro, *J. Am. Chem. Soc.*, **108**, 2444–2445 (1986).
12. N. J. Turro, C.-C. Cheng, X.-G. Lei and E. M. Flanigen, *J. Am. Chem. Soc.*, **107**, 3739–3941 (1985).
13. V. Ramamurthy, D. R. Corbin and D. F. Eaton, *J. Org. Chem.*, **55**, 5269–5278 (1990).
14. V. P. Rao, N. Han and N. J. Turro, *Tetrahedron Lett.*, **31**, 835–838 (1990).

5. Fragmentation reactions

5.1. INTRODUCTION

The bond-breaking reactions involving carbon–carbon and carbon–hydrogen bonds, described earlier as Norrish type I and Norrish type II processes, are so important and so characteristic in the photochemistry of carbonyl compounds that they merit a separate chapter. Other examples of photochemically induced carbon–carbon bond cleavage have been noted, such as the ring opening of cyclobutanes, producing two alkenes, or that of cyclopropanes, yielding isomeric alkenes. It is important to remember that even reversible bond-opening steps can be useful synthetically, as they may allow stereochemical changes difficult to achieve by ionic means. The epimerization of steroids described in Chapter 4 is an example. The isomerization of a *trans*-penem (e.g. (1)) to its *cis* isomer is another.[1]

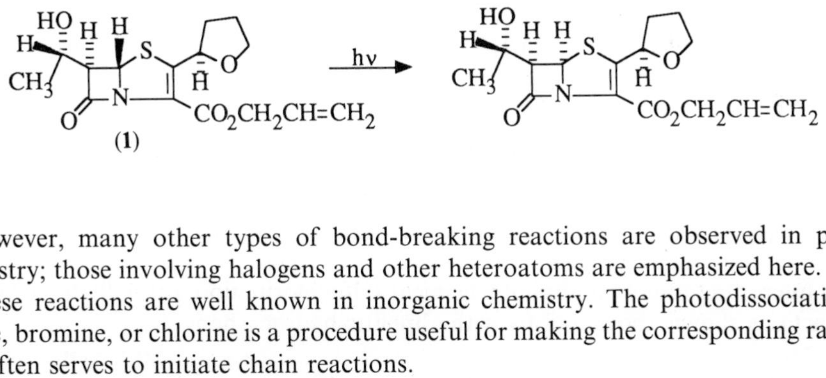

(1)

However, many other types of bond-breaking reactions are observed in photochemistry; those involving halogens and other heteroatoms are emphasized here. Some of these reactions are well known in inorganic chemistry. The photodissociation of iodine, bromine, or chlorine is a procedure useful for making the corresponding radicals and often serves to initiate chain reactions.

As shown in Chapter 1, the bonds between carbon and halogens are quite weak. Likewise, it is easy to cleave the bond between carbon and a heteroatom that is attached to atoms with which it forms a good leaving group. This is the case for the carbon–nitrogen bond, when nitrogen is part of an azide, azo, or diazonium ion grouping.

5.2. PHOTOHALOGENATION

The synthesis of insecticides has provided some of the best known examples of photochlorination reactions. For instance, the photochlorination of benzene to a mixture of diastereomeric hexachlorocyclohexanes (2) can be essentially quantitative. Note that the chlorination of benzene catalyzed by Lewis acids yields products of substitution rather than addition.

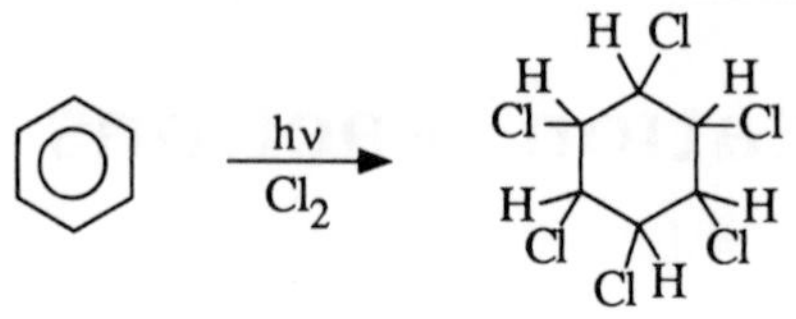

(2)

Several more complex chlorination reactions may be accomplished. For example, photochlorination may initiate the conversion of a sulfonyl chloride into the corresponding chloride in good yield. In the photochlorination of benzenesulfonyl chloride (3), chlorine is regenerated, so that less than 1 equivalent is needed at the start.

$$C_6H_5SO_2Cl \xrightarrow[\text{Cl}_2 \ (0.25 \ \text{eq})]{hv} C_6H_5Cl + SO_2 + Cl_2$$

(3)

In the more complex examples shown in Fig. 5.1, the overall processes amount to substitution reactions, but a fragmentation step must have been involved in these quantitative conversions.

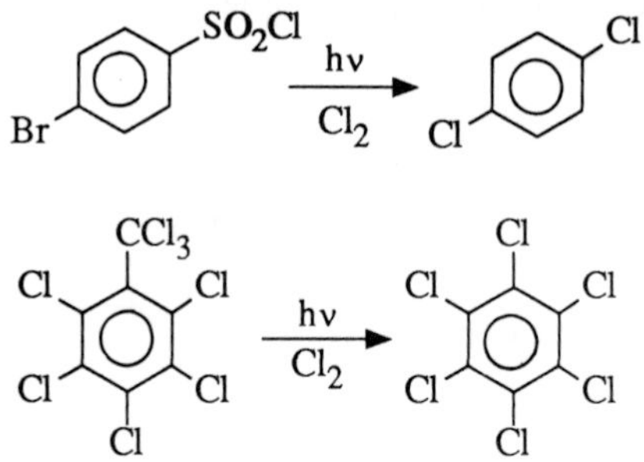

Fig. 5.1 Photochlorination reactions accompanied by fragmentation reactions.

It is interesting to note that photochemical sulfochlorination, formally the reverse of the reactions shown above, occurs with aliphatic hydrocarbons:

$$RH + SO_2 + Cl_2 \xrightarrow{hv} RSO_2Cl + HCl$$

It can also be achieved using SO_2Cl_2 instead of $SO_2 + Cl_2$.

5.3. PHOTOLYSIS OF OTHER INORGANIC CHLORIDES

Nitrosyl chloride is probably the most important example in this category. Photolyzed in the presence of cyclohexane (4), it produces nitrosocyclohexane (5), tautomeric with cyclohexanone oxime (6). Further acid-catalyzed Beckmann rearrangement of the oxime leads to a seven-membered ring lactam, ε-caprolactam (7), which can be polymerized to a polyamide (Nylon-6). This process is used on an industrial scale.

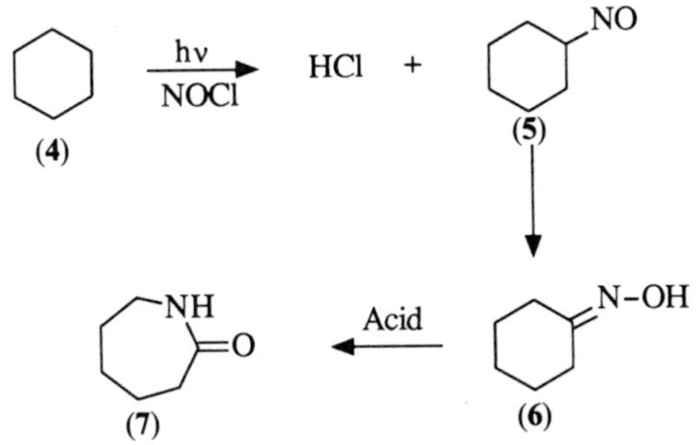

Photolysis of BBr_3 or BI_3 in benzene solution yields the corresponding phenyl boron dihalides:[2]

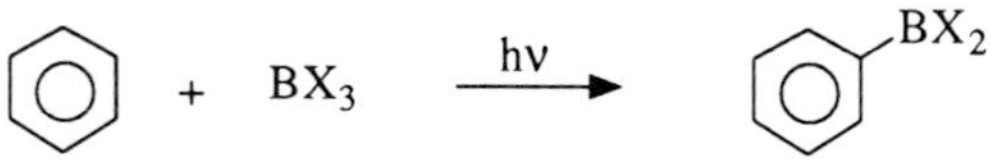

5.4. PHOTOLYSIS OF ALKYL HALIDES

Photoinduced homolytic cleavage of the carbon–halogen bond initially produces an alkyl radical and a halogen radical:

$$R-X \xrightarrow{h\nu} \left[R\cdot + X\cdot \right] \longrightarrow R^+ + X^-$$

As indicated, however, this pair of radicals can subsequently transfer an electron from the less electronegative carbon to the more electronegative halogen. In the presence of a good nucleophile Y^-, reaction with R^+ yields the product of the overall substitution of X^- by Y^- in the starting material. A particularly good example of this reaction has been described in the cubane series. There, the backside attack required in an S_N2 reaction is impossible, and S_N1 reactions are prohibited because the cubyl

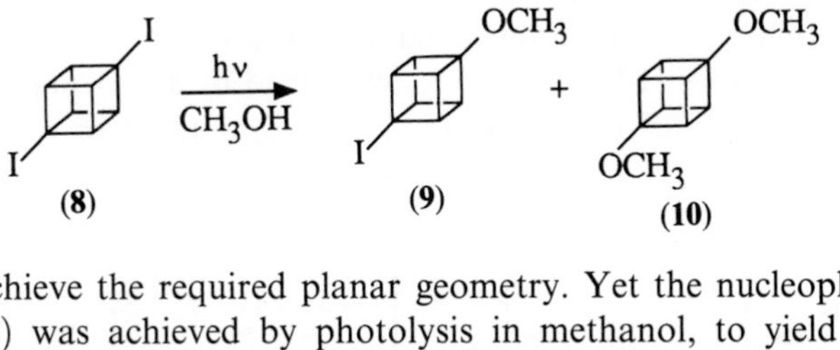

cation cannot achieve the required planar geometry. Yet the nucleophilic reaction of diiodocubane (8) was achieved by photolysis in methanol, to yield the mono- (9) (4%) and the dimethoxycubane (10) (52%).[3] The photosubstitution of the cubane derivative (8) is an extension of earlier work in which photosubstitution of 1-norbornyl iodide was converted into its 1-methoxy derivative. There is no substitution under ionic conditions, even when Ag^+ is present.

In cases with a proton at a position adjacent to the reaction site, elimination could be a major reaction path as shown in the norbornyl series (Fig. 5.2). This reaction was not reported in the cubane case, where it would have generated a very strained bridgehead double bond (anti-Bredt structure).

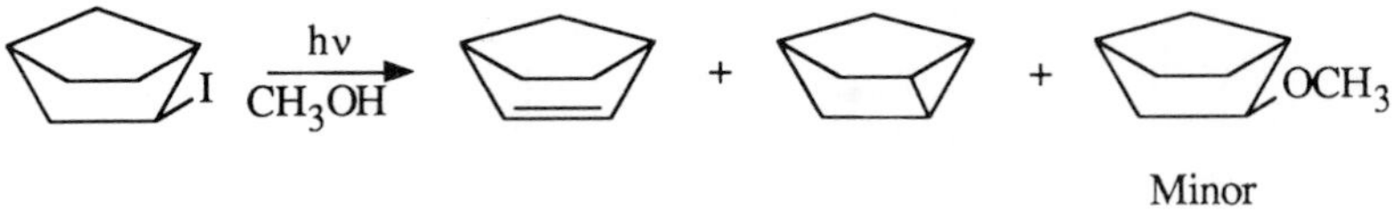

Fig. 5.2 The products of photolysis of 2-iodonorbornane in methanol. In this case the methoxy derivative is the minor reaction product.

A related reaction generated benzocyclobutadiene (**13**).[4] The *cis*-1,2-diiodocyclo-butane (**11**) underwent competing isomerization to the *trans*-diiodo compound (**12**) and loss of iodine to yield the cyclobutadiene.

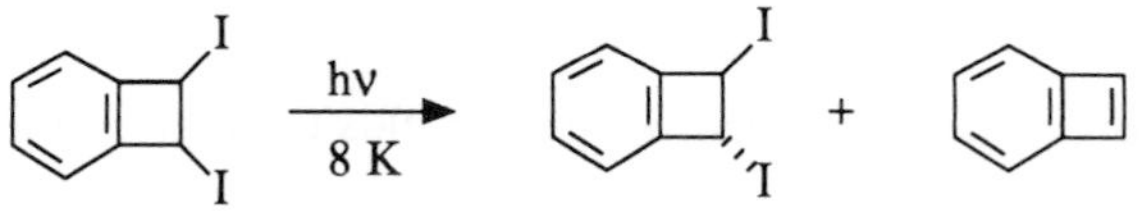

The photolysis of *gem*-diiodo derivatives is also quite interesting, because it yields reagents capable of addition to double bonds, even hindered ones, producing cyclo-propanes in excellent yield (Fig. 5.3). The cyclopropane shown in Fig. 5.3 was obtained in 97% yield, and its formation has been suggested to involve a carbocation intermediate. In molecules having two different double bonds, the selectivity of the photochemical reaction may be different from that observed with other cyclopropanation reagents.

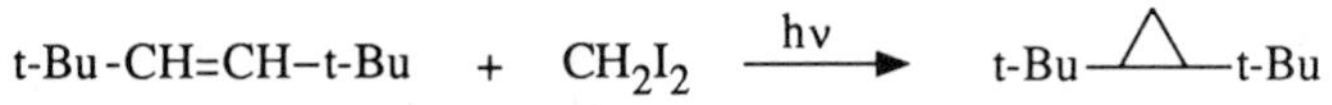

$$\text{t-Bu-CH=CH--t-Bu} \quad + \quad \text{CH}_2\text{I}_2 \quad \xrightarrow{h\nu} \quad \text{t-Bu}—\triangle—\text{t-Bu}$$

Fig. 5.3 Photocyclopropanation with diiodomethane.

Although the above fragmentation reactions were most successful with alkyl iodides, conditions for successful utilization of alkyl bromides have been developed.[5]

The same type of reaction occurs when alkenyl halides are irradiated.[6] An interesting example is shown here, where the same product (**16**) was obtained by irradiation of the two different vinyl iodides (**14**) and (**15**), as explained by a rearrangement of the vinyl cation intermediate obtained from (iodomethylene)cyclopentane.

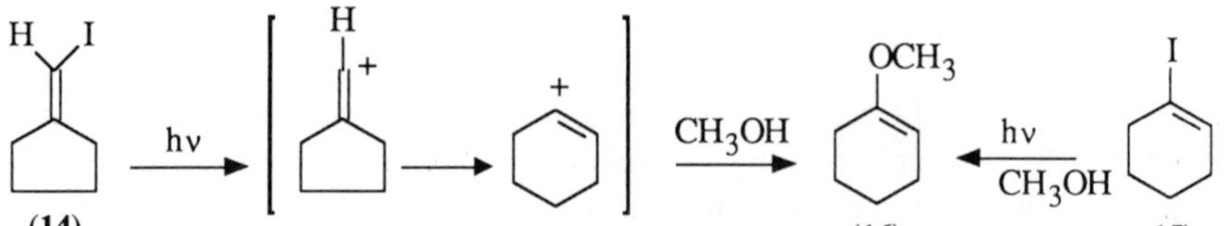

It is important to note that the radical initially formed in the photolysis of an organic halide may be prevented from conversion into the cation, for example by operating in a glassy medium at low temperature. This was illustrated in the synthesis of the triplet ground state biradical molecules shown in Fig. 5.4, in which the corresponding dichlorides were irradiated at 77 K in ethanol.[7] *O*-Xylylene (**17**) and its analog (**18**), on the other hand, were obtained in their singlet state by irradiation of the dihalides.

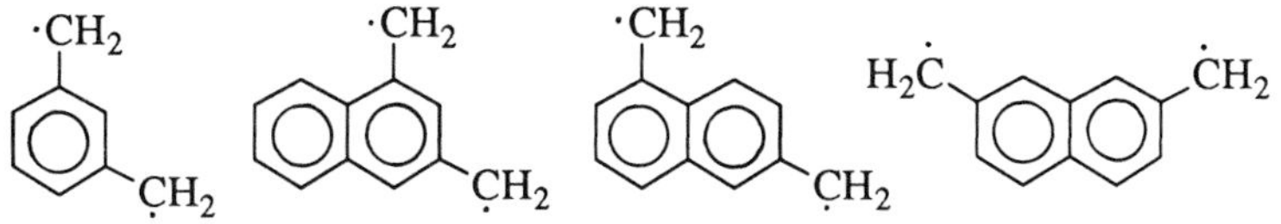

Fig. 5.4 Photochemically generated triplet ground state diradicals.

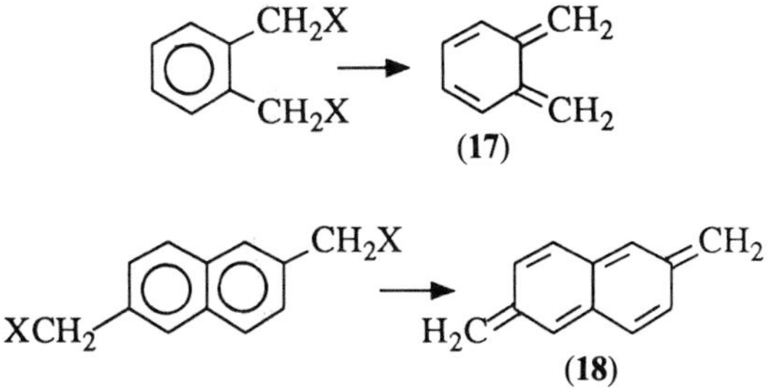

(17)

(18)

5.5. PHOTOLYSIS OF N-HALOGENATED COMPOUNDS

N-Bromosuccinimide (NBS) is probably the best known example of a reagent capable of undergoing photoinduced cleavage of a nitrogen–halogen bond, a useful reaction for the bromination of allylic, benzylic, and heteroaromatic compounds. While the initial step is undoubtedly the fission of the nitrogen–bromine bond to succinimyl and bromine radicals, the subsequent steps are mechanistically complex, particularly in relation to the organic fragment(s) generated. The actual brominating reagent is Br_2, generated through the reaction of bromine radicals with unreacted NBS.

5.6. PHOTOLYSIS OF NITRITE ESTERS

Usually called the Barton reaction, after its inventor, the photolysis starts with a homolytic fission, generating a nitrosyl and an alkoxyl radical. A number of reactions may ensue. For example, the alkoxyl radical can abstract a hydrogen atom intramolecularly and generate, through a six-membered ring transition state, a new carbon-centered radical at a position often difficult to functionalize directly. Figure 5.5 illustrates the application of this approach to functionalization of a methyl group in a steroid.

A good demonstration of the Barton reaction's utility for functionalization at what appears to be a distant site (but which is accessible through a six-membered ring transition state) is in the synthesis of perhydrohistrionicotoxin. Competing functionalization of the butyl side-chain was also observed in this step of the synthesis.[8]

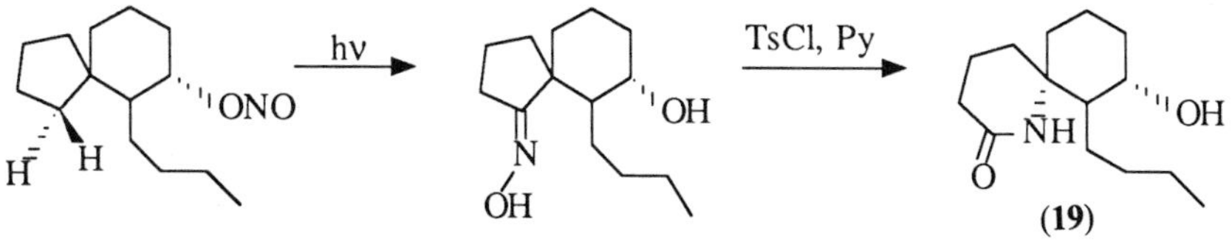

(19)

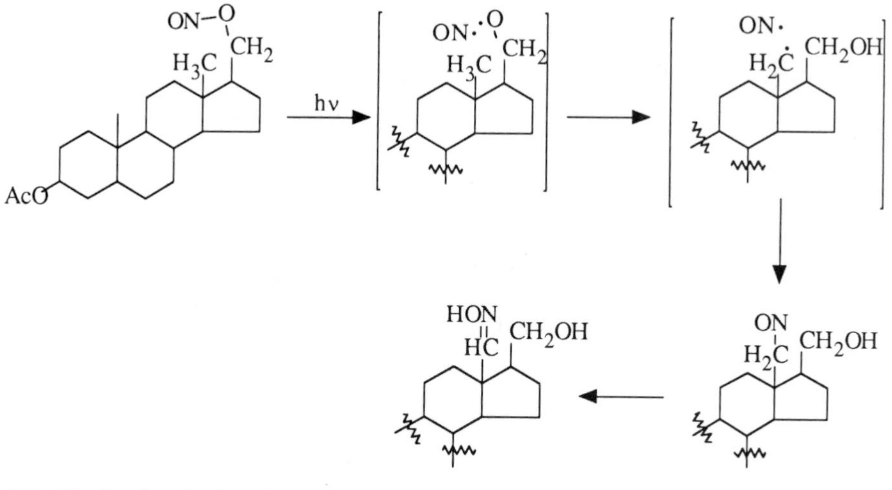

Fig. 5.5 The functionalization of a methyl group in a steroid by the Barton reaction.

5.7. PHOTOLYSIS OF N-CHLORAMINES

A very closely related reaction, the photolysis of N-chloramines, also allows functionalization at distant, unactivated, sites. It is illustrated here with steps used in the
synthesis of dihydroconessine (**20**).[9]

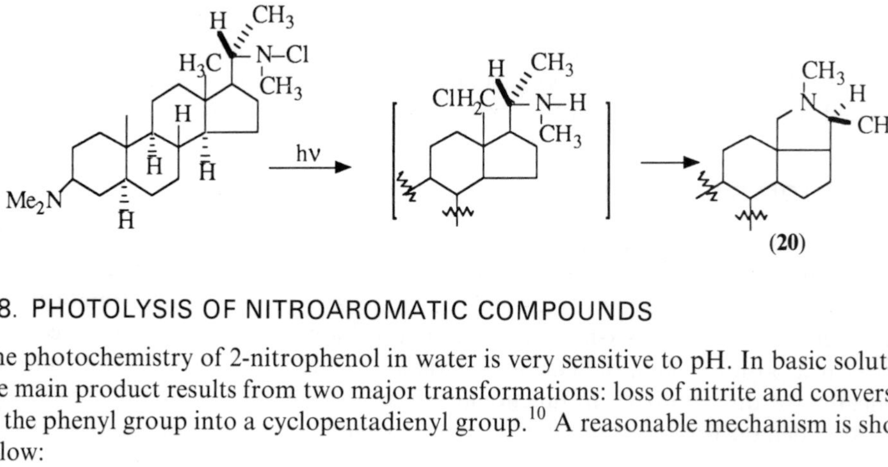

5.8. PHOTOLYSIS OF NITROAROMATIC COMPOUNDS

The photochemistry of 2-nitrophenol in water is very sensitive to pH. In basic solution,
the main product results from two major transformations: loss of nitrite and conversion
of the phenyl group into a cyclopentadienyl group.[10] A reasonable mechanism is shown
below:

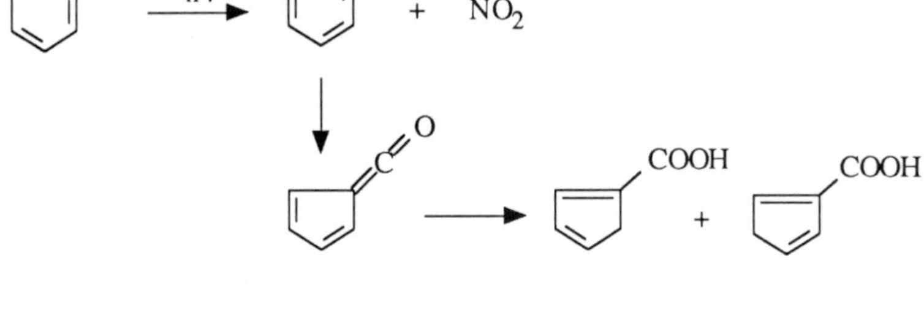

5.9. PHOTOLYSIS OF AZO, DIAZO, AND AZIDE DERIVATIVES

Molecular nitrogen is an excellent leaving group. Upon irradiation, it can be easily expelled from organic molecules. Whether the diradical intermediate formed reacts from a singlet or a triplet state is beyond the scope of this discussion.

5.9.1. Photolysis of diazonium ions

The photochemical decomposition of diazonium ions was incorporated into a process used for the reproduction of documents (blueprints) long before Xerox and related reproducing machines became available (Fig. 5.6). The irradiation of a diazonium salt leads to the loss of N_2 and formation of a neutral product (in aqueous medium, the aryl cation may also be trapped by the solvent to give a phenolic product). To the extent that decomposition of the diazonium ion has occurred, the normal coupling reaction with a phenolic reagent no longer takes place. Therefore, the areas of a paper impregnated with the diazonium ion which are protected from light by a mask (or lines on a document) will produce a colored azo dye after treatment with a phenol in basic solution, while the areas not protected from light will be colorless.

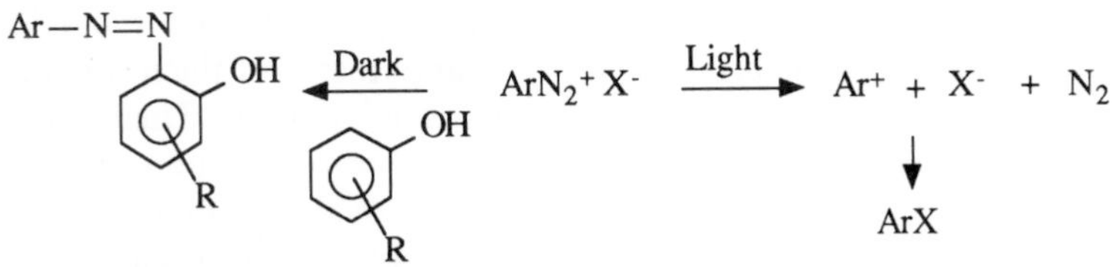

Fig. 5.6 Reactions used in an early process for the reproduction of documents.

5.9.2 Photolysis of diazo derivatives

The photolysis of diazo derivatives is a classic method for forming carbenes. For example, diazomethane (CH_2N_2) gives N_2 and $:CH_2$ upon irradiation. The photochemical decomposition of diazo compounds can occur at more than one site in a molecule, such as (21),[11] where a quintet ground state molecule was generated at low temperature (this is a dicarbene where each carbene is in the triplet state; see definition in Chapter 1).

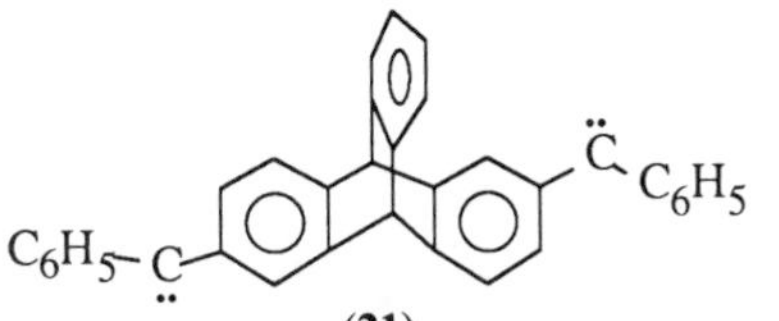

(21)

Diazoketones can give products of carbene reactions (addition to olefins, ether formation with alcohols), but they mostly yield products of Wolff rearrangements, as illustrated by the synthesis of D-norsteroids (22).[12] A Wolff reaction gives a ketene. Further reaction with a nucleophilic solvent such as water, an alcohol or an amine yields a carboxylic acid, ester, or amide. When the ketene reacts with water and converts into a carboxylic acid, the starting material changes from a water-insoluble mole-

(22)

cule to a water-soluble molecule, particularly in the presence of a base. This transformation has found important practical applications. After the diazoketone has been spread over a suitable surface, irradiation through a selected pattern in a mask will reproduce that pattern on the surface. Washing the surface with basic aqueous solution will remove the acidic organic product from the areas which have been exposed to light, while the areas kept in the dark (where the mask shielded the light) will keep their coating. Further etching of the exposed surface with an acid, and deposition of a metal, gives an integrated circuit. The process may be repeated, each time coating the sample with diazoketone and changing the mask; this is one of the basic steps in the formation of photo-resists used in the electronic industry.

Another approach consists of having resin possessing groups (hydroxyl groups, for example) which react with the ketene to be formed. The product becomes more insoluble in organic solvents than the starting material, and it is now the unreacted diazoketone which can be washed away. This modern approach to the old lithography process is aptly called *photolithography*. A clever application of the technique has been directed at the planned synthesis of peptides on a solid support. There, the pattern of exposure to light through a mask determines the regions of the support which become activated for subsequent chemical reaction upon removal of a photolabile protecting group. After reaction with a reagent, such as a suitably protected aminoacid containing a photolabile group, the process can be repeated through many cycles. Such a high degree of miniaturization can be achieved that pentapeptides have been synthesized on $50 \times 50\,\mu m$ squares on a solid support. It has been claimed that 40 000 different compounds could thus be created side by side on a $1\,cm^2$ support.[13]

Diazo fragmentation reactions become even more interesting when two diazo groups flank the carbonyl group, as in the synthesis of acenaphthyne (23). This is more strained than in a benzyne.[14] The irradiation of bisdiazocyclohexanone at 8 K showed that ketene was initially formed, and that it was the precursor to a cyclopropenone. The ketene was not observed under the conditions (15 K) in which acenaphthyne was produced. The decarbonylation occurring in the last step is common in the photochemistry of cyclic ketones (Chapter 4).

(23)

A similar reaction, performed at 254 nm in an argon matrix, led to new ketene analogs, as shown in Fig. 5.7.[15] It is not known whether cyclopropenones are intermediates, as none were detected. The linear product lost CO upon further irradiation at longer wavelengths.

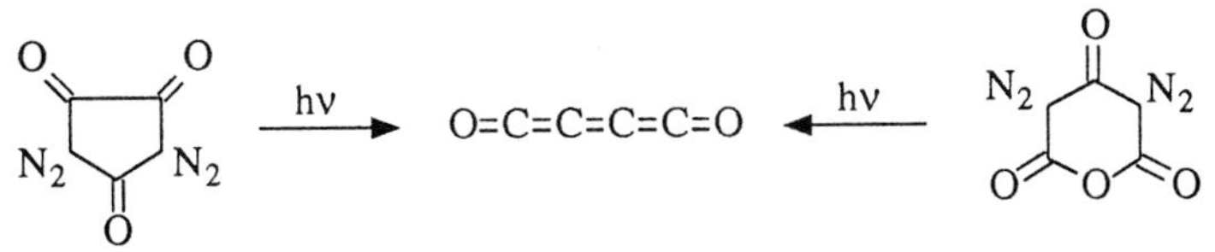

Fig. 5.7 Photolysis of bisdiazoketones.

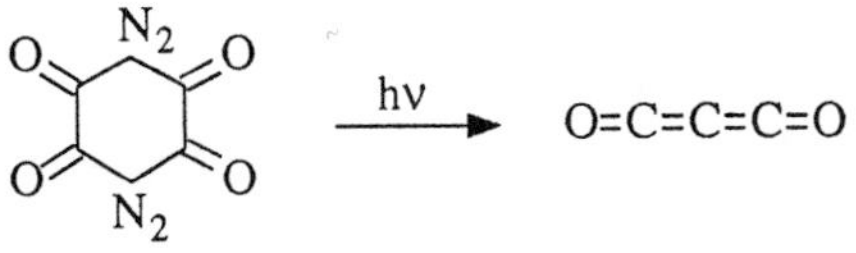

Fig. 5.8 Photochemical synthesis of carbon suboxide.

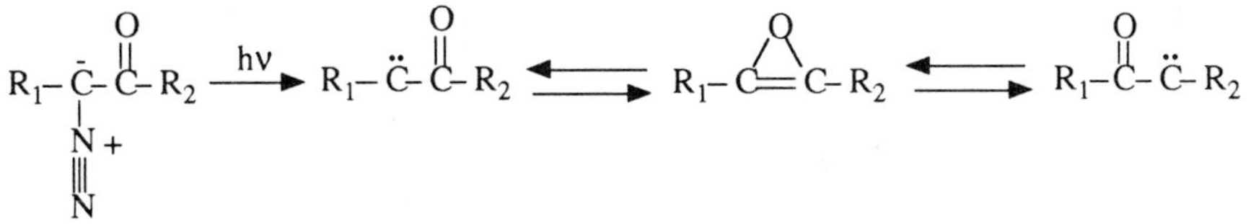

Fig. 5.9 Oxirene formation from ketocarbenes.

A similar reaction reported in this study involved additional carbon–carbon bond fragmentation (Fig. 5.8), although it is not known whether this step occurred before or after the loss of nitrogen.

Even in the simplest cases, the photolysis of a diazoketone may lead to several products. The first is a ketocarbene, which may then produce a ketene or an unsaturated ketone. Among the other possibilities, the carbene may seek stabilization by reacting with the oxygen, reversibly forming the long elusive oxirene, which can revert to either one of two isomeric ketocarbenes (Fig. 5.9). The oxirene can be detected spectroscopically below 25 K.[16]

Vinylcarbenes and cyclopropenes are interconvertible:

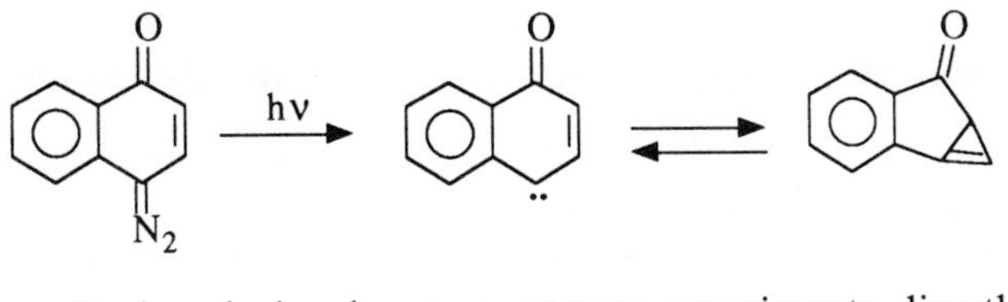

Spectroscopic monitoring during low-temperature experiments directly revealed this interconversion.[17] The cyclization appears to take place mostly from the singlet carbene. Related triplet carbenes can react with O_2:

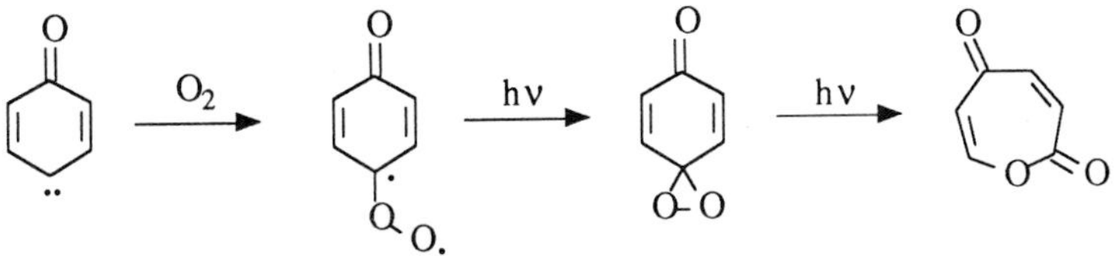

As shown, the diradical formed initially can be cyclized photochemically, and the dioxirane produced may be isomerized into a lactone. A reaction similar to the Wolff rearrangement provided a convenient synthesis of a 1-oxygenated carbapenem (25) from a readily available cephalosporin derivative (24).[18] The mechanism suggested is not fully known, as the proposed sulfine intermediate (26) could not be isolated.

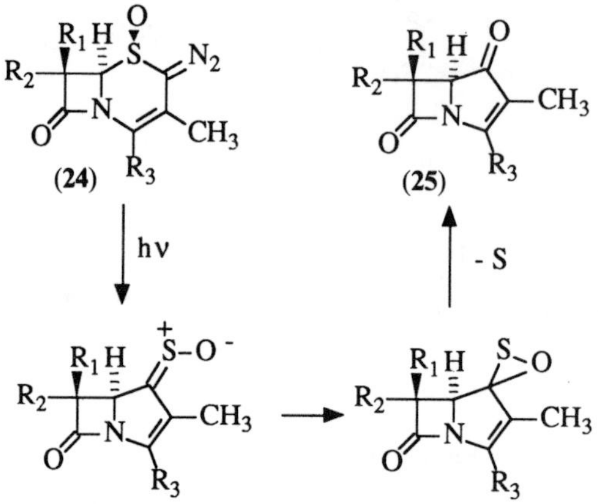

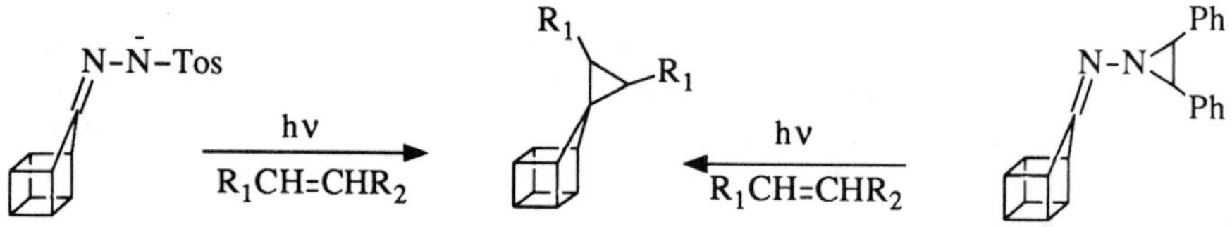

5.9.3. Photolysis of tosylhydrazone salts and aziridylimines

The same mixture of products (29) is obtained upon photolysis of the two carbene precursors (27) and (28).[19] The carbene was shown, through analysis of the products of CH_3OD addition, to be in equilibrium with the bridgehead alkene:

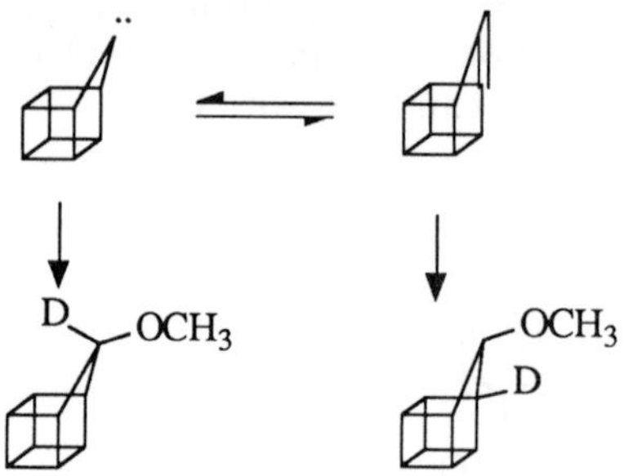

5.9.4. Photochemistry of azo derivatives

The loss of N_2 from azo derivatives to produce a pair of radicals which recombine is a general reaction (Fig. 5.10).

$$R_1-N=N-R_2 \xrightarrow{\ h\nu\ } [R_1{}^{\cdot} + N_2 + R_2{}^{\cdot}] \longrightarrow R_1-R_2 + N_2$$

Fig. 5.10 Photochemical decomposition–recombination of azo derivatives.

The recombination is seldom a quantitative process, and individual radicals may react with other molecules. The example of 2,2′-azobisisobutyronitrile (**30**), a standard initiator for radical chain reactions, is well known. Photoinitiators of radical chain reactions are used in many circumstances; an example is in the curing of paints.

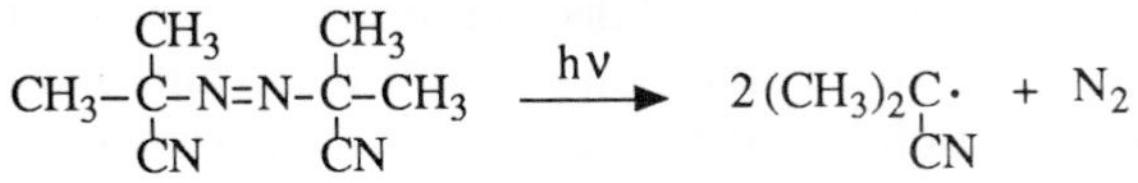

$$CH_3-\underset{\underset{CN}{|}}{\overset{\overset{CH_3}{|}}{C}}-N=N-\underset{\underset{CN}{|}}{\overset{\overset{CH_3}{|}}{C}}-CH_3 \xrightarrow{\ h\nu\ } 2\,(CH_3)_2\underset{\underset{CN}{|}}{C}{\cdot} + N_2$$

(**30**)

Diazirines present a special case of decomposition of azo compounds as they can serve as photoaffinity labels (Section 9.2), where they are reportedly more reactive than diazo groups.[20] Dialkylcarbenes are usually difficult to intercept because they undergo facile intramolecular rearrangement to alkenes. However, the photochemical laser decomposition of dimethyldiazirine (**31**) produced singlet dimethyl carbene (**32**), which could be intercepted with pyridine, producing the ylide (**33**) characterized spec-

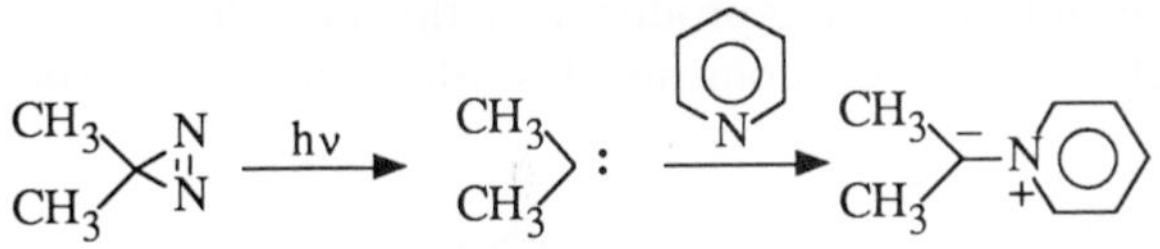

(**31**)   (**32**)   (**33**)

troscopically.[21] This proved that the carbene was a reaction intermediate. A recent example of diazirine irradiation proved that (**34**) (a tritiated analog of 11-*cis*-retinal) cross-linked the visual photoreceptor rhodopsin.[22] The homocubyl carbene generated in Section 5.93 by two different methods was also obtained by photolysis of the diazirine precursor (**35**), and its reaction products were identical.

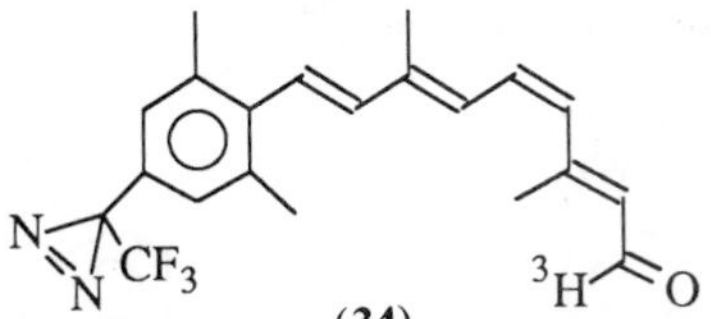

(**34**)

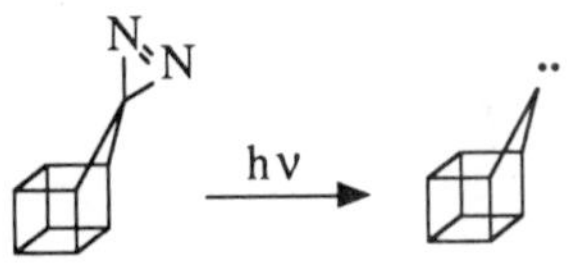

(**35**)

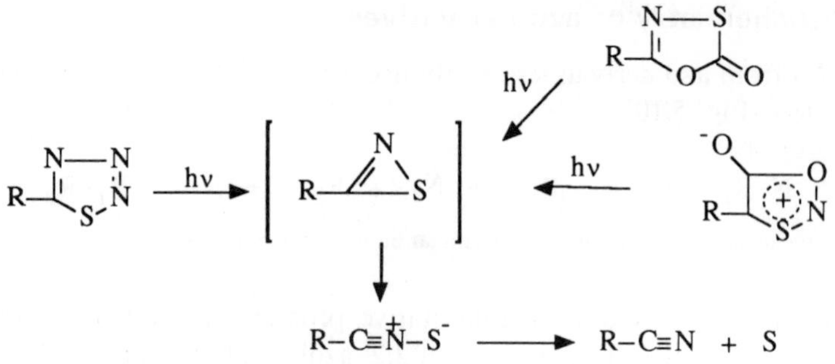

Fig. 5.11 Photofragmentation reactions leading to a nitrile sulfide (R=C₆H₅).

Many examples of this photochemical decomposition are known, and the loss of N_2 also takes place in complex molecules containing nitrogen–nitrogen double bonds. For example, photofragmentation of thiatriazoles leads to the synthesis of nitrile N-sulfides (Fig. 5.11).

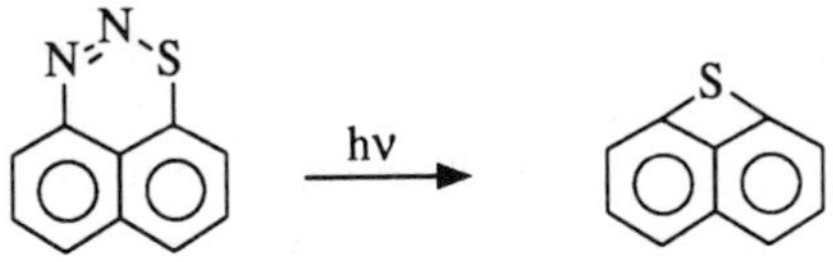

(36)

Another interesting synthetic application of photofragmentation with loss of N_2 occurs quantitatively in benzene and produces the thietane (36).[23] The loss of N_2 can be combined with cycloaddition reactions, as illustrated here with carbon disulfide:[23]

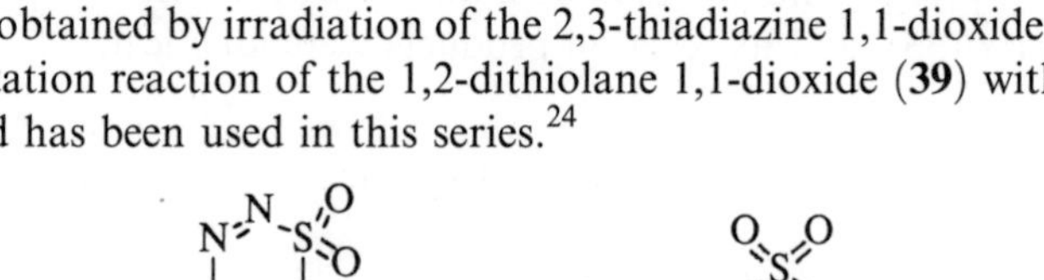

Some interesting reactions are closely related. For example, the S, S-dioxide (38) of the thietane (37) is obtained by irradiation of the 2,3-thiadiazine 1,1-dioxide (31). The analogous fragmentation reaction of the 1,2-dithiolane 1,1-dioxide (39) with loss of SO_2 is also known and has been used in this series.[24]

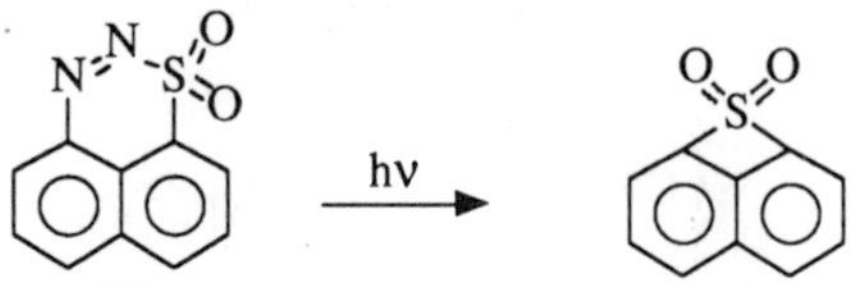

(37)   (38)

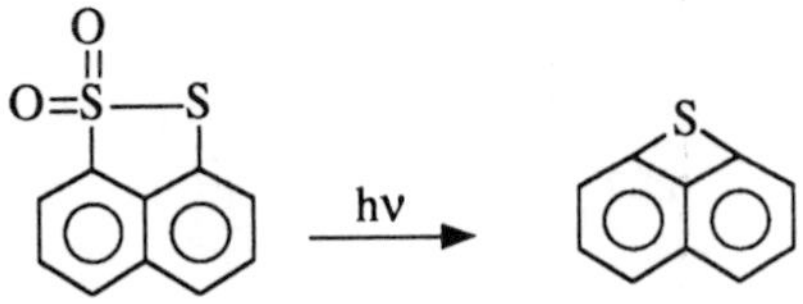

(39)

5.9.5. Photolysis of azides

The decomposition of azides is one of the easiest syntheses of nitrenes. Nitrenes have been very useful for making three-membered rings containing nitrogen, as well as amine derivatives.

$$R-\bar{N}-\overset{+}{N}{\equiv}N \xrightarrow{h\nu} R-N: + N_2$$
$$\text{nitrene}$$

The chemistry of ketonitrenes is very similar to that of ketocarbenes mentioned above (Fig. 5.12). Isocyanates, rather than ketenes, are the usual products. This reaction is called a Curtius rearrangement.

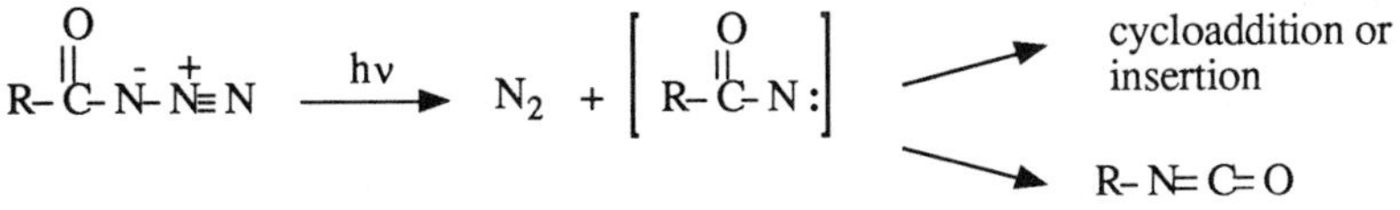

Fig. 5.12 Photochemical decomposition of acyl azides.

Usually, the products of rearrangement predominate over the products of direct nitrene insertion. These, however, may be important, as shown in a key step of the total synthesis of the alkaloid garryine (in which a photochemical allylic oxidation step was also included later in the scheme).[25]

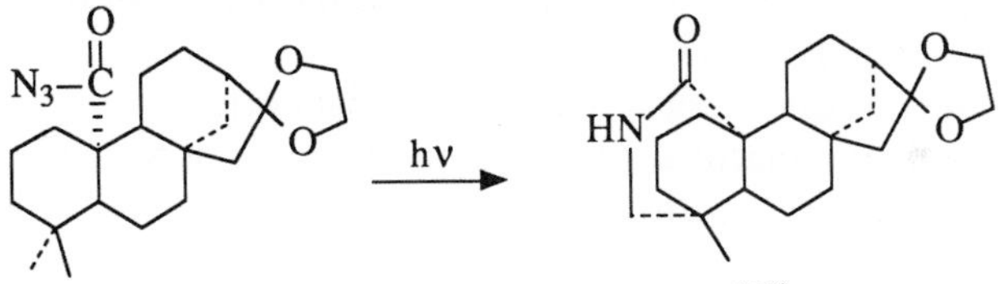

(**40**)

Azides have frequently been used in biochemical studies, as part of the general approach of *photoaffinity labeling*. In a typical case, a substrate to an enzyme is substituted with an azido group. When the substrate has formed a complex with the enzyme, and before conversion into a product, the azide moiety is obviously near the active site of the enzyme. Photolysis produces covalent bonds between the substrate and the enzyme. By determining the location of these points of attachment, the three-dimensional shape of the activity enzyme–substrate complex can be revealed.

A phenyl azide moiety has often been used for affinity labeling. However, the photochemistry turns out to be more complex than anticipated, because an intramolecular reaction of the phenyl nitrene easily takes place (Fig. 5.13).[26] Two pathways are shown

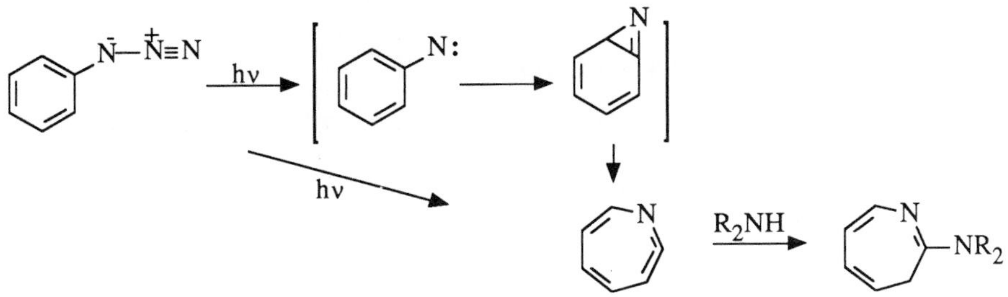

Fig. 5.13 Ring enlargement of phenyl azide.

in Fig. 5.13. In one, a nitrene intermediate undergoes cycloaddition to produce an azirine. A subsequent $2\pi + 2\pi + 2\sigma$ electrocyclic reaction isomerizes the bicyclic azirine into the seven-membered product. The other pathway generates the product directly by coupling the departure of nitrogen with the ring enlargement. The final step, reaction with a nucleophilic reagent, reduces the strain in the ring. In related work, the interconversion of phenylnitrene (41) and isomeric pyridylmethylenes (42) has been demonstrated.[27]

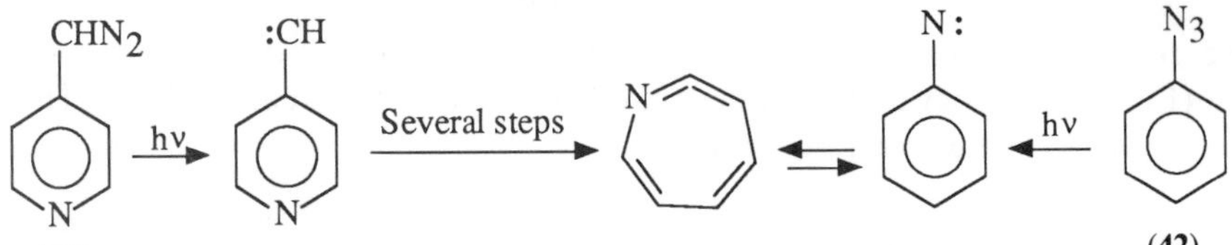

When present during the photolysis of azides, oxygen adds to the nitrene formed, generating a nitroso oxide able to isomerize into a dioxaziridine and the nitro isomer, and to act as an agent for oxygen atom transfer.[28]

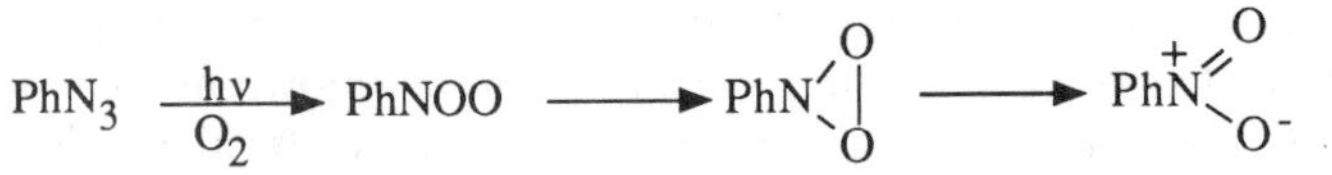

The photochemistry of azides is not limited to molecules with the azido substituent attached to carbon. Similar rearrangements have been described in molecules in which the azide is attached to silicon, germanium, or phosphorus.[29]

The fragmentation of molecules containing more than one photosensitive group may be selective. One example is with iodinated tyrosine derivatives in which the amino group was (4-azido)benzoylated. Irradiation in the presence of an alcohol or an amine led to loss of nitrogen from the azide group, yielding azepines which still contained an iodine substituent.[30] This approach may be useful in photoaffinity labeling experiments with radioactive iodine (cf. Chapter 9.12).

Another example concerns a molecule substituted with an azo and an azide moiety, both groups capable of losing nitrogen. In a pretty reaction, it was shown that the overall transformation appeared to be an intramolecular trapping of the initially formed nitrene by the azo group, generating a triaziridine ring:[31]

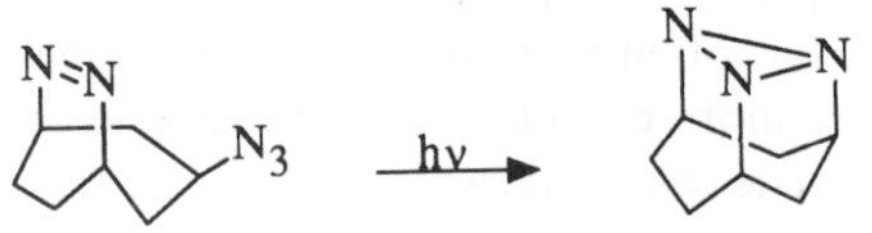

5.10. OTHER FRAGMENTATION REACTIONS

Innumerable photofragmentation reactions have been recorded in the literature. The most common leaving groups, besides N_2, are CO_2 and CO, particularly when included in cyclic structures. The loss of these groups has led to the synthesis of elusive compounds, such as cyclobutadiene or benzyne.

The reaction conditions are often critical in determining the products formed upon fragmentation. For example, while a pyrazoline-3,5-dione is a logical precursor capable of losing one molecule of N_2 and two of CO, only one of each is lost thermally, yielding a ketene. Irradiation in a matrix at very low temperature, in contrast, leads to the loss of the three molecules, and generates a carbene which reacts intramolecularly to yield an unsaturated product:[32]

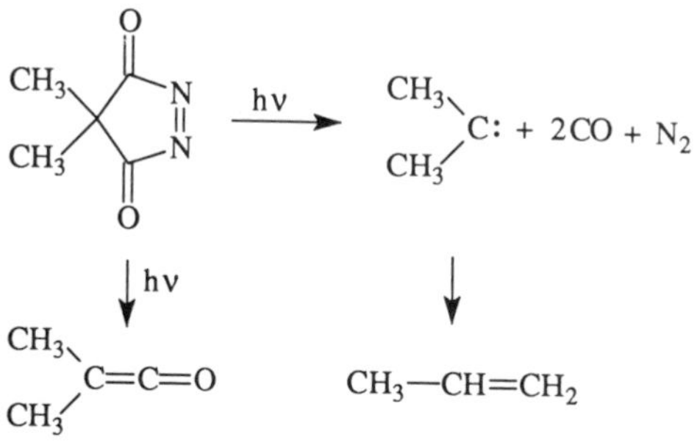

The decarboxylation of simple carboxylic acids is often an important step in organic synthesis, but it may be difficult. Photochemically, however, it can be performed easily in the presence of a nitrogen heterocyclic base, such as acridine, and t-butyl mercaptan acting as a hydrogen donor (Fig. 5.14(a)). The same procedure converts α-hydroxy-acids into carbonyl compounds (Fig. 5.14(b)).[33]

The combined decarboxylation and decarbonylation of anhydrides was used in the synthesis of benzyne at 4–10 K. The complete scheme, which is wavelength-dependent, includes the photoisomerization of benzyne and its reaction with CO trapped in the matrix,[34] and is shown in Fig. 5.15.

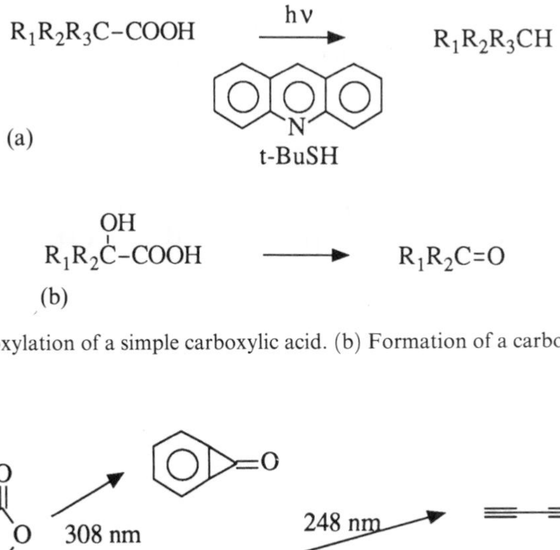

Fig. 5.14 (a) Decarboxylation of a simple carboxylic acid. (b) Formation of a carbonyl compound from an α-hydroxyacid.

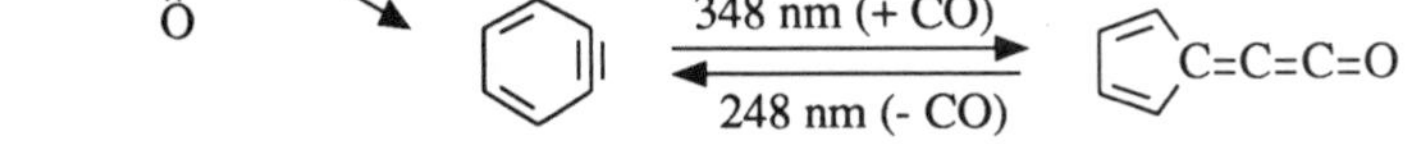

Fig. 5.15 The photochemical decarboxylation/decarbonylation of an anhydride. Benzyne formed in this reaction can undergo the photoisomerization reactions shown.

An unusual leaving group is the oxygen atom. It can be created in the triplet ground state ($^3P_1$) by photosensitized decomposition of N_2O (of course, since the other fragment is N_2, this reaction could be viewed as an example of N_2 elimination, belonging to the previous section). The reaction, of great importance in atmospheric chemistry, provides a method for creating ozone by reaction of oxygen atoms with O_2. It can be carried out in the laboratory at 253.7 nm in the gas phase using mercury vapor as a sensitizer. This interesting synthetic method is suitable for mechanistic studies since O_2 can be excluded from the system at the start of the experiment and cannot be directly implicated in the formation of any of the products. Atomic oxygen is an extremely reactive species which was shown to add to alkenes to produce 'hot' oxiranes, which further isomerized or fragmented into CO and other products.[35] One example of reaction is with a cyclopropene, as shown:

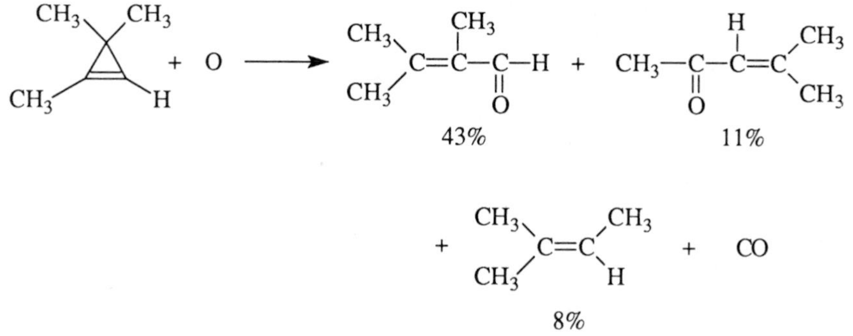

2-Methyl-2-butene (**44**) is assumed to result from the decomposition of the cyclopropanone (**43**) produced from diradicals formed by addition of atomic oxygen.

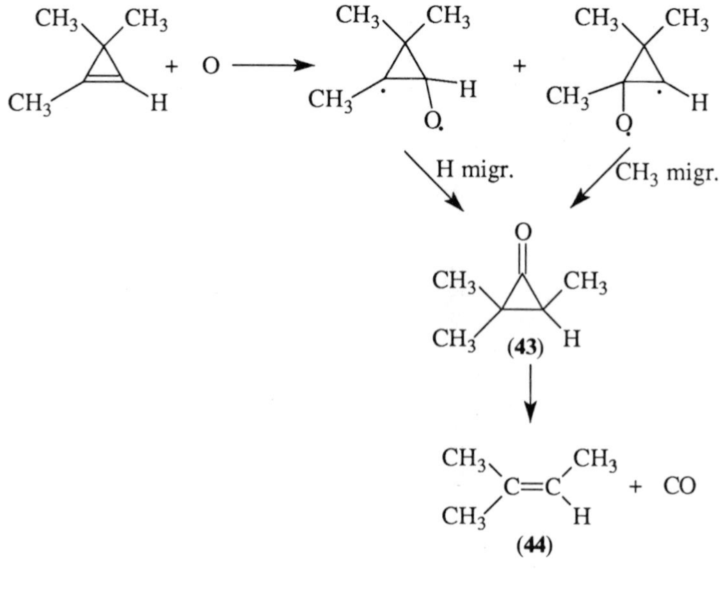

REFERENCES

1. H. Iwata, R. Tanaka, S. Imajo, Y. Oyama and M. Ishiguro, *J. Chem. Soc. Chem. Commun.*, 285–287 (1991).
2. R. A. Bowie and O. C. Musgrave, *Proc. Chem. Soc.*, 15 (1964).
3. D. S. Reddy, G. P. Sollott and P. E. Eaton, *J. Org. Chem.*, **54**, 722–723 (1989).
4. O. L. Chapman, C. C. Chang and N. R. Rosenquist, *J. Am. Chem. Soc.*, **98**, 261–262 (1976).
5. P. J. Kropp and R. L. Adkins, *J. Am. Chem. Soc.*, **113**, 2709–2717 (1991).
6. P. J. Kropp, *Acct. Chem. Res.*, **17**, 131–137 (1984).
7. M. C. Biewer, C. R. Biehn, M. S. Platz, A. Despres and E. Migirdicyan, *J. Am. Chem. Soc.*, **113**, 616–620 (1991).
8. E. J. Corey, J. F. Arnett and G. N. Widiger, *J. Am. Chem. Soc.*, **97**, 430–431 (1975).
9. E. J. Corey and W. R. Hertley, *J. Am. Chem. Soc.*, **81**, 5209–5212 (1959). E. J. Corey and W. R. Hertley, *Chem. Soc.*, **80**, 2903–2904 (1958).
10. A. Alif, J. F. Pilichowski and P. Boule, *J. Photochem. Photobiol. A*, **59**, 209–219 (1991).
11. H. Tukada and K. Mutai, *J. Chem. Soc. Chem. Commun.*, 35–37 (1991).
12. M. P. Cava and E. Moroz, *J. Am. Chem. Soc.*, **84**, 115–116 (1962). J. Meinwald, G. G. Curtin and P. G. Gassman, *J. Am. Chem. Soc.*, **84**, 116–117 (1962).
13 S. P. A. Fodor, J. L. Read, M. C. Pirrung, L. Stryer, A. T. Lu and D. Solas, *Science*, **251**, 767–773 (1991).
14. O. L. Chapman, J. Gano, P. R. West, M. Regitz and G. Maas, *J. Am. Chem. Soc.*, **103**, 7033–7036 (1981).
15. G. Maier, H. P. Reisenauer, H. Balli, W. Brandt and R. Janoscheck, *Angew. Chem. Int. Ed. Engl.*, **29**, 905–908 (1990).
16. C. Bachmann, T. Y. N'Guessan, F. Debu, M. Monnier, J. Pourcin, J.-P. Aycard and H. Bodot, *J. Am. Chem. Soc.*, **112**, 7488–7497 (1990).
17. G. Bucher and W. Sanders, *J. Org. Chem.*, **57**, 1346–1351 (1992).
18. R. L. Rosati, L. V. Leilani, P. Morrissey and J. A. Retsema, *J. Med. Chem.*, **33**, 291–297 (1990).
19. N. Chen, M. Jones, Jr, W. R. White and M. S. Platz, *J. Am. Chem. Soc.*, **113**, 4981–4992 (1991).
20. H. Bayley and J. R. Knowles, *Meth. Enzymol.*, **46**, 69–114 (1977).
21. D. A. Modarelli and M. S. Platz, *J. Am. Chem. Soc.*, **113**, 8985–8986 (1991).
22. T. A. Nakayama and H. G. Khorana, *J. Org. Chem.*, **55**, 4953–4956 (1990).
23. J. Nakayama, T. Fukushima, E. Seki and M. Hoshino, *J. Am. Chem. Soc.*, **101**, 7684–7687 (1979).
24. J. Meinwald and S. Knapp, *J. Am. Chem. Soc.*, **96**, 6532–6534 (1974).
25. S. Masamune, *J. Am. Chem. Soc.*, **86**, 290–291 (1964).
26. O. L. Chapman and J. P. LeRoux, *J. Am. Chem. Soc.*, **100**, 282–285 (1978).
27. O. L. Chapman, R. S. Sheridan and J. P. LeRoux, *J. Am. Chem. Soc.*, **100**, 6245–6247 (1978).
28. S. Ishikawa, S. Tsuji and Y. Sawaki, *J. Am. Chem. Soc.*, **113**, 4282–4288 (1991).
29. G. Bertrand, J. P. Majoral and A. Baceiredo, *Acc. Chem. Res.*, **19**, 17–23 (1986).
30. U. Henriksen and O. Buchardt, *Tetrahedron Lett.*, **31**, 2443–2444 (1990).
31. W. Merterer, O. Klingler, R. Thiergardt, E. Beckmann, H. Fritz and H. Prinzbach, *Chem. Ber.*, **124**, 621–633 (1991).
32. G. Maier, M. Heider and D. Sierakowski, *Tetrahedron Lett.*, **32**, 1961–1962 (1991).
33. K. Okada, K. Okubo and M. Oda, *J. Photochem. Photobiol. A*, **57**, 265–277 (1991).
34. J. G. Radziszewski, B. A. Hess, Jr. and R. Zahradnik, *J. Am. Chem. Soc.*, **114**, 52–57 (1992).
35. R. J. Cvetanovic, *Adv. Photochem.*, **1**, 115–182 (1963).

6. Photochemistry with oxygen

6.1. INTRODUCTION

Oxygen, one of our major environmental components, is required in nature for the survival of most organisms. In the absence of light, most of the chemistry in which oxygen participates is based on radical reactions. Molecular oxygen is a diradical, one of the few simple molecules having a triplet ground state. It reacts with organic radicals to produce species having a peroxide bond:

$$R\cdot + \cdot O{-}O\cdot \rightarrow R{-}O{-}O\cdot$$

Standard radical chain reactions may lead to the formation of a variety of peroxidic products, such as $R{-}O{-}O{-}H$, $R{-}O{-}O{-}R$, or $R{-}O{-}O{-}R'$.

The peroxide bond, being very weak, can undergo homolytic fission (to give two fragments, each bearing an oxygen radical) under a variety of conditions. Fortunately, these oxidation reactions involving molecular oxygen with molecules in nature tend to be slow. The oxidation reactions often proceed in poor yield because natural antioxidants, such as tocopherol (vitamin E), protect many molecules in their environment.

In contrast, the presence of light can create much more reactive oxygen species. Singlet oxygen ($^1O_2$) and the superoxide anion radical (O_2^{-}, often called simply the superoxide ion) are the primary derivatives, from which hydrogen peroxide and hydroxyl radicals ($\cdot$OH) can be formed. Here again, living cells are often protected by natural components such as carotenoid pigments (which quench $^1O_2$) and superoxide dismutase (an enzyme which specifically converts O_2^{-} into hydrogen peroxide and water). Photobiologists describe light-dependent reactions involving oxygen as *photodynamic reactions* and further divide them into reactions of type I, which have a free radical component because of an electron transfer step, and type II, which produce $^1O_2$ (cf. section 10.3)[*,1]

6.2. ELECTRONICALLY EXCITED FORMS OF OXYGEN

Three electronically excited molecules are formed from triplet ground state oxygen (called $^3\Sigma_g^-$): one triplet excited state ($^3\Sigma_u^-$) and two singlet excited states ($^1\Sigma_g^+$ and $^1\Delta_g$). Unfortunately, the exact spin assignments for the electronic states of molecular oxygen are not simple.[2]

6.3. DETECTION OF $^1O_2$

The energy difference between the ground state $^3\Sigma_g^-$ and the excited state $^1\Sigma_g^+$ corresponds to 762.1 nm radiation, and the difference between $^3\Sigma_g^-$ and $^1\Delta_g$ to 1268.7 nm.

---

[*]The term 'photodynamic' has also been used to indicate light-dependent transport phenomena.

The energy difference between two interacting $^1\Delta_g$ molecules and two molecules in the ground state corresponds to 634.3 nm (one-half of 1268.7 nm). These three emissions from electronically excited molecular oxygen have been observed experimentally.

The detection and characterization of $^1O_2$ is not an easy task. The proper choice of solvent is critically important, since it affects the lifetime of $^1O_2$. The maximum lifetime, at the limit of zero pressure in the gas phase, is 45 minutes. However, it is only 4.4×10^{-6} seconds in water, but 10 000 times longer, 4.2×10^{-2} seconds, in chlorofluorocarbon-113.[3]

Photochemically generated $^1O_2$ is usually detected by one of two means. Either $^1O_2$ itself is observed, using spectroscopic techniques, or its presence is inferred from the formation of characteristic products in the presence of specific reagents.

The direct method for detecting $^1O_2$ measures the near-infrared light emission at 1270 nm, associated with the return of the excited molecule to the ground state. Unfortunately, this emission is very weak; therefore it requires rather sophisticated and expensive equipment, even for qualitative measurement.[*]

Experiments for indirect measurement, while more easily performed, may be less reliable. One approach is based on the reaction of $^1O_2$ with specific reagents, most often furan derivatives, such as 2,5-dimethylfuran, furfuryl alcohol, or diphenylisobenzofuran. Other possible reagents include heterocyclic derivatives, such as imidazole or histidine, sulfides, hindered olefins (such as adamantylideneadamantane), or vinyl ethers. In all cases, the presence of $^1O_2$ is inferred either from the disappearance of the added reagent or from the formation of known reaction products. Chromatographic techniques, either high-performance liquid chromatography (HPLC) or gas chromatography, are particularly convenient for obtaining kinetic data in conjunction with the chemical assays. One convenient spectroscopic method which has often been used involves the trapping of $^1O_2$ by imidazole or histidine; the bleaching of p-nitrosodimethylaniline by the intermediate formed can be followed by measuring the decrease in absorption at 440 nm.[5]

A second indirect approach to the detection of $^1O_2$ takes advantage of the large differences in $^1O_2$ lifetime in different solvents. If the light-dependent disappearance of a standard $^1O_2$ trap (such as a furan derivative) has been measured in water, for example, the rate is expected to increase in deuterated water, in which the lifetime of $^1O_2$ is known to be longer. Other solvent couples that might be used are CH_3OH vs CD_3OD, CD_3OH or CH_3OD, C_6H_6 vs C_6D_6, or $CHCl_3$ vs $CDCl_3$. Conversely, any added reagents which compete for $^1O_2$ with the detecting reagent should reduce the rate of reaction with the latter. Sodium azide (NaN_3), and DABCO (1,4-diazabicyclooctane) are two $^1O_2$ quenchers used frequently in aqueous solutions. In organic solvents, β-carotene is perhaps the most efficient $^1O_2$ quencher.

The use of quenchers is not without pitfalls. Since a quencher must react with an electronically excited molecule, one must be sure that it is $^1O_2$ and not the electronically excited sensitizer that has been quenched. Either process would reduce the amount of diagnostic product formed. In practice, one must measure the products formed in the absence of quencher, as well as in the presence of various concentrations of quencher, in order to ascertain that the normal target and the quencher are indeed competing for $^1O_2$. This is best expressed graphically in the form of a Stern–Volmer plot (see Chapter 1, Section 1.22.1).

---

[*]For a review of instrumentation, see ref. 4.

Analyzing for reaction products derived from the quencher itself is not always necessary and would often be impractical. When NaN_3 is used as the quencher, for example, more than 99% of the reactions involve physical rather than chemical quenching.[6]

If $^1O_2$ is suspected to be produced in a photochemical reaction, leading to an oxidation product, the course of the reaction may be confirmed by generating $^1O_2$ independently by non-photochemical means, and comparing the two results.

The formation of singlet oxygen may also be determined by electron spin resonance (ESR) spectroscopy, utilizing either nitrones,[7] 2,2,6,6-tetramethyl-4-piperidone (1) in neutral solution, or 2,2,6,6-tetramethylpiperidine in basic solution.[8] The product (2) is detected when (1) is used, but the detailed steps leading to its formation are not known.

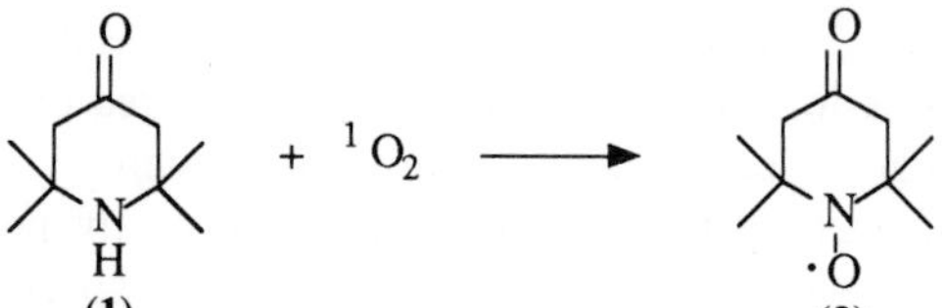

Alternatively, the formation of $^1O_2$ in solution may be analyzed by following the change in ESR spectrum associated with the conversion of an aromatic nitroxide into its endoperoxide, as in the reaction of (3).[9]

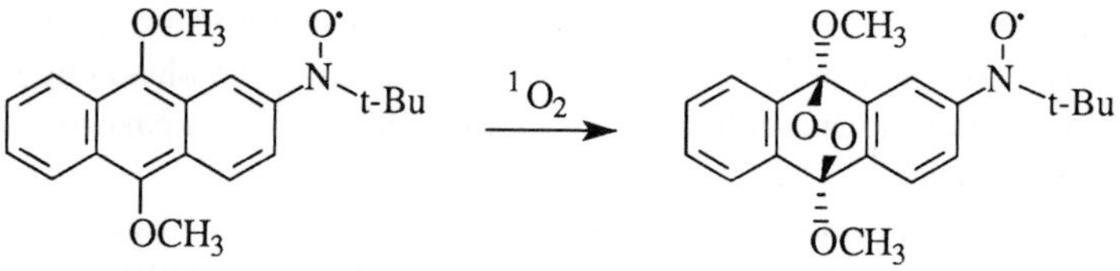

6.4. PHOTOCHEMICAL SYNTHESES OF $^1O_2$

Dye-sensitized reactions were used to generate $^1O_2$ long before the nature of the oxidizing species was known. A number of sensitizers have been popular. Xanthene dyes such as rose bengal or erythrosin B are among the best known; methylene blue, acridine orange, and porphyrin derivatives are also frequently used.

As mentioned earlier, the extremely short lifetime of $^1O_2$ creates a fundamental problem, since pure $^1O_2$ is normally not readily available as a reagent in the absence of its precursor(s) and of the eventual reaction product(s) arising from the precursor(s). In other words, it is very difficult to study clean bimolecular reactions between $^1O_2$ and a substrate when $^1O_2$ is generated in solution by classical methods. Recent progress in separating $^1O_2$ from its precursors and their eventual reaction products has ensured cleaner reactions between $^1O_2$ and a substrate. Two approaches have been described:

1. Using a sensitizer which is polymeric and insoluble, only the reactions that occur in solution are studied. Rose bengal covalently bound to a polymer is commercially

available for this type of application (Sensitox I). After irradiation, it can be filtered off and reused if desired. One disadvantage of this technique is that $^1O_2$ is generated in a heterogeneous environment. Because it is almost impossible to exactly reproduce the experimental conditions, the approach is poorly suited for kinetic studies. Also, the sensitizer is bleached with time. Other immobilized photosensitizers have been described, such as $Ru(bpy)_3^{2+}$ in a zeolite, which has high catalytic turnover and high stability.[10] One practical application of immobilized sensitizers is in water disinfection, where the photodynamic inactivation of water-borne bacteria has been demonstrated.[11]

2. Producing $^1O_2$ continuously in the gas phase, and studying its reactions in solution. The short liftetime of $^1O_2$ in the gas phase creates limitations. Two approaches have been described. In the method of Wang and Midden, the sensitizer is immobilized as a very thin layer on a glass plate, which is placed about 0.1 mm above the surface of the solution containing the substrate. Light shining through the sensitizer-coated plate produces $^1O_2$, which can diffuse toward the aqueous phase.[12,13] Although rose bengal was initially recommended for this technique, the nature of the photosensitizer should be immaterial. Fluoranthene[14] and tetraphenylporphines[15] have been successfully used. A plant leaf has even been substituted successfully for the sensitizer, leading to the suggestion that natural components of the plant might create $^1O_2$ as a defense mechanism against predators.[16]

In principle, this technique keeps the substrate from reacting with $O_2^{\cdot-}$, which does not diffuse through the gas phase. However, the procedure suffers from severe limitations. First, the sensitizer is temporarily held on the lower side of a glass plate extremely close to the liquid layer containing the substrate. Great skill is required to set up the experiment without accidental contact between the sensitizer and the solution. Should particles of sensitizer fall into the solution, they would induce normal sensitized reactions. One must also avoid contact between sensitizer and solution during irradiation. The temperature usually increases during these experiments. As the solution undergoes thermal expansion, the already small gap between the separated phases decreases and could lead to temporary contact between the sensitizer and substrate. Stringent temperature control during the irradiations is therefore required. A critical control consists of running one experiment with the substrate omitted. After the sensitizer plate is removed, the substrate should be added and photolyzed. If none of the transformations usually attributed to the $^1O_2$ reactions are detected, one may have greater faith in the results of similar, complete, experiments.

In a more rigorous interpretation of the experimental results obtained with Midden and Wang technique, Parker concluded that $O_2(^1\Delta_g)$ was indeed the active oxidizing agent, and that the experiments agreed completely with the theory. For experiments using bacterial targets, theory and experiments did not agree because the kill rate did not depend linearly on the $O_2(^1\Delta_g)$ concentration in the immediate vicinity of the bacteria.[17] From the observed quadratic dependence on $O_2(^1\Delta_g)$, the cytotoxic species was claimed to be $O_2(^1\Sigma_g^+)$, formed by pooling energy between two $O_2(^1\Delta_g)$ molecules.

In an alternative approach for generating $^1O_2$ in the gas phase, a glass tube internally coated with an efficient sensitizer is irradiated while a rapid stream of oxygen flows through. The effluent carrying $^1O_2$ is bubbled into the solution containing the substrate.[18] This technique is not suitable for substrates such as erythrocytes or liposomes, which are too susceptible to hydrodynamic damage.

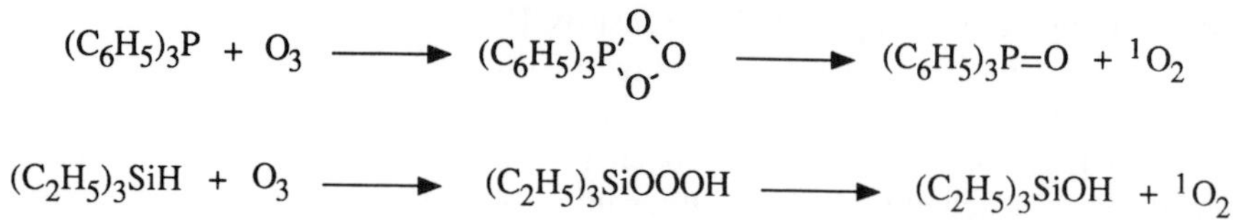

$$(C_6H_5)_3P + O_3 \longrightarrow (C_6H_5)_3P{\overset{O}{\underset{O^-}{\diagdown}}}O \longrightarrow (C_6H_5)_3P{=}O + {}^1O_2$$

$$(C_2H_5)_3SiH + O_3 \longrightarrow (C_2H_5)_3SiOOOH \longrightarrow (C_2H_5)_3SiOH + {}^1O_2$$

Fig. 6.1   Generation of ${}^1O_2$ from ozone with triphenylphosphine and triethylsilane.

6.5. NON-PHOTOCHEMICAL SYNTHESES OF ${}^1O_2$

${}^1O_2$ has been prepared by a variety of non-photochemical procedures:

1. From molecular oxygen—the gas is allowed to flow through a tube submitted to electrodeless discharge at 6.7 MHz and is bubbled into a solution of reactant. Products typical of ${}^1O_2$ reactions are observed,[19] but, because oxygen atoms and ozone are formed simultaneously, it is not a clean method for testing the effect of ${}^1O_2$ in biological systems or for investigating specific oxidation pathways.

2. From ozone—the approach uses reagents such as triphenylphosphine or a trialkylsilane, which form very stable oxidation products, as shown in Fig. 6.1.[20,21] The initial ozone adduct decomposes thermally near room temperature to produce ${}^1O_2$.

3. From hydrogen peroxide—by reaction with sodium hypochlorite or with chlorine gas in alkaline medium:

$$H_2O_2 + ClO^- \rightarrow H_2O + Cl^- + {}^1O_2$$

This reaction is stoichiometric. Many other elements can induce the generation of ${}^1O_2$ from hydrogen peroxide, often catalytically.[22] One example is sodium molybdate, which produces a quantitative conversion of hydrogen peroxide in basic medium:[23]

$$2\,H_2O_2 + MoO_4^{2-} \rightleftharpoons 2\,H_2O + MoO_6^{2-}$$

$$MoO_6^{2-} \rightarrow MoO_4^{2-} + {}^1O_2$$

4. From organic endoperoxides. While most reactions of ${}^1O_2$ give products irreversibly, some are reversible. A typical case is the reaction with polycyclic aromatic hydrocarbons (Fig. 6.2), exemplified with a naphthalene derivative. The choice of substituents can affect the physical properties of the molecules. Introducing at least one carboxylic acid group in a side-chain can make both the starting material and its deoxygenated product water soluble. This is a useful tool for generating ${}^1O_2$ in aqueous solution.[24]

As one example of the benefits which may be gained by generating ${}^1O_2$ non-photochemically, the emission at 1930 nm corresponding to ${}^1\Sigma_g^- \rightarrow {}^1\Delta_g$ could be observed when a sensitizer such as benzophenone was irradiated in the presence of oxygen, but only the emission at 1270 nm corresponding to ${}^1\Delta_g \rightarrow {}^3\Sigma_g^+$ was detected when

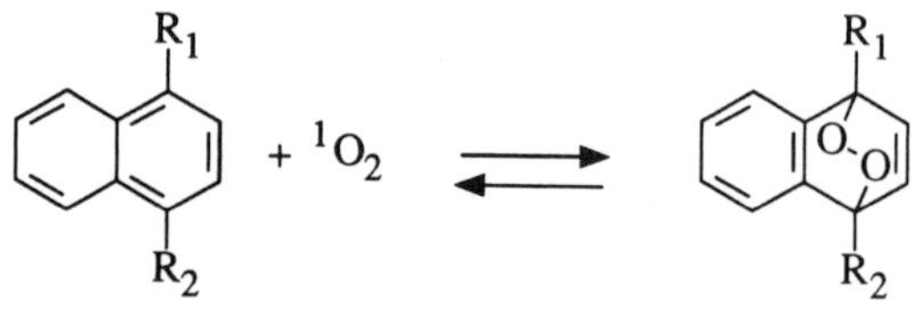

Fig. 6.2   The reaction of a naphthalene derivative with ${}^1O_2$ (forward direction), and the synthesis of ${}^1O_2$ from an endoperoxide (reverse direction).

1,4-dimethylnaphthalene endoperoxide was decomposed either thermally or with a laser at 266 nm.[25]

5. From organic hydroperoxides—the oxidation of hydroperoxides, for example by cerium(IV), produces hydroperoxy radicals which dimerize and decompose with the formation of $^1O_2$:

$$2 \; RR'R''COO\cdot \; \rightarrow \; RR'R''COOOOCRR'R''$$

$$RR'R''COOOOCRR'R'' \; \rightarrow \; RR'R''COOCRR'R'' + O_2$$

If $R'' = H$:

$$RR'HCOOOOCHRR' \; \rightarrow \; RR'CO + RR'CHOH + O_2$$

Organic hydroperoxides are several orders of magnitude better than hydrogen peroxide for generating $^1O_2$.[26]

6. From the disproportionation of $O_2^{\;\bar{}}$:

$$2 \; O_2^{\;\bar{}} + 2 \; H^+ \; \rightarrow \; H_2O_2 + O_2$$

There has been much debate as to whether molecular oxygen is produced in its ground state or in an excited state. It appears that a tiny proportion of the oxygen produced is in the singlet excited state, but most is in the ground state.[27]

7. From pulse radiolysis. Recently, $^1O_2$ has been generated by pulse radiolysis, and its detection by a time-resolved method has been achieved.[28] In this approach, electronically excited benzene is initially produced by reaction with the electrons generated in the reactor. The triplet excited state of benzene reacts with an oxygen molecule to produce $^1O_2$, and the luminescence is measured with a detector system connected to the reactor with a fiber optic cable.

6.6. REACTIONS OF $^1O_2$ WITH UNSATURATED MOLECULES

The best known reactions of $^1O_2$ in organic chemistry are with alkenes (Fig. 6.3). A unifying reaction mechanism involves the initial formation of a perepoxide (which is not isolated). Further transformations include: (1) the formation of an allylic hydroperoxide (an overall ene reaction which produces a migration of the double bond); (2) a

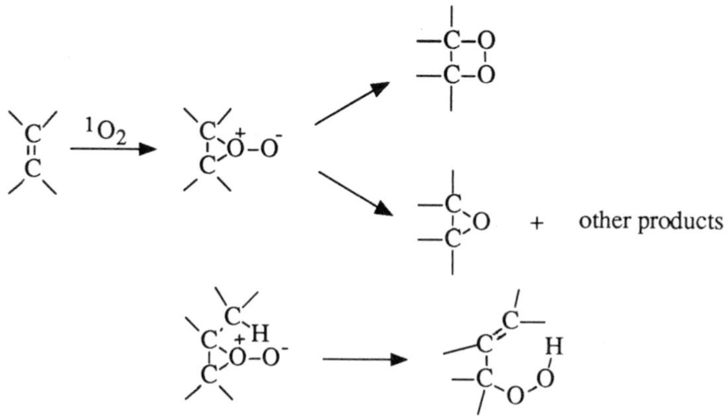

Fig. 6.3 Typical reactions of $^1O_2$ with alkenes.

ring expansion to a dioxetane; and (3) a redox process which creates an epoxide. Unfortunately, it is still impossible to generate at will any one of these products from the same starting material. The more substituted alkenes tend to produce epoxides and dioxetanes. The less substituted alkenes which have allylic hydrogens tend to produce allylic hydroperoxides. These can be reduced chemically to allylic alcohols, or they can subsequently decompose via free radical mechanisms. The latter are particularly important in lipids.

Despite the apparent complexity of the reactions between $^1O_2$ and organic molecules, selectivity is often observed. Hindered alkenes, for example, may be converted into reasonably stable dioxetanes in good yield. A good example is adamantylideneadamanane (4) which can be used for titrating $^1O_2$ by monitoring the adamantanone (5) formed upon thermal decomposition.

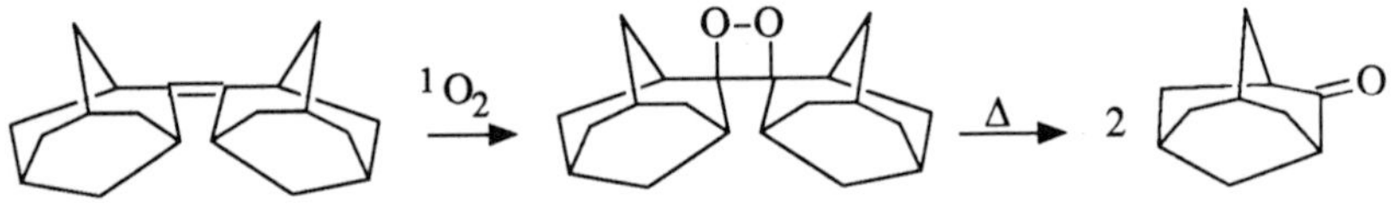

(4)   (5)

As indicated in Chapter 1, one of the carbonyl fragments is generated in an electronically excited state. It should be noted that this cleavage of a double bond into two carbonyl fragments by reaction with $^1O_2$ is equivalent to the well-known ozonolysis method.

$^1O_2$ also typically reacts with 1,3-dienes, producing cyclic peroxides:

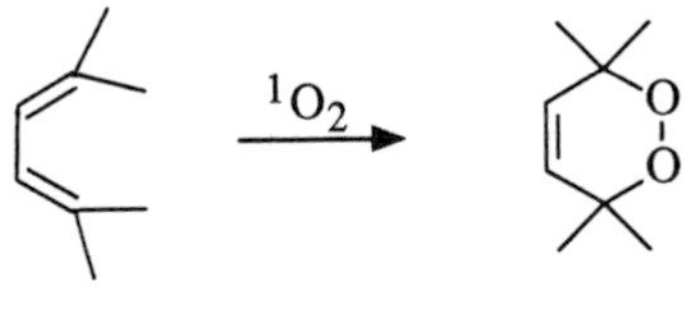

Depending on the nature of the substituents and reagents, these cyclic peroxides can undergo several different reactions, usually associated with breaking of the peroxidic linkage. The formation of cyclic peroxides, also known as endoperoxides, is particularly important in the case of fused aromatic compounds, as it can be reversed and, as explained earlier, it provides an effective method for storing $^1O_2$ via endoperoxides.

Although endoperoxides from many polycyclic aromatic hydrocarbons were prepared without difficulty, benzene 1,4-endoperoxide was not known until a clever indirect synthetic method was developed. It features $^1O_2$ addition to the benzene dimer (6), followed by photochemical cleavage (300 nm) of the cyclobutane ring at low temperature, generating benzene and benzene endoperoxide (7). Photolysis at 365 nm rearranged the peroxide into the diepoxide (8) without cleaving the cyclobutane ring.[29]

Note that some stable endoperoxides do not decompose to produce $^1O_2$. They decompose thermally into oxygen-centered radicals, and are useful for inducing radical chain reactions. The naturally occurring terpene derivative ascaridole (9) is a typical example.

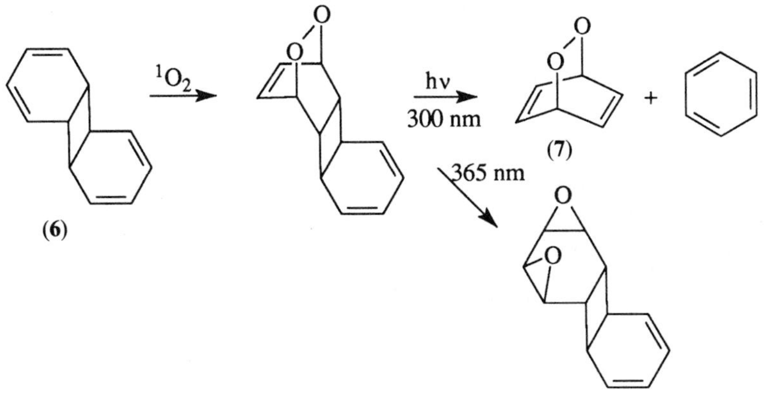

(6)   (7)   (8)

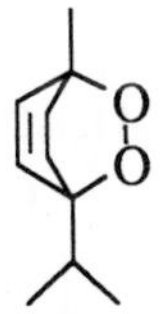

(9)

Despite the possibility of many competing reactions taking place with $^1O_2$, selectivity is often observed. This is demonstrated in Woodward's synthesis of chlorophyll, where only one double bond was affected (in the south-west direction of the structure) in preparing (10), in a manner which suggested the intermediacy of a dioxetane.[30]

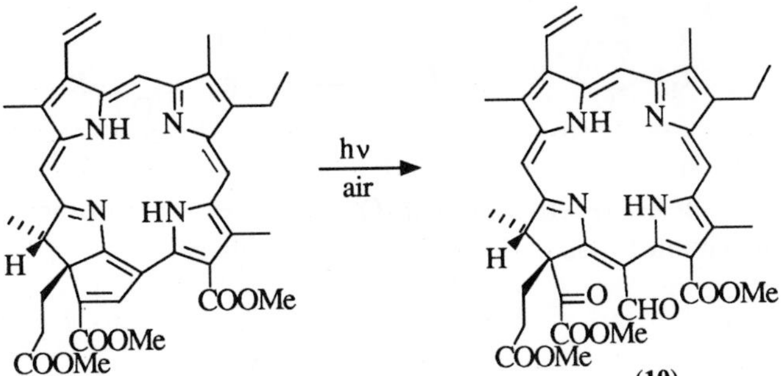

(10)

6.7. REACTIONS OF $^1O_2$ WITH SULFIDES

Sulfides are easily oxidized by $^1O_2$, initially producing a peroxidic intermediate, which leads to the formation of sufoxides and sulfones. In protic solvents the dipolar structure has been assigned to the initial intermediate, but the cyclic isomer is formed in aprotic solvents.[31] Both routes are illustrated below:

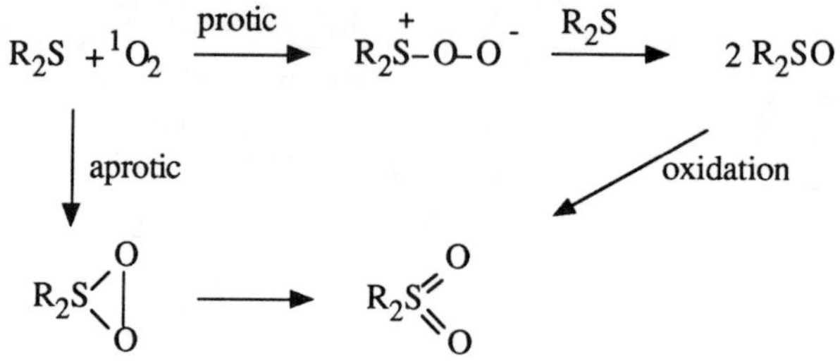

$$R_2S + {}^1O_2 \xrightarrow{\text{protic}} R_2\overset{+}{S}\text{-O-O}^{-} \xrightarrow{R_2S} 2\,R_2SO$$

with branches labeled "aprotic", "oxidation", showing structures R_2S with O and O leading to R_2S with O and O.

6.8. MISCELLANEOUS REACTIONS OF ${}^1O_2$

Photochemically generated ${}^1O_2$ oxidizes diazo compounds to produce carbonyl ylides, perhaps through a five-membered ring intermediate:[32]

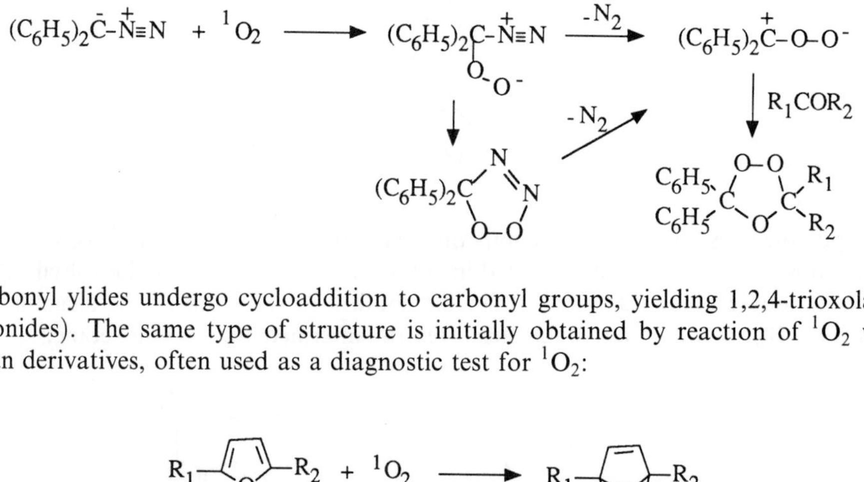

$$(C_6H_5)_2\overset{-}{\underset{}{C}}\text{-}\overset{+}{N}\text{≡}N + {}^1O_2 \longrightarrow (C_6H_5)_2\overset{+}{C}\text{-}\overset{+}{N}\text{≡}N \xrightarrow{\text{-}N_2} (C_6H_5)_2\overset{+}{C}\text{-O-O}^{-}$$

with intermediate $(C_6H_5)_2C$ five-membered ring with N, N, O-O labeled, $\xrightarrow{-N_2}$, and $\xrightarrow{R_1COR_2}$ giving the trioxolane structure $\begin{smallmatrix}C_6H_5 \\ C_6H_5\end{smallmatrix}$ with O-O and R_1, R_2.

Carbonyl ylides undergo cycloaddition to carbonyl groups, yielding 1,2,4-trioxolanes (ozonides). The same type of structure is initially obtained by reaction of ${}^1O_2$ with furan derivatives, often used as a diagnostic test for ${}^1O_2$:

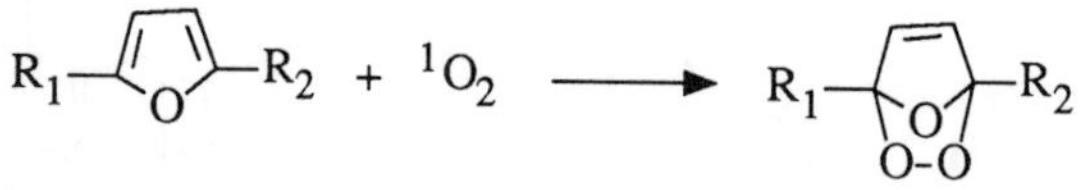

$$R_1\text{—}\underset{O}{\boxed{}}\text{—}R_2 + {}^1O_2 \longrightarrow R_1\text{—}\underset{\underset{O\text{-}O}{O}}{\boxed{}}\text{—}R_2$$

Thiophenes are much less reactive toward ${}^1O_2$. Oxazoles, on the other hand, undergo a quantitative conversion to triamides via 1,2,4-trioxolanes:[33]

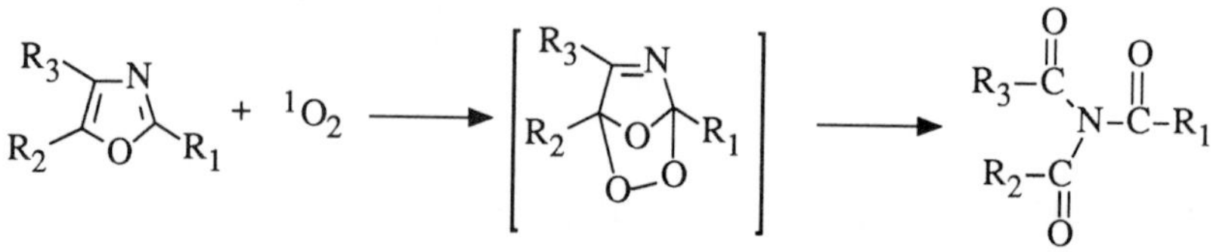

$$\underset{R_2}{\overset{R_3}{\diagup}}\underset{O}{\overset{N}{\diagdown}}R_1 + {}^1O_2 \longrightarrow \left[\underset{R_2}{\overset{R_3}{\diagup}}\underset{O\text{-}O}{\overset{N}{\diagdown}}R_1\right] \longrightarrow \begin{matrix}R_3\text{-}\overset{O}{\overset{\|}{C}} & \overset{O}{\overset{\|}{C}}\text{-}R_1 \\ & N \\ R_2\text{-}\overset{}{\underset{\|}{C}} \\ O\end{matrix}$$

The triamide (11) has been used for macrocyclic lactonization, where the alcohol intramolecularly attacks the carbonyl of the alkylacylamide.

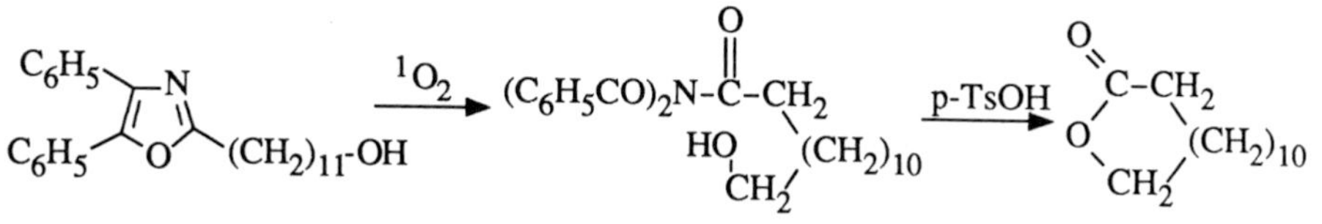

(11)

$^1O_2$ is known to react with a large number of organic molecules, including proteins, lipids, and nucleic acids. It is therefore highly cytotoxic, a property of major importance in the photodynamic therapy of cancers, which is becoming increasingly valuable (Chapter 11).

6.9. BIOLOGICAL ACTIVITY OF $^1O_2$

The diversity of chemical reactions known for $^1O_2$ readily explains the toxicity of this compound, which is generated photochemically from a large variety of sensitizers. Most molecules absorbing light in the visible or ultraviolet regions are candidates for making $^1O_2$. Therefore, it is not surprising to find that cosmetic products have long been known to lead to skin reactions when wearers are exposed to sunlight. This is particularly well documented in the case of citrus extracts containing furocoumarins related to those used in puvatherapy and other aspects of photodynamic therapy (cf. Chapter 11).

It is important to note that a sensitizer may be involved with the formation of $^1O_2$ in two different manners: (1) the sensitizer acts as a catalyst for forming $^1O_2$ whenever light strikes, and it is $^1O_2$ which is the toxic agent, or (2) the sensitizer helps generate $^1O_2$ but is itself then modified into a toxic product by the $^1O_2$ formed. This is occasionally the case with polycyclic aromatic hydrocarbons. In a similar manner, aflatoxin B_1 may become more mutagenic by reacting with $^1O_2$ produced through photosensitization by aflatoxin B_2.[34]

Another interesting example is that of the psoralens, which are very important sensitizers used in medicine (Chapter 11). These molecules are known to interact with different kinds of biomolecules, particularly DNA, and to undergo cycloaddition reactions, which have been studied extensively. However, psoralens (e.g. (12)) also produce $^1O_2$, with which they react to form dioxetanes (e.g. (13)), which are stable at low temperatures. Dioxetanes have a high mutagenic acticity in *Samonella typhumurium* strain TA 100.[35]

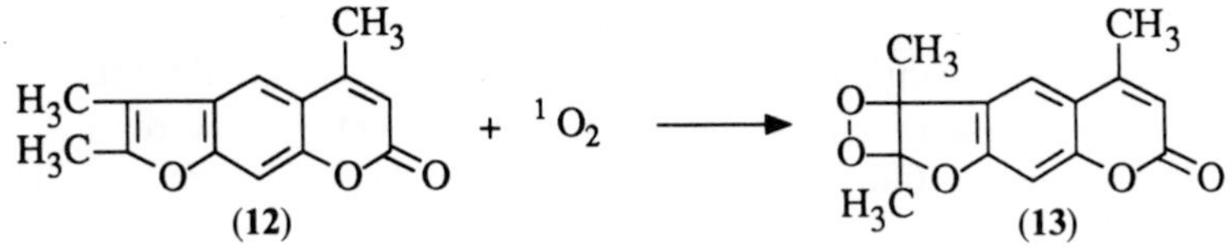

(12)   + $^1O_2$ →   (13)

It is interesting to note that furanocoumarins were earlier known to undergo opening of the furan ring, following what appeared to be a [4 + 2]cycloaddition reaction with

$^1O_2$. It is reasonable to believe that the initial attack is a $[2 + 2]$cycloaddition reaction, as shown, and that a rearrangement follows in which there is a 1,3 suprafacial shift of one oxygen atom.

It is important to point out that many sensitizers have demonstrable biological activity, acting for example as herbicides, insecticides, or fish poisons. The fact that a compound generates $^1O_2$ with high quantum yields in the laboratory cannot be taken as proof that the biological activity *in vivo* is necessarily caused by $^1O_2$. Because of the effect of the cellular environment and possibly of rapid metabolic transformations, other processes (superoxide formation, radical reactions, etc.) may actually be responsible for the lethal biological damage. Mechanistic studies *in vivo* are difficult.

6.10. $^1O_2$ IN A CHEMICAL LASER

An intense yellow chemiluminescence develops on addition of I_2 to a flow of oxygen gas containing $O_2(^1\Delta_g)$. A strong atomic line is also emitted at 1315 nm, and laser action can therefore be achieved by using a chemical source of $^1O_2$. This process was thought to have potential military applications.

6.11. THE SUPEROXIDE RADICAL ANION

The name given by photobiologists to the photochemical reactions involving $^1O_2$ which were described above is *photodynamic reactions of type II*. Another class, called *photodynamic reactions of type I*, involves one electron transfer step to oxygen, either from the excited sensitizer or from a molecule with which it has reacted. Two products are created, the superoxide anion radical O_2^- and a cation radical derived from the electron donor.

Most of the photochemistry involving $^1O_2$ has been performed in organic solvents. The quantum yield for the formation of $^1O_2$, in particular, has been determined for many photosensitizers, but almost always in organic solvents, where a lifetime of $^1O_2$ is usually much longer than in aqueous solutions. Remember that *the quantum yields for $^1O_2$ production are different in different solvents.*

The electron transfer from a donor D to molecular oxygen initially produces a 'tight' pair of species, the radical cation D^+ and the radical anion O_2^-. In order to separate the two, the electrostatic attraction between them must be overcome. The charge separation is difficult in a non-polar organic solvent, but is greatly assisted by the presence of water or some other solvent which is either protic or has a high dielectric constant:

$$\text{Donor}^* + O_2 \rightarrow [\text{Donor}^+ + O_2^-]_{solv} \rightarrow \text{Donor}^+{}_{solv} + O_2^-{}_{solv}$$

Therefore, one may expect to observe O_2^- in aqueous solution even with sensitizers which produce $^1O_2$ with a quantum yield close to 1 in organic solvents. Experiments have confirmed this expectation.

The superoxide ion (as O_2^- is usually called) is a very good nucleophile, as well as a base which can be protonated to $HOO\cdot$. The stability of O_2^- is highly pH-dependent, since the protonated form, $HOO\cdot$, disproportionates:

$$2\ HOO\cdot \rightarrow H_2O_2 + O_2$$

This reaction is also called a dismutation, and any enzyme which catalyzes this conversion is called a superoxide dismutase (some superoxide dismutases are commercially available). At least a small part of the oxygen created in the dismutation process may be in the singlet state.

Hydrogen peroxide, produced in the dismutation of O_2^-, is also a well known oxidizing agent. Biochemically, hydrogen peroxide is decomposed by the enzyme catalase into water and oxygen:

$$H_2O_2 + Catalase \rightarrow H_2O + \frac{1}{2}O_2$$

Non-enzymatically, transition metal ions convert hydrogen peroxide into hydroxyl radicals, probably the most reactive species produced in the photochemistry of oxygen. This is called the Fenton reaction.

The Fenton reaction is usually carried out in the laboratory with Fe^{2+}, which is also widely distributed in living cells:

$$H_2O_2 + Fe^{2+} \rightarrow HO\cdot + OH^- + Fe^{3+}$$

One should also note that O_2^- may be involved in a related redox process, the Haber–Weiss reaction.[36] In the presence of a metal ion, Fe^{3+} for example, O_2^- is initially converted to hydrogen peroxide, which is then decomposed into $\cdot OH$ by Fe^{2+}:

$$2\ O_2^- + 2\ H^+ \rightarrow H_2O_2 + O_2$$

$$Fe^{3+} + O_2^- \rightarrow Fe^{2+} + O_2$$

$$Fe^{2+} + H_2O_2 \rightarrow Fe^{3+} + OH^- + \cdot OH$$

$$\overline{3\ O_2^- + 2\ H^+ \rightarrow 2\ O_2 + OH^- + \cdot OH}$$

The complexity of O_2^- chemistry is now becoming apparent. This species can react either as an oxidant or a reducing reagent. One must cope with the reactions of the radical anion and its conjugate acid, with hydrogen peroxide and its possible reactions, including those with transition metal ions or with the enzyme catalase which lead to $\cdot OH$, and with the reactions of this highly reactive hydroxyl radical.

6.12. DETECTION OF O_2^-

O_2^- can be detected by ESR spectroscopy, following spin trapping by nitrones either of O_2^- itself, or of $\cdot OH$ (from secondary reactions):[37]

O_2^- can also be detected by luminescence, using 9-acridone-2-sulfonic acid as a reagent.[38]

The reduction of well-known substrates such as nitro blue tetrazolium (NBT) or ferricytochrome c by O_2^- provides convenient methods of detection. The electron transfer is observed spectroscopically, by monitoring the increase in absorbance at 550 nm associated with the reduced species. While a solution of cytochrome c remains homogeneous as the reduction proceeds, the reduced form of NBT is highly insoluble. Light scattered by the precipitate formed during the reduction contributes also to the increased absorbance at 550 nm. When the transfer is observed, one must then determine whether the electron is transferred to the reagent from the sensitizer itself or from O_2^-. The selectivity of the enzyme superoxide dismutase (SOD) facilitates the distinction. If the reduction of SOD or cytochrome c is decreased markedly when the enzyme is added at the start of the photolysis, O_2^- must have been an intermediate in the photoinduced reduction. Since this test assesses the outcome of a competition experiment (SOD and cytochrome c compete for the O_2^- generated), the extent of cytochrome c reduction must be inversely related to the concentration of SOD present.

When the effect of SOD cannot be firmly established, a control experiment should be performed to be sure that SOD was not inactivated by $^1O_2$ which might have been produced in the photolysis. Since O_2^- can be generated in the absence of light, it is possible to test the activity of SOD which has been irradiated aerobically in the presence of a sensitizer in the conditions of the experiment.

The easiest way to generate O_2^- without irradiation is with the enzyme xanthine oxidase (XO, also commercially available) in the presence of xanthine or acetaldehyde. The activity of SOD is determined by establishing the rate of reduction of cytochrome c in the presence of known concentrations of substrate and XO, and by measuring how the addition of SOD—either irradiated or not irradiated—affects this rate in the presence of the sensitizer of interest.

Comparing the extent of a reaction in the presence and in the absence of oxygen does not always directly demonstrate the participation of O_2^-. The reason is that many diagnostic reactions of O_2^- involve an electron transfer to a substrate. When cytochrome c is photoreduced, for example, it is the reduction of Fe^{3+} to Fe^{2+} which is eventually detected. However, in the presence of oxygen, the initial electron transfer takes place from an electronically excited donor to oxygen, which ultimately transfers it to cytochrome c. The absence of oxygen does not necessarily prevent the excited donor from transferring the electron directly to cytochrome c.

A further complication is that cytochrome c can be photoreduced in the absence of any photosensitizers (a ligand or component of the protein backbone supplies the electron). While almost totally absent in the presence of oxygen, this photoreduction is quite important in an inert atmosphere and is further magnified by the presence of other potential electron donors, such as EDTA, a common component of biochemical media.

The oxygen requirement in many photodynamic reactions can be demonstrated in part by observing that photochemical transformations do not occur in the presence of an inert gas such as argon, nitrogen, carbon monoxide, or carbon dioxide. This is not always easy, as oxygen often must be removed completely for the difference to be noticeable. While chemists can speed up gas exchanges by freeze–thawing techniques, repeatedly evacuating the reaction vessel and flushing it with the inert gas, the approach does not always work in biological studies with cells which can be irreversibly damaged under high vacuum. For example, the rate of gas exchange in agar plates bearing microorganisms can be infuriatingly slow. Furthermore, one never knows when the exchange is complete. It is not surprising to find, therefore, that reactions once considered to be non-photodynamic, because they occurred even after extensive deoxygenation with an inert gas, eventually turned out to be oxygen-dependent. Such was the case of phenylheptatriyne, a natural product found in the plant family Asteraceae (Compositae). First described as an antibiotic compound, it actually has no such activity in the dark. Mechanistic studies originally led to the conclusion that the phototoxicity did not depend on the presence of oxygen, a conclusion which was later proved to be erroneous.[39] There are other examples of toxic molecules actually deriving their activity from light activation, such as α-terthienyl and the antibiotic gilvocarcin V.

6.13. ATOMIC OXYGEN

Fragmentation reactions of suitable precursors may lead to the formation of atomic oxygen, which is very reactive. This reaction has been mentioned at the end of Chapter 5 (Section 5.10).

6.14. PHOTOLYTIC OZONATION

Photolytic ozonation is used for the destruction in water of organic compounds which do not degrade when treated with ozone alone. Irradiation of ozone in water at 253.7 nm is now believed to produce hydrogen peroxide directly:

$$O_3 + H_2O \xrightarrow{h\nu} O_2 + H_2O_2$$

Complex reactions may ensue, including the rapid reaction of ozone with the hydrogen peroxide anion, and eventual formation of hydroxyl radicals which then react with the organic molecules.[40]

6.15. DIATOMIC SULFUR

It should be mentioned that the well-known chemical similarities between oxygen- and sulfur-containing molecules in organic chemistry are less obvious in the photochemical field. The equivalent of $^1O_2$ is $^1S_2$, which is obtained only with the greatest difficulty, for example by decomposition of cyclic disulfide precursors such as (**14**) and (**15**).[41,42] One must note that both the cycloaddition step, converting two thiocarbonyl groups into a 1,2-dithietane, and its decomposition to an alkene and diatomic sulfur (this can be either photochemical or thermal), are different from the usual chemistry of carbonyls and dioxetanes. In the chemistry of the oxygen analogs, alkenes are the precursors to dioxetanes and the carbonyls are the end-products; the reverse has been shown above with the sulfur compounds.

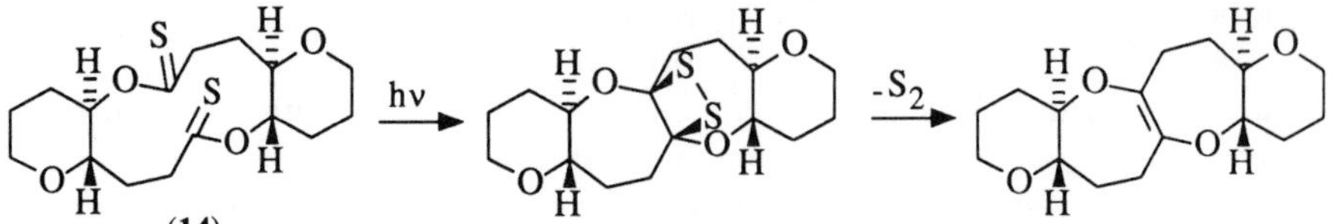

(14)

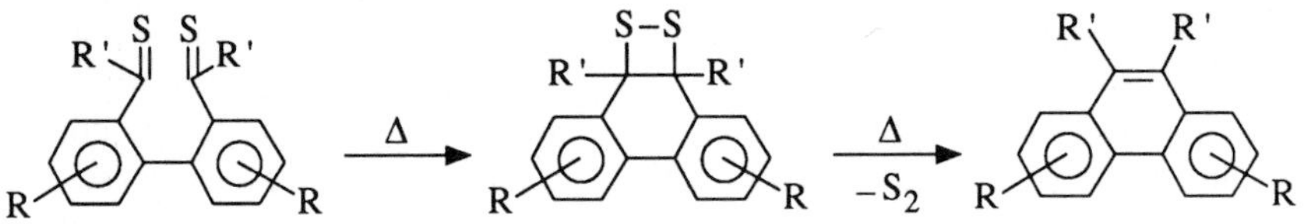

(15)

The chemistry of S_2, however, is somewhat similar to that of $^1O_2$, as it adds to 1,3-dienes to produce unstable six-membered cyclic disulfides:

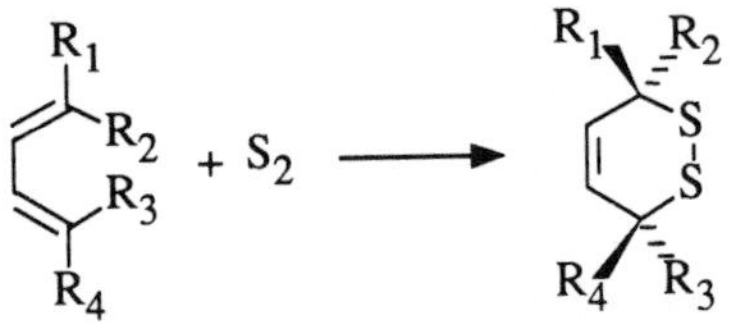

The situation is more complex with strained or cyclic dienes, which yield trithiolanes, products containing a five-membered ring with *three* sulfur atoms. They probably result from the loss of one sulfur atom from adducts with 2 mol of S_2.

S_2 adds to hindered alkenes, yielding episulfides,[43] but the equivalent of the allylic hydroperoxydation reaction does not seem to have been reported.

6.16. SULFUR MONOXIDE

The formation of sulfur monoxide has been observed in many photochemical reactions, particularly in the photolysis of sulfur dioxide, sulfur trioxide, carbon disulfide + molecular oxygen, carbon disulfide + nitrogen dioxide, and in the photolysis and thermolysis of sulfoxides. One example of reactions claimed to have generated sulfur monoxide photochemically is shown in Fig. 6.4. Depending upon the nature of the substituent R, varying amounts of the second products which revealed delocalization of a charge (radical or anion) on carbon and oxygen atoms were obtained.[44] Little work comparable to the studies with singlet oxygen has been recorded with sulfur

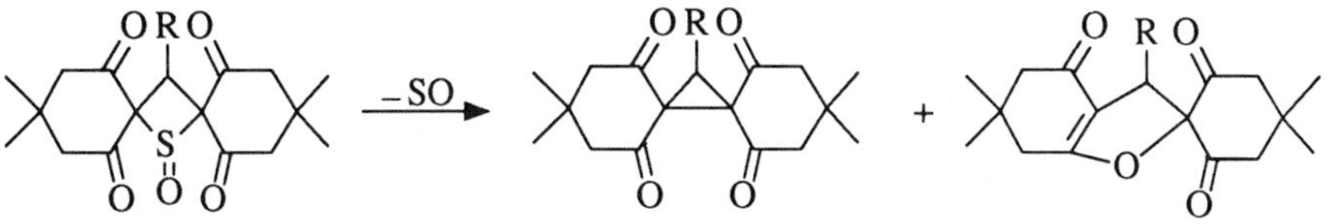

Fig. 6.4 An example of a photochemical reaction involving the generation of sulfur monoxide.

monoxide. Although some of the fragmentation reactions probably generate sulfur monoxide in a singlet state, the ground state of the molecule is triplet.

6.17. PHOTOCHEMISTRY WITHOUT OXYGEN

After having provided a rather extensive discussion of the photochemistry which involves oxygen, it is fitting to stress that organic chemists almost routinely exclude oxygen from their samples during irradiation (unless they try to achieve photooxidation reactions). There are two reasons: (1) to avoid the production of reactive oxygen species ($^1O_2$ and O_2^-), which could react with either starting materials or products and lower the yields of desired products; (2) to avoid physical quenching, which reduces the concentration of excited molecules and, consequently, reduces the rates of product formation. The quantum yield for the quenching of triplet excited states by oxygen, to form $^1O_2$, is not independent of the sensitizer concentration. Below a threshold, it gradually decreases as the concentration of sensitizer decreases.[45]

REFERENCES

1. J. D. Winkler, K. Deshayes and B. Shao, *J. Am. Chem. Soc.*, **111**, 769–770 (1989).
2. M. Kasha, In A. A. Frimer (ed.) *Singlet O_2*, vol. 1, pp. 1–12. CRC Press, Boca Raton (1985).
3. B. M. Monroe, In A. A. Frimer (ed.) *Singlet O_2*, vol. 1, pp. 177–224. CRC Press, Boca Raton (1985).
4. M. A. J. Rodgers, In R. V. Bensasson, E. J. Jori, E. J. Land and T. J. Truscott (eds.) *Primary Photoprocesses in Biology and Medicine, Nato ASI Series A*, vol. 85, pp. 1–23. Plenum Press, New York (1984).
5. I. Kraljic and S. El Mohsni, *Photochem. Photobiol.*, **28**, 577–581 (1978).
6. W. R. Haag and T. Mill, *Photochem. Photobiol.*, **45**, 317–321 (1987).
7. J. R. Harbour, S. L. Issler and M. L. Hair, *J. Am. Chem. Soc.*, **102**, 7778–7779 (1980).
8. J. Moan and E. Wold, *Nature*, **279**, 450–451 (1979).
9. F. W. Keana, V. S. Prabhu, S. Ohmiya and C. E. Klopfenstein, *J. Org. Chem.*, **51**, 3456–3462 (1986).
10. L. T. Pettit and M. A. Fox, *J. Phys. Chem.*, **90**, 1353–1354 (1986).
11. S. A. Bezman, P. A. Burtis, T. P. J. Izod and M. A. Thayer, *Photochem. Photobiol.*, **28**, 325–329 (1978).
12. W. R. Midden and S. Y. Wang, *J. Am. Chem. Soc.*, **105**, 4129–4135 (1983).
13. C. M. Krishna, Y. Lion and P. Riesz, *Photochem. Photobiol.*, **45**, 1–6 (1987).
14. T. P. Wang, J. Kagan, S. Lee and T. Keiderling, *Photochem. Photobiol.*, **52**, 753–757 (1990).
15. A. V. Borisov, V. I. Tsivenko, I. A. Myasnikov and E. A. Venediktov, *Zh. Fiz. Khim.*, **64**, 1712–1714 (1990).
16. M. R. Berenbaum and R. A. Larson, *Experientia*, **44**, 1030–1032 (1988).
17. T. A. Dahl, W. R. Midden and P. E. Hartman, *Photochem. Photobiol.*, **46**, 345–352 (1987).
18. W. C. Eisenberg, A. Snelson, J. Veltman and R. W. Murray, *Tetrahedron Lett.*, **22**, 1949–1952 (1981).
19. E. J. Corey and W. C. Taylor, *J. Am. Chem. Soc.*, **86**, 3881–3882 (1964).
20. R. W. Murray and M. L. Kaplan, *J. Am. Chem. Soc.*, **91**, 5358–5364 (1969).
21. E. J. Corey, M. M. Mehrotra and A. U. Khan, *J. Am. Chem. Soc.*, **108**, 2472–2473 (1986).
22. J. M. Aubry, *La Recherche*, **17**, 971–973 (1986).
23. J. M. Aubry, B. Cazin and F. Duprat, *J. Org. Chem.*, **54**, 726–728 (1989).
24. I. Saito, T. Matuura and K. Inoue, *J. Am. Chem. Soc.*, **105**, 3200–3206 (1983).
25. P. T. Chou and H. Frei, *Chem. Phys. Lett.*, **122**, 87–92 (1985).
26. J. R. Kanofsky, *J. Org. Chem.*, **51**, 3386–3388 (1986).
27. J. R. Kanofsky, *J. Am. Chem. Soc.*, **108**, 2977–2979 (1986).
28. A. A. Gorman, I. Hamblett and E. J. Land, *J. Am. Chem. Soc.*, **111**, 1876–1877 (1989).

29. T. Noh and N. C. Yang, *J. Am. Chem. Soc.*, **113,** 9412–9414 (1991).
30. R. B. Woodward, *Angew. Chem.*, **72,** 651–662 (1960).
31. Y. Watanabe, N. Kuriki, K. Ishiguro and Y. Sawaki, *J. Am. Chem. Soc.*, **113,** 2677–2682 (1991).
32. R. W. Murray and M. M. Morgan, *J. Org. Chem.*, **56,** 684–687 (1991), and references cited therein.
33. H. H. Wasserman, K. E. McCarthy and K. S. Prowse, *Chem. Rev.*, **86,** 845–856 (1986).
34. A. A. Stark and D. F. Liberman, *Mutation Res.*, **247,** 77–86 (1991).
35. W. Adam, H. Hauer, T. Mosandl, C. R. Saha-Moeller, W. Wagner and D. Wild, *Liebigs Ann. Chem.*, 1227–1236 (1990).
36. F. Haber and J. Weiss, *Proc. R. Soc. Ser. A*, **147,** 332–351 (1934).
37. R. J. Carmichael, M. M. Mossoba and P. Riesz, *FEBS Lett.*, **164,** 401–405 (1983).
38. N. Suzuki, T. Itagaki, A. Goto, B. Yoda, I. Mizumoto, T. Nomoto, M. Kobayashi and H. Inaba, *Chem. Express*, **6,** 25–28 (1991).
39. J. Kagan and R. W. Tuveson, *Bioactive Molecules*, **7,** 71–84 (1988).
40. G. R. Peyton and W. H. Glaze, *ASC Symp. Ser.*, **327,** 76–88 (1987).
41. K. Steliou, *Acc. Chem. Res.*, **24,** 341–350 (1991).
42. K. Steliou, P. Salama and X. Yu, *J. Am. Chem. Soc.*, **114,** 1456–1462 (1992).
43. W. Ando, H. Sonobe and T. Akasaka, *Tetrahedron Lett.*, **28,** 6653–6656 (1987).
44. S. Ito and J. Mori, *Bull. Chem. Soc. Japan*, **51,** 3403–3404 (1978).
45. M. Terazima, M. Tonooka and T. Azumi, *Photochem. Photobiol.*, **54,** 59–66 (1991).

7. Electron transfers in photochemistry

7.1. INTRODUCTION

Electronically excited molecules are normally in the vicinity of other molecules, with which they may react chemically or to which they may transfer energy. The complex formed by one molecule in an excited state and another molecule in the ground state is called an *exciplex*. In a special case the two molecules are the same: this excited dimeric exciplex is called an *excimer*.

Physical methods can detect the formation of excimers or exciplexes. The most convenient is the observation of fluorescence. Any changes in fluorescence maxima and/or intensity following a change in concentration of either the starting material or an added chemical confirm important intermolecular interactions. If there are no immediate chemical reactions, an exciplex must be involved.

Fluorescence quenching may be just a physical process of energy exchange. However, chemical reactions may also ensue. One process which may start chemical reactions is *the exchange of one electron between one molecule in the excited state and one molecule in the ground state*:

$$D^* + M \rightarrow [D^* \, M] \rightarrow D^{\dot{+}} + M^{\dot{-}}$$

The initial exciplex may also be viewed as a charge transfer complex. Knowing the excited state redox potential of the donor in relation to the redox potential of the acceptor provides an important factor in predicting both whether an electron transfer is feasible and its direction. Quantitative data have been discussed and tabulated elsewhere.[1,2]

An increasingly large field of organic photochemistry involves electron transfer reactions started by an active participant, which may even operate catalytically. One category of photo-redox reactions uses electron-poor sensitizers, for example aromatic molecules bearing electron-withdrawing substituents such as cyano or ester groups. Another category uses electron-rich sensitizers, such as aromatic molecules substituted with alkoxyl groups.

7.2. ELECTRON-POOR SENSITIZERS: CYANOAROMATIC SENSITIZERS

Many cyanoaromatic compounds derived from benzene, naphthalene, or anthracene are electron acceptors. They have been useful in generating cation radicals from organic molecules via single-electron transfer (SET) reactions. Such photoinduced electron transfer reactions are also abbreviated to PET.

Many reactions involving cation radicals are distinctive, because they look neither like typical radical reactions nor like typical ionic reactions. An example is the photochemical reaction of an alkene with an alcohol. As mentioned earlier, a radical usually

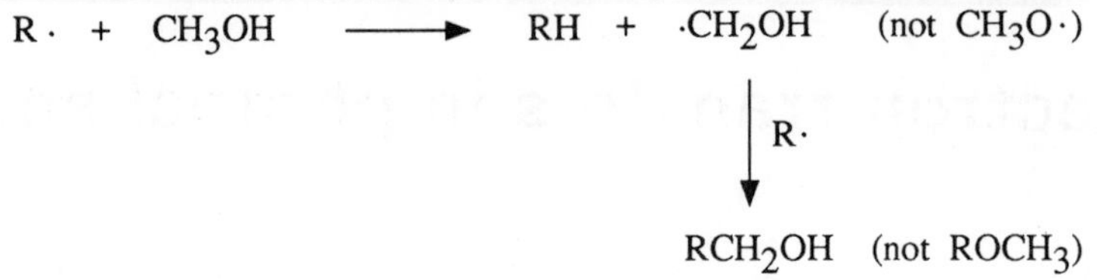

$$R\cdot \; + \; CH_3OH \longrightarrow RH \; + \; \cdot CH_2OH \quad (not \; CH_3O\cdot)$$

$$\Big\downarrow R\cdot$$

$$RCH_2OH \quad (not \; ROCH_3)$$

Fig. 7.1. Typical reaction of a radical with CH_3OH.

abstracts one hydrogen from an alcohol such as CH_3OH to produce a carbon-centered radical. The combination of the two types of radicals therefore produces a new alcohol rather than an ether (Fig. 7.1).

An ether, rather than an alcohol, is formed in good yield in the photochemical reaction of 1,1-diphenylethylene with CH_3OH in the presence of 1,4-dicyanobenzene, suggesting that the product is formed through an ionic rather than a radical reaction. Even more interesting, the product appears to result from an anti-Markovnikov addition reaction. One explanation is illustrated in Fig. 7.2. The first step is the formation of a charge transfer complex, resulting from electron transfer. Nucleophilic attack of the cation radical by the solvent generates a radical species. Finally, the electron which was

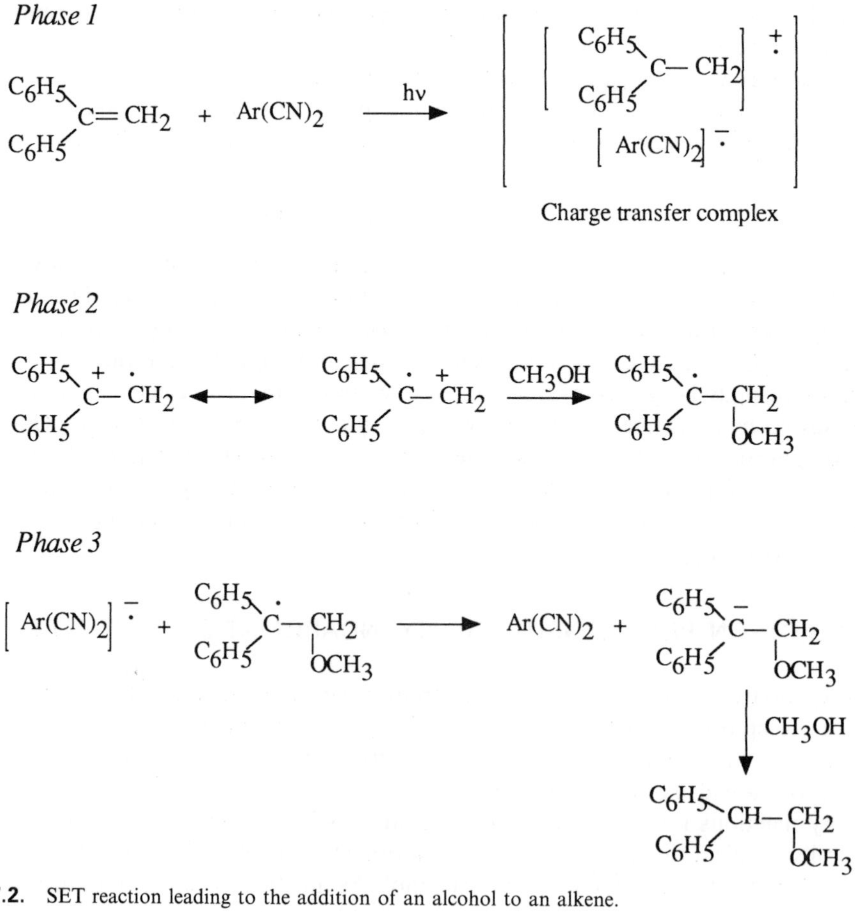

Phase 1

Charge transfer complex

Phase 2

Phase 3

Fig. 7.2. SET reaction leading to the addition of an alcohol to an alkene.

initially accepted by the sensitizer is returned to the newly formed radical, converting it into a carbanion which is then protonated by the solvent. The catalysis may also occur intramolecularly, since this electron transfer scheme does not require that the sensitizer and substrate be separate molecules.

The photoamination of electron-rich alkenes is another example of useful reaction sensitized by p-dicyanobenzene (DCNB).[3] In general, little regioselectivity was observed, unless a methoxyl was present at the 4-position in one ring, the amino group becoming predominantly attached to the benzylic position of this ring.

$$Ar_1-CH=CH-Ar_2 + NH_3 \xrightarrow[DCNB]{h\nu} Ar_1-CH_2-\underset{NH_2}{CH}-Ar_2 + Ar_1-\underset{NH_2}{CH}-CH_2-Ar_2$$

The electron initially transferred to the sensitizer does not necessarily come from a carbon–carbon double bond. Strained carbon–carbon single bonds (from three-membered rings, for example) have been photochemically broken with electron-poor sensitizers. Quadricyclene (1), for example, can be isomerized into norbornadiene (2) by photosensitization with chloranil, an electron-poor quinone (A represents chloranil in the scheme below).

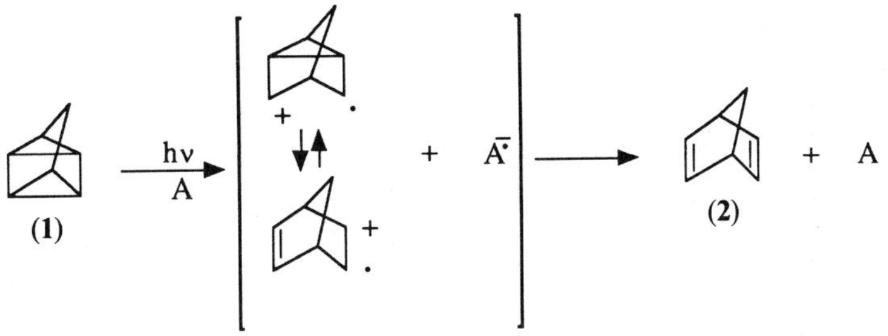

Photochemical electron transfer between a sensitizer and reagents probably does not take place over long distances. In some photosensitized Diels–Alder reactions, for example, a ternary complex involving the sensitizer, the diene, and the dienophile has been detected. Further support is provided by the observation that an optically active bis-dicyanoanthracene sensitizer (3) induced enantioselectivity in the cycloaddition reaction.[4] The product (4) was obtained with 15% enantiomeric excess in toluene from either enantiomeric sensitizer, the actual values depending both on solvent and temperature.

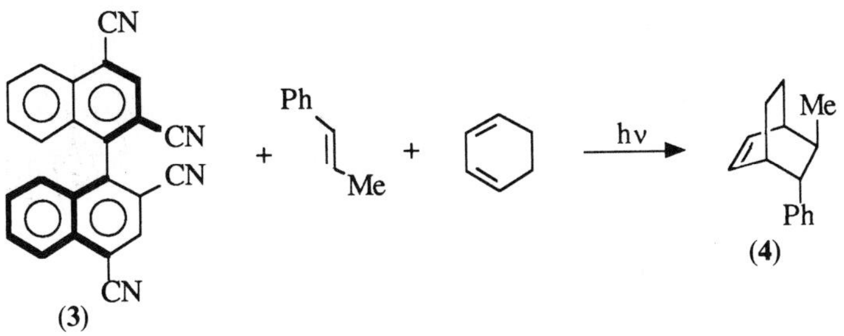

Consistent with the observation that cyanoaromatic sensitizers are involved in electron transfer reactions over short distances, steric hindrance about the cyano group in the sensitizer has a significant effect on the quantum and chemical yields of product formation. The reaction investigated was the lactonization shown below:

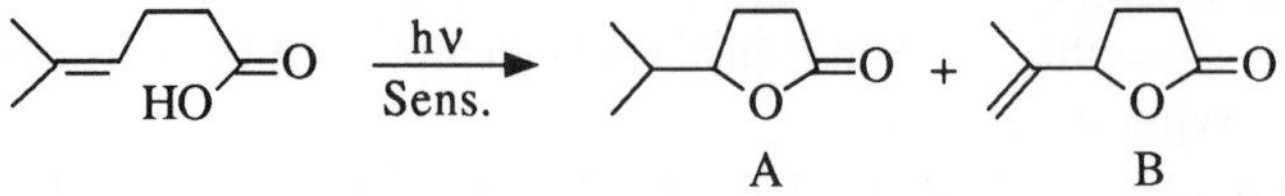

$$A \qquad\qquad B$$

The total yield increased from 36 to 89% and the yield of the major product A increased from 31 to 83% in going from 1,4-dicyano- to 1,4-dicyano-2,3,5,6-tetraethylbenzene. This gain was ascribed to hindrance of the back electron transfer by the presence of the bulky substituents on the sensitizer.[5]

Although cyanoaromatic molecules have often been used to induce SET reactions, these molecules do not always survive the reactions intact. They may undergo substitution, addition, or reduction. For example, 38% of 1,4-dicyanobenzene but only 5% of the more hindered sensitizer were consumed in the lactonization shown above. The course of such photosensitized reactions is markedly affected by the medium and particularly by the presence of salts, which lengthen the lifetime (and reduce the reactivity) of radical anions in the absence of oxygen.[6]

7.3. PHOTOINDUCED GENERATION OF RADICAL IONS IN SOLVENTS OF LOW POLARITY

High-polarity solvents are used in most of the successful reactions yielding new products from an initial photochemical electron transfer. They allow separation of the initially formed pair of a radical cation with a radical anion, and prevent the reverse electron transfer, back to the initial reagents. Since the use of low-polarity solvents would greatly extend the range of electron transfer reactions, it is interesting to find that cationic acceptors such as N-methylacridinium (MA^+) (5) have successfully served as sensitizers in solvents such as benzene, chloroform, or methylene chloride.[7]

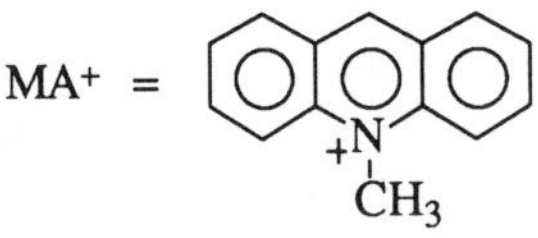

(5)

7.4. ELECTRON-RICH SENSITIZERS

An electron transfer in the opposite direction (from the sensitizer to the substrate) is possible when *electron-rich* sensitizers are used. Returning to the case of 1,1-diphenylethylene, photolysis in the presence of 1-methoxynaphthalene converts the alkene into a radical anion, which is then protonated. The radical produced can transfer one electron back to the radical cation, regenerating the sensitizer while producing a cation which reacts with the solvent. Therefore, the overall addition appears to be ionic in this case, with an overall Markovnikov addition of the alcohol to the alkene (Fig. 7.3).

The reactions outlined above illustrate the general principles. In some cases, the mechanisms are more complex. One hint is when the sensitizer no longer operates catalytically and cannot be recovered quantitatively. Figure 7.4 illustrates an intramolecular trapping reaction sensitized with 1-cyanonaphthalene, where the major product resulted from the anti-Markovnikov lactonization reaction, but where 70% of the sensitizer was consumed.[8] An example of a ring-opening reaction with an electron-

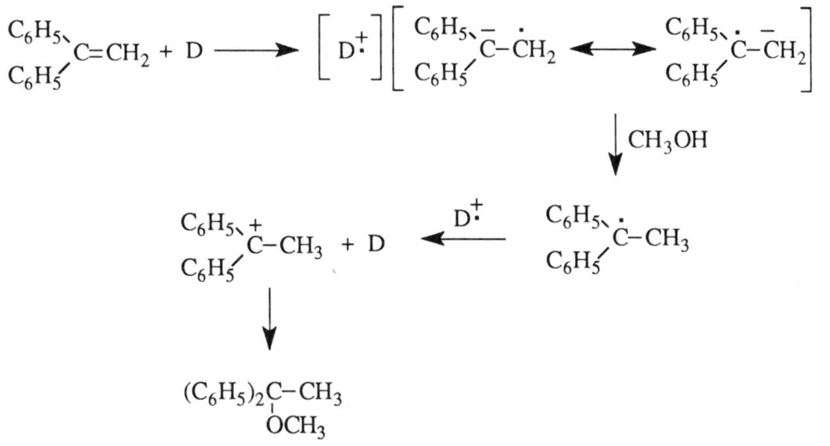

Fig. 7.3. Addition of methanol to an alkene catalyzed by an electron-rich sensitizer.

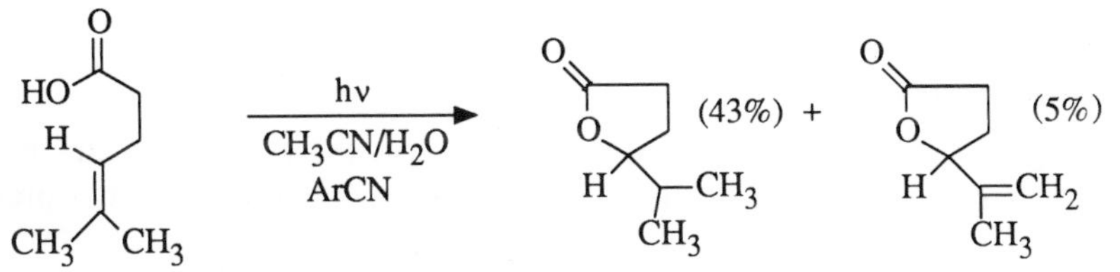

Fig. 7.4. Photosensitized lactonization in the anti-Markovnikov direction.

poor sensitizer was described earlier (Section 7.2). Cyclobutane rings formed by cyclo-addition reaction at pyrimidines are produced when DNA is irradiated (Chapter 9). The repair of this damage can be mediated by the enzyme photolyase (which utilizes a photochemical step), or it can be performed photochemically with an aromatic amine acting as a reducing reagent through a SET reaction:[9]

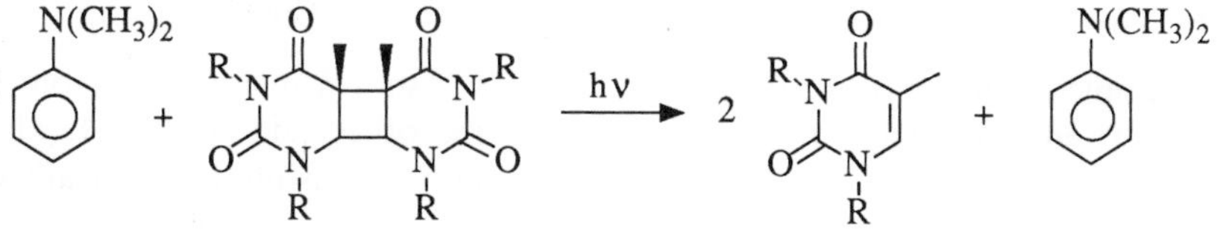

7.5. OTHER EXAMPLES OF ELECTRON TRANSFER REACTIONS IN PHOTOCHEMISTRY

Besides electron transfer reactions to O_2, leading to O_2^-, there are many other cases where one reagent in a photolyzed mixture transfers one electron. Amines are notoriously good electron donors, and there is an extensive literature on the photoreduction reactions they mediate, particularly with alkenes and carbonyl compounds.

In a typical amine reaction, one electron is transferred to the electronically excited sensitizer, creating an amine cation radical. If a hydrogen is attached to an adjacent atom, its transfer to the sensitizer produces a radical, which after donating one electron is converted into an imine or iminium ion derivative (Fig. 7.5). The iminium ion is expected to hydrolyze readily to CH_3CHO and $(C_2H_5)_2NH$. In this scheme the sensitizer

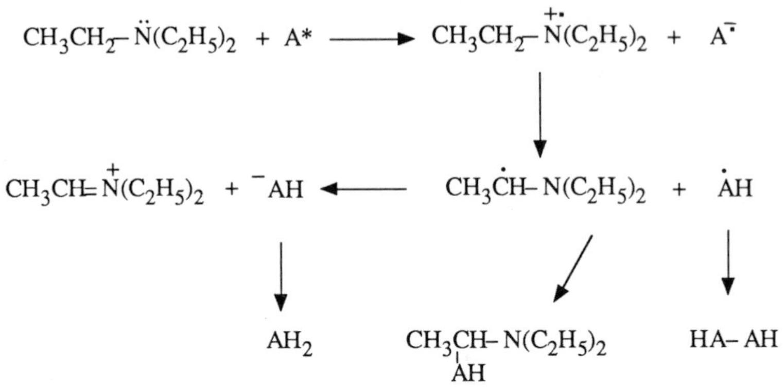

$$CH_3CH_2-\overset{..}{N}(C_2H_5)_2 + A^* \longrightarrow CH_3CH_2-\overset{+\cdot}{N}(C_2H_5)_2 + A^{\bar{\cdot}}$$

$$CH_3CH=\overset{+}{N}(C_2H_5)_2 + {}^-AH \longleftarrow CH_3\overset{\cdot}{C}H-N(C_2H_5)_2 + \overset{\cdot}{A}H$$

$$AH_2 \qquad \underset{AH}{CH_3CH-N(C_2H_5)_2} \qquad HA-AH$$

Fig. 7.5. General scheme showing the fate of triethylamine as an electron donor interacting with the electronically excited acceptor A.

A is reduced to HA−AH and AH$_2$. The photochemistry of alkenes and of carbonyl compounds provides examples of all these reactions.

7.5.1. Amine-alkenes

The scheme described in Fig. 7.5 applies perfectly to the photochemistry of stilbene, $C_6H_5CH=CHC_6H_5$. The products formed from irradiation in the presence of $N(C_2H_5)_3$ are shown in Fig. 7.6, identified as in Fig. 7.5.

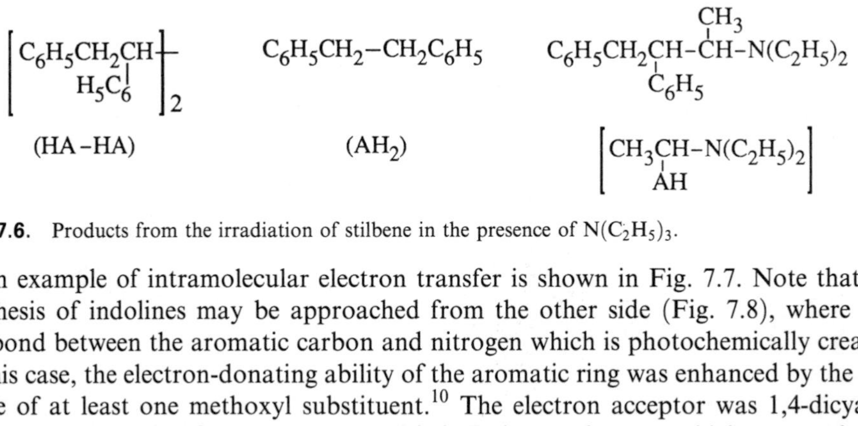

$$\left[\underset{H_5C_6}{C_6H_5CH_2CH} \right]_2 \qquad C_6H_5CH_2-CH_2C_6H_5 \qquad \underset{C_6H_5}{\overset{CH_3}{C_6H_5CH_2CH-CH-N(C_2H_5)_2}}$$

$$(HA-HA) \qquad\qquad (AH_2) \qquad\qquad \left[\underset{AH}{CH_3CH-N(C_2H_5)_2} \right]$$

Fig. 7.6. Products from the irradiation of stilbene in the presence of $N(C_2H_5)_3$.

An example of intramolecular electron transfer is shown in Fig. 7.7. Note that the synthesis of indolines may be approached from the other side (Fig. 7.8), where it is the bond between the aromatic carbon and nitrogen which is photochemically created. In this case, the electron-donating ability of the aromatic ring was enhanced by the presence of at least one methoxyl substituent.[10] The electron acceptor was 1,4-dicyano-naphthalene, but its fate was not revealed. It is not known which was reduced, oxygen or the sensitizer.

Interestingly, this sequence could be extended through another photoinduced electron transfer reaction, as shown in Fig. 7.9. The irradiating wavelengths were selected

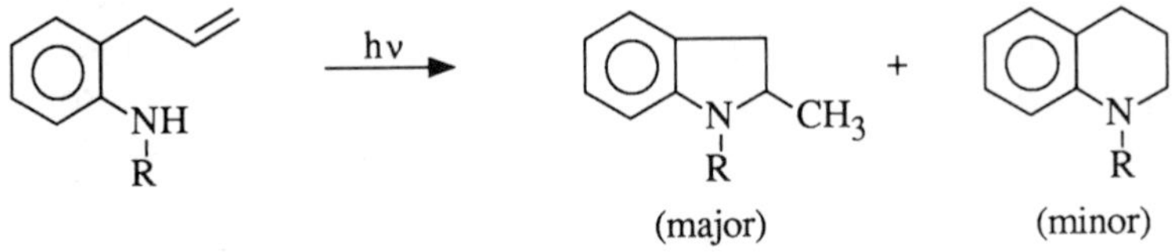

(major)   (minor)

Fig. 7.7. A photo-redox synthesis of indolines and tetrahydroquinolines.

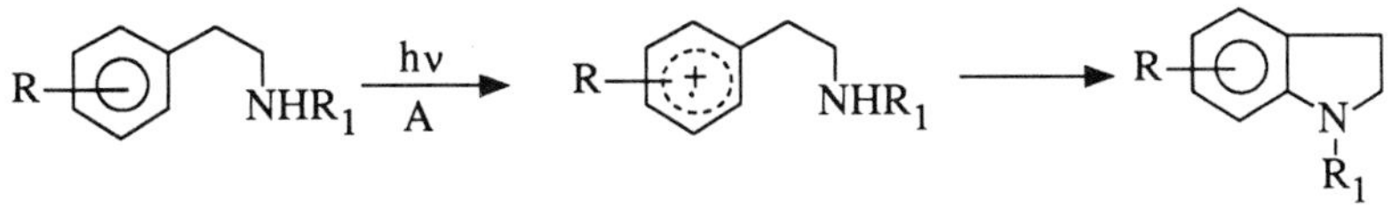

Fig. 7.8. Alternative photochemical synthesis of indolines.

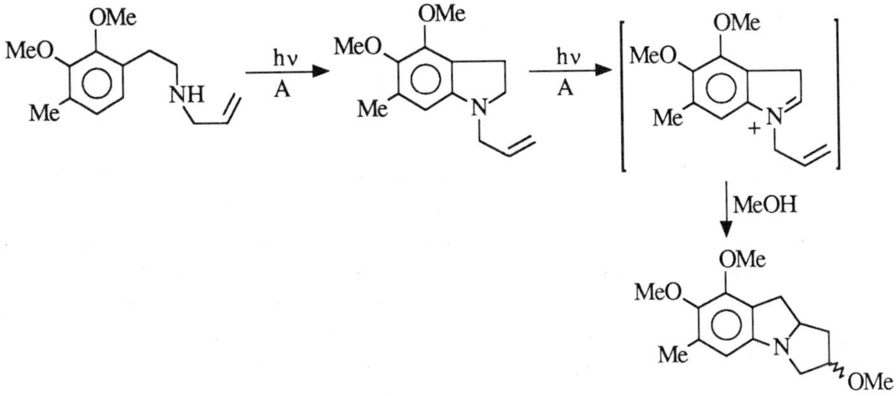

Fig. 7.9. Synthesis and further reactions of an indoline.

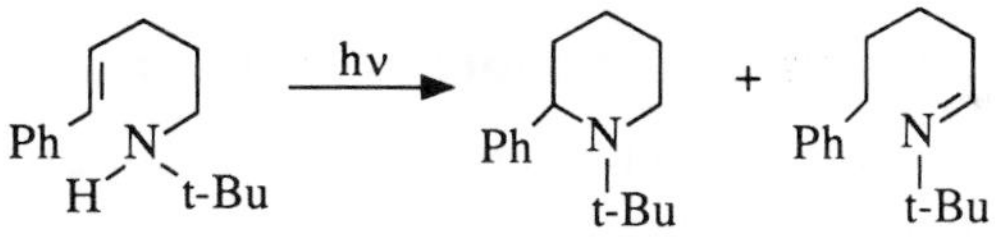

Fig. 7.10. Photocyclization of a *t*-butylamine into a tetrahydroquinoline without formation of a five-membered ring product.

in such a way that the electron acceptor A was directly excited in the second cyclization, but not in the first.

Increasing the bulk of the *N*-alkyl substituent in systems where competition may be possible affects the ratio of products. It increasingly favors six-membered ring products and disproportionation over five-membered ring products (Fig. 7.10).[11]

7.5.2. Photochemical reactions of amines and carbonyl compounds

In this case AH· is a familiar intermediate in the photochemistry of the carbonyl group (Chapter 4), as it can be formed upon hydrogen abstraction by the excited carbonyl. The same types of products may therefore be expected here, namely the pinacol reduction product, an alcohol, and the mixed coupling product.

When the cation radical of a tertiary amine has been formed photochemically, another type of reaction may be found. A carbon–carbon bond cleaves, leading to an iminium cation and a radical (Fig. 7.11).[12]

7.5.3. Oxygenation reactions

Probably the most important photooxygenation reaction involving an electron transfer

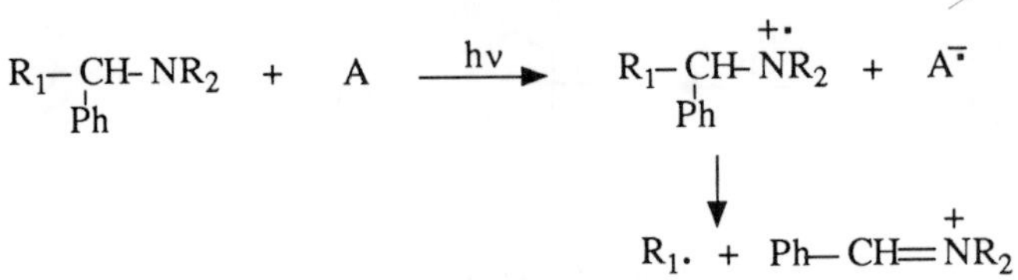

Fig. 7.11. General scheme showing cleavage in a SET reaction.

step is the process described as a type I photodynamic reaction (Chapter 6). The key step is an electron transfer between the electronically excited sensitizer (or some other molecule) and oxygen, to form a radical cation and the superoxide ion $O_2^{\cdot-}$, the oxygen radical anion. $O_2^{\cdot-}$ is a powerful nucleophile, as well as an oxygen radical. In addition to the combination of the radical cation/$O_2^{\cdot-}$ pair, to produce a peroxide, a multitude of reactions are possible, particularly in a complex biological environment. Perhaps most important are those in which radical chain reactions are started. At this point, any radical formed may react with *molecular* oxygen, to produce peroxidic species. This type of chemistry is particularly important in lipids, where allylic hydroperoxides are readily formed, and serve as the sources of oxygen-centered radicals able to abstract hydrogens from other positions and further oxygenate the substrate. It accounts also for the cleavage of nucleic acids under conditions where $O_2^{\cdot-}$ can be formed. These may be photochemical, but may also involve simple redox chemistry, for example with Fe^{2+} or Cu^+ ions.

7.5.4. Supported sensitizers for photooxidation reactions

Sensitizers supported on solids may be convenient since they can be removed by simple filtration from a reaction mixture in which the products are in solution. The availability of solid-supported rose bengal was mentioned in Chapter 6. 2,4,6-Trinitrobenzene and other nitroaromatic molecules are well-known electron acceptors. Covalently linked to silica gel, they compete favorably with homogeneous sensitizers for the benzylic oxidation of a methyl group into an aldehyde (Fig. 7.12).[13]

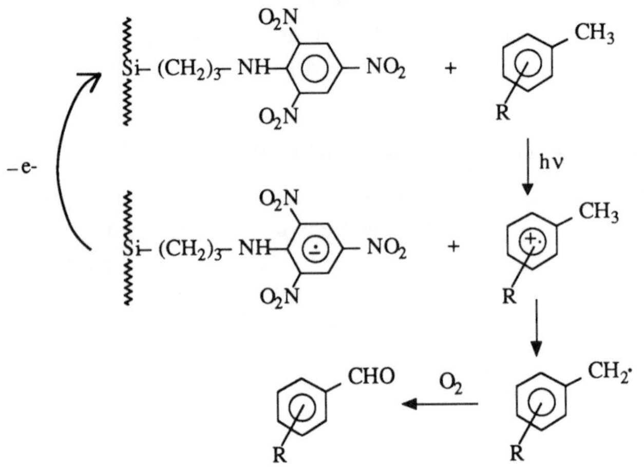

Fig. 7.12. Photooxidation of a methyl group into an aldehyde using 2,4,6-trinitrobenzene supported on silica gel as a sensitizer.

7.5.5. Deoxygenation reactions

While photoreactions leading to alcohols are very common (from carbonyl compounds, for example), deoxygenation reactions are much more difficult to achieve. One general method uses electron transfer from the common solvent hexamethylphosphoric tri-amide (HMPTA, $[(CH_3)_2N]_3PO$) to an ester (e.g. (**6**)) derived from the alcohol.[14]

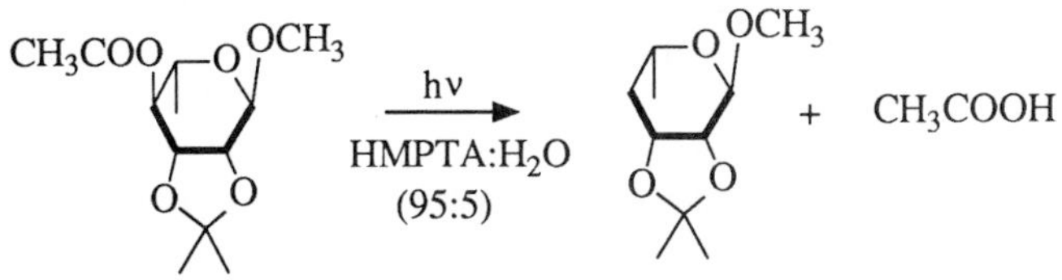

7.5.6. Aromatic substitution reactions ($S_{RN}1$ mechanism)

Aromatic substitution reactions can occur by a mechanism involving an initial electron transfer which creates an aromatic radical anion derivative, later dissociating into an aromatic radical and a departing anion. The aromatic radical may then add a different anion to form a new radical anion, which, after transfer of one electron, yields a pro-duct of an overall aromatic substitution reaction (Fig. 7.13). This mechanism is called $S_{RN}1$ (where the abbreviations stand for Substitution, Radical, Nucleophilic, 1st order). The further conversion of ArX into ArY occurs through a chain reaction.

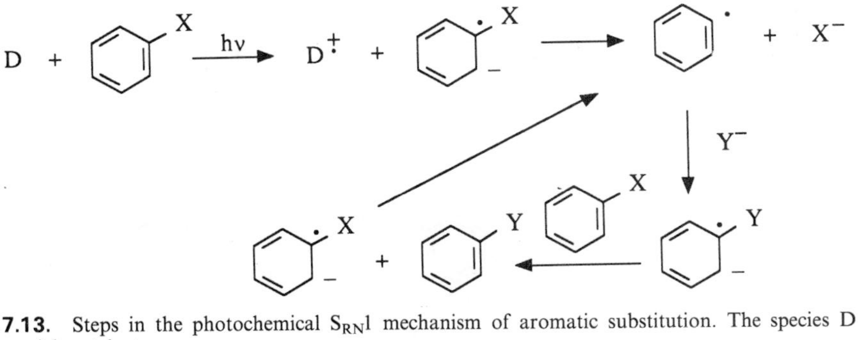

Fig. 7.13. Steps in the photochemical $S_{RN}1$ mechanism of aromatic substitution. The species D is an electron-rich catalyst.

Many nucleophiles have been found to participate in photochemical aromatic substi-tution reactions, as illustrated with enolate ions[15] (Fig. 7.14(a)) and carbanions (Fig. 7.14(b)). $S_{RN}1$ reactions with aromatic nucleophiles (such as $ArNH_2$, $ArOH$, or $ArSH$) do not necessarily form a new bond at the heteroatom. In the example below, product A is the only one formed when Y = O, and it predominates (9 : 1) when Y = NH. The ratio is reversed with Y = S (2 : 8), and only B is formed when Y = Se.[16]

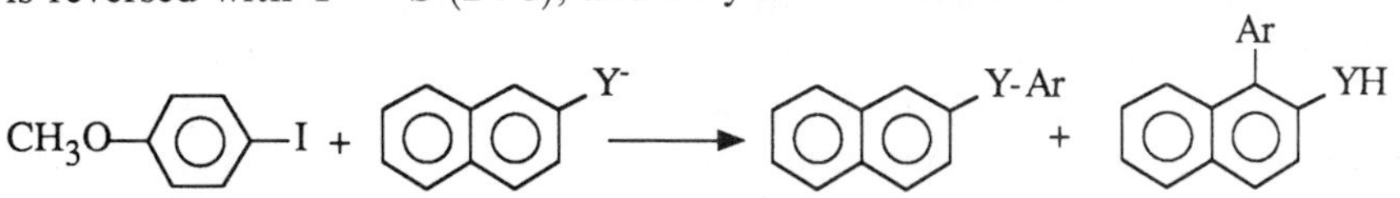

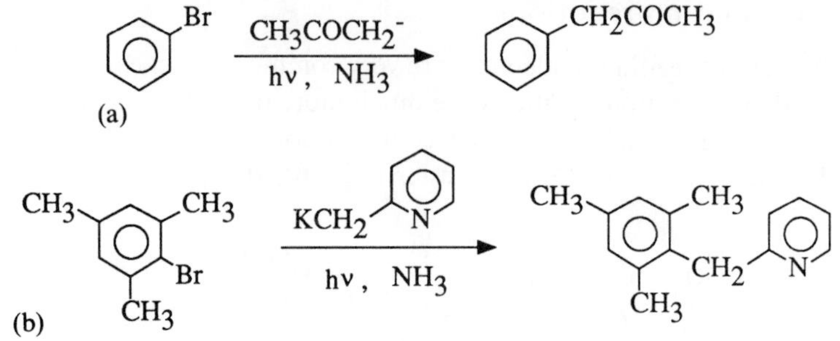

Fig. 7.14.   (a) Aromatic photosubstitution with an enolate ion. (b) Aromatic photosubstitution with a carbanion.

7.6. OTHER REACTIONS WITH AROMATIC COMPOUNDS

A benzylic position can be activated through photoinduced electron transfer, using an *electron-poor* reagent. The initially formed aromatic radical cation loses a proton from the benzylic position, creating a benzylic radical. It reacts, in turn, with the radical anion which has been created, or with other reagents, such as oxygen etc. In Fig. 7.15, only the resonance forms of the intermediates which are convenient for the argument are shown. A typical reagent A would be a carbonyl compound, for example a ketone: the initial electron transfer from the phenyl ring produces the corresponding phenyl radical cation and a ketyl radical anion. Upon loss of a proton, this radical cation is converted into a benzylic radical which reacts with the ketyl radical anion and is protonated to a tertiary alcohol.

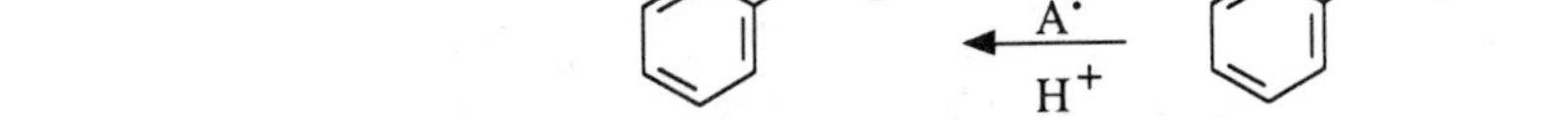

Fig. 7.15.   Typical photobenzylation reaction with an electron-poor reagent (A).

Figure 7.16 summarizes the reactions with an ester, which produce an intermediate similar to that obtained when a benzylic Grignard reacts with an ester. This intermediate loses CH_3O^-, to form a ketone, which can abstract an electron in a new photochemical step and form a new ketyl radical anion which will either dimerize, giving a pinacol, or react with another benzylic radical and produce a tertiary alcohol (Fig. 7.16).

Many drugs are known to produce photosensitization. For example, non-steroidal anti-inflammatory drugs (NSAIDs) are designed to inhibit inflammation, but nevertheless lead to inflammation in the presence of light. Many of these compounds are halogenated aromatic compounds, of which diclofenac is an example. The detailed mechanism of photoinduced inflammation is still not known, but the photochemistry of the drug is summarized in Fig. 7.17. The mechanism of the first reaction is not known,

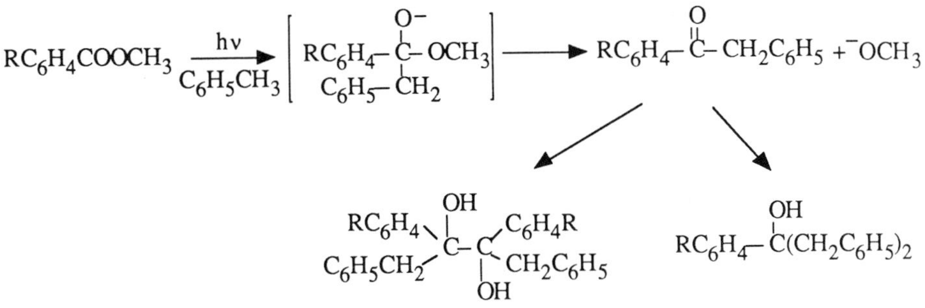

Fig. 7.16. Photobenzylation of a carboxylic ester.

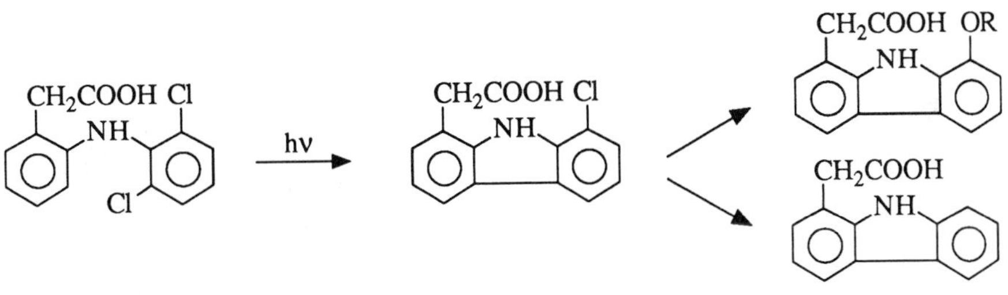

Fig. 7.17. Photochemical transformation of diclofenac.

but is believed not to involve an aryl radical, since its formation would be endothermic. The photogeneration of a chloride ion or loss of HCl from a protonated exciplex has been postulated. This step is followed either by photosolvolysis (R = H in water or R = CH_3 in methanol), or by photoreduction. In the former case, it is probably an $S_{RN}1$ reaction. In the latter, protonation followed by loss of the chloride ion has been postulated on the basis of experiments in deuterated solvents, which yield deuterated products.[17]

Electron transfer reactions may take place intramolecularly. The following example is a model explaining why replacement of thymine by 5-bromouracil in DNA enhances photosensitivity, leading to DNA–protein cross-linking, strand breaks, and creation of alkali-labile sites. The self-complementary hexanucleotide d(GCA^{Br}UGC)$_2$ was shown to undergo cleavage upon irradiation. The mechanism postulates an intramolecular electron transfer from adenine to an adjacent 5-bromouracil. Following loss of Br$^-$, intramolecular hydrogen abstraction by the uracil radical created, followed by reaction with water, produce a hemiacetal which releases adenine and yields the lactone shown (Fig. 7.18).[18]

7.7. PHOTOREDUCTION OF BENZENE TO DIHYDROBENZENE

Once an aromatic ring has transferred an electron to a suitable acceptor, the radical cation generated can be readily reduced by a hydride ion, producing a radical which can then pick up an electron from the original acceptor. After protonation, a dihydro-benzene is generated (Fig. 7.19).

In more complex cases, it is a ring bearing an electron-*donating* substituent which is predominantly reduced, because such a substituent helps stabilize the intermediate

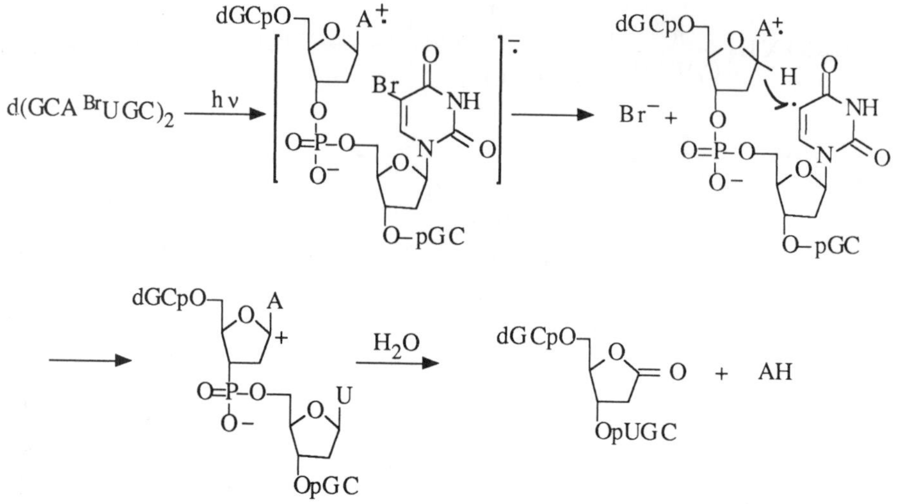

Fig. 7.18. Intramolecular electron transfer reaction in a hexadeoxynucleotide containing 5-bromouracil.

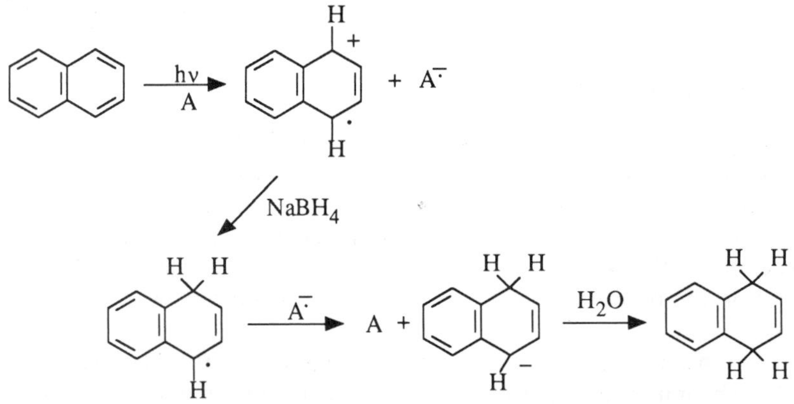

Fig. 7.19. Photoreduction of naphthalene in the presence of $NaBH_4$.

radical cation. This is just the opposite of the traditional reduction method, with dissolving metals in liquid NH_3, which initially forms a radical anion. In that case it is a ring containing electron-withdrawing substituents which is reduced predominantly.

A complementary method for the photoreduction of aromatic rings starts with the formation of a radical anion, using an electron-rich sensitizer or reagent. Here, the first intermediate produced is the same as in the Birch reduction. Protonation by a protic solvent or an acidic reagent, followed by electron transfer to the radical formed and protonation of the resulting anion, yield the dihydroaromatic product (Fig. 7.20). The chemistry associated with $N(C_2H_5)_3$ has been described earlier (see Fig. 7.5).

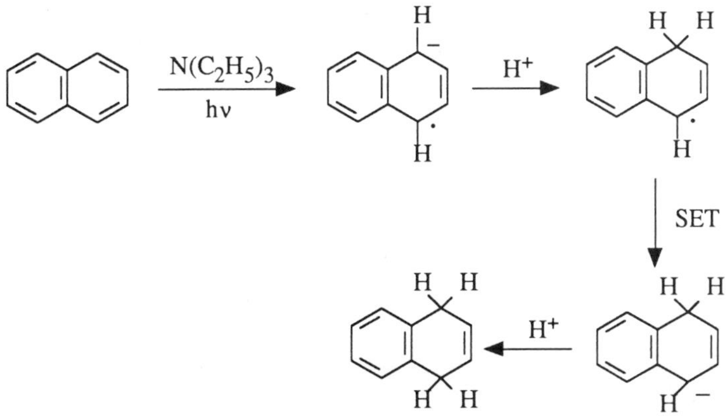

Fig. 7.20. Photoreduction of naphthalene in the presence of triethylamine, an electron-rich reagent.

7.8. PHOTOINDUCED ELECTRON TRANSFER REACTIONS INVOLVING INORGANIC REAGENTS

The general principles governing the photoreactions which include electron transfer steps accommodate inorganic reagents, either as donors or as acceptors. Indeed, there is a rich photochemistry involving either simple inorganic reagents (such as $FeCl_3$, $TiCl_4$, $Ru(bpy)_3^{2+}$, or $AgOSO_2CF_3$), or a variety of semiconductors.

One recent example demonstrated the one-pot synthesis of the optically active cyclic iminoacids proline and pipecolinic acid over semiconductors in aqueous medium. Reasonable steps are outlined here for the conversion of l-ornithine (7) into proline over TiO_2 loaded with PtO_2.[19] The first is an oxidative process on the surface of TiO_2, while the last is a reduction.

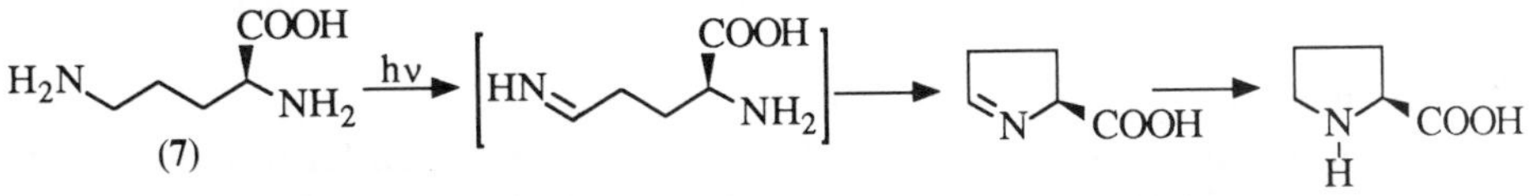

The ready availability of TiO_2 and other semiconductor photocatalysts should make the search for practical applications rewarding. One of the promising areas is in waste water treatment. A number of industrially important chlorinated solvents can be photocatalytically degraded over TiO_2.[20] The overall degradation produces CO_2 and Cl^-, and the addition of H_2O_2 enhances the rate of the process. Another useful application is in the photooxidation in water of organophosphorous compounds, such as pesticides.[21] Here again, adding strong oxidizing species enhances the rate of catalytic activity of TiO_2. In another example of photoreaction over semiconductors, CO_2 is reduced to the formate ion by irradiation in the presence of a suspension of zinc sulfide in aqueous 2,5-dihydrofuran.[22] The addition of platinum or silver can change the photocatalytic activity of TiO_2 significantly.[23] In view of the number of photoreactions taking place over TiO_2, it is surprising that this chemical has been advocated as a constituent of sunscreens for protection of human skin. A recent patent application even combines TiO_2 and ZnO in sunscreen compositions.[24]

Photochemical conversion of simple molecules into biological molecules may have implications in the field of chemical evolution. For example, irradiating a mixture of NH_3, CH_4 and H_2O in the presence of a platinized TiO_2 suspension produced a mixture of aminoacids, notably glycine, alanine, serine, aspartic acid and glutamic acid.[25] Photolysis in the presence of hydrous ferrous oxide or of an aqueous suspension of montmorillonite clays containing hydrous ferrous oxide converted N_2 into NH_3.[26]

7.9. PHOTOINDUCED ELECTRON TRANSFER REACTIONS IN BIOLOGICAL SYSTEMS

Photochemical reactions in biological systems are very difficult to study from a mechanistic standpoint, because the complexity of the medium and the presence of different microenvironments defy accurate analysis. However, electron transfer reactions involving isolated biological molecules are readily detected. The syntheses of O_2^- has already been mentioned. The possibility of producing photo-redox reactions is clearly demonstrated when ferricytochrome c is irradiated *in vitro* and its absorbance monitored at 550 nm. No obvious reaction takes place when oxygen is present. However, in the absence of oxygen, the photoreduction of the metal ion is observed. The electron donated to Fe^{3+} must come from the organic moiety in the molecule, although its precise origin has not been firmly established. Adding EDTA increases the rate of reduction, but it is not clear whether this reagent operates as a source of electrons, or protons, or both.

The example of phototoxicity of TiO_2 in malignant HeLa cells is interesting because it illustrates the potential biological activity of inorganic semiconductors. Ultrafine particles which had no cytotoxicity in the dark completely killed the cells in a 10 minute irradiation at a concentration of 0.1 mg/ml. Electron microscopy revealed that the TiO_2 was in the cytoplasm as well as on the cell membranes. The photodamage to the cells was believed to come from $\cdot OH$ and $HOO\cdot$ radicals, as well as from photogenerated holes in TiO_2.[27]

Many different types of reactions have been observed in the photochemistry of nucleic acids and their components. Photo-induced electron transfer reactions logically account for the addition of protic species (H_2O, C_2H_5OH) to pyrimidines, as in the photohydration of cytosine (**8**).

(**8**)

DNA–protein crosslinks and purine photoproducts (e.g. (**10**), (**11**), and (**12**) from thymine (**9**)) are also known. The photochemistry of proteins with sunlight involves mostly the aminoacids which absorb at the longest wavelengths, namely phenylalanine, tyrosine, and tryptophan. Some of this photochemistry is discussed in Chapter 9.

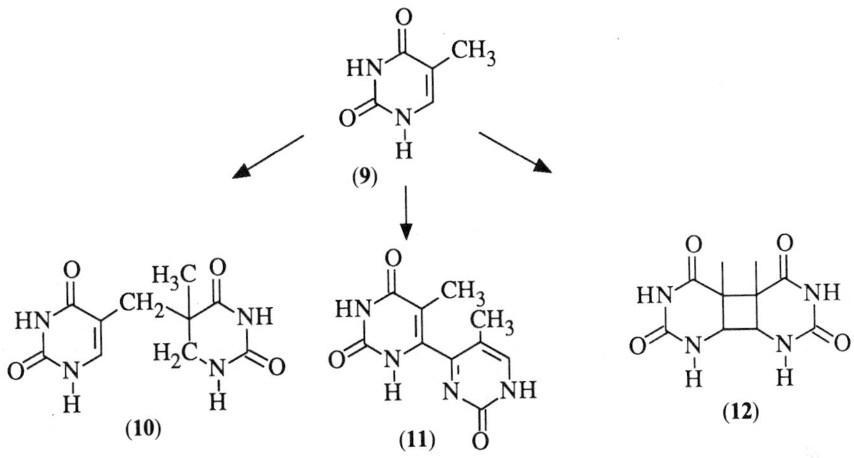

7.10. PHOTOELECTROCHEMICAL CELLS

One important goal in the design of these cells is the conversion of solar energy into chemical or electrical energy. Photogalvanic cells produce electricity when a photochemical reaction produces both reduced and oxidized species, and when the electrons can be carried away through metal electrodes without allowing back reaction of the photogenerated species. The practical upper limit for solar conversion in such cells is only 4%.[26]

Cells in which a semiconductor is in contact with an electrolyte which can undergo redox chemistry minimize the back reactions, but numerous variables need to be optimized in order to achieve both good solar energy conversion and stability.[29] Many different electrodes have been used. The present state of the art is a solar cell continuously operating for about 1 year, producing an open circuit potential of 1.2 V while converting solar energy to electrical energy with 16.4% efficiency.[30] The electrode is a n-CdSe semiconductor, and the photoreaction in the electrolyte, to which cyanide is added, is as follows:

$$Fe(CN)_6^{4-} + h\nu \rightarrow Fe(CN)_6^{3-} + e^-$$

Future applications of this technology may include production of chemical and clinical sensors, and decontamination of air and water streams by oxidizing organic pollutants with illuminated semiconductors.[31]

REFERENCES

1. P. S. Mariano and J. L. Stavinoha, In W. M. Horspool (ed.) *Synthetic Organic Chemistry*, pp. 145–257. Plenum Press, New York (1984).
2. M. Julliard and M. Chanon, *Chem. Rev.*, **83**, 425–506 (1983).
3. M. Yasuda, T. Isami, J. Kubo, M. Mizutani and T. Yamashita, *J. Org. Chem.*, **57**, 1351–1354 (1992).
4. J.-I. Kim and G. B. Schuster, *J. Am. Chem. Soc.*, **112**, 9635–9637 (1990).
5. P. G. Gassman and S. A. De Silva, *J. Am. Chem. Soc.*, **113**, 9870–9872 (1991).
6. M. A. Kellett, D. G. Whitten, I. R. Gould and W. R. Bergmark, *J. Am. Chem. Soc.*, **113**, 358–359 (1991).

7. W. P. Todd, J. P. Dinnocenzo, S. Farid, J. L. Goodman and I. R. Gould, *J. Am. Chem. Soc.*, **113**, 3601–3602 (1991).
8. P. G. Gassman and K. J. Bottorff, *J. Am. Chem. Soc.*, **109**, 7547–7548 (1987).
9. S.-R. Yeh and D. E. Falvey, *J. Am. Chem. Soc.*, **113**, 8557–8558 (1991).
10. G. Pandey, M. Sridhar and U. T. Bhalerao, *Tetrahedron Lett.*, **31**, 5373–5376 (1990).
11. F. D. Lewis and G. D. Reddy, *Tetrahedron Lett.*, **31**, 5293–5296 (1990).
12. L. Y. C. Lee, X. Ci, C. Giannotti and D. G. Whitten, *J. Am. Chem. Soc.*, **108**, 175–177 (1986).
13. M. Julliard, C. Legris and M. Chanon, *J. Photochem. Photobiol. A*, **61**, 137–152 (1991).
14. C. Portella, H. Deshayes, J. P. Pete and D. Scholler, *Tetrahedron*, **40**, 3635–3644 (1984).
15. R. A. Rossi and J. F. Bunnett, *J. Org. Chem.*, **38**, 1407–1410 (1973).
16. A. Pierini, M. T. Baumgartner and R. A. Rossi, *J. Org. Chem.*, **56**, 580–586 (1991).
17. D. E. Moore, S. Roberts-Thomson, D. Zhen and C. C. Duke, *Photochem. Photobiol.*, **52**, 685–690 (1990).
18. H. Sugiyama, Y. Tsutsumi and I. Saito, *J. Am. Chem. Soc.*, **112**, 6720–6721 (1990).
19. B. Ohtani, S. Tsuru, S. Nishimoto and T. Kagiya, *J. Org. Chem.*, **55**, 5551–5553 (1990).
20. T. Hisanaga, K. Harada and K. Tanaka, *J. Photochem. Photobiol. A*, **54**, 113–118 (1990).
21. C. K. Gratzel, M. Jirousek and M. Gratzel, *J. Mol. Catal.*, **60**, 375–387 (1990).
22. H. Kirsch and G. Twardzik, *Chem. Ber.*, **124**, 1161–1162 (1991).
23. A. Sclafani, M.-N. Mozzanega and P. Pichat, *J. Photochem. Photobiol. A*, **59**, 181–189 (1991).
24. C. A. Cole, M. K. O. Lindemann, E. R. Lukenbach and R. C. Stutzman, Sunscreen compositions comprising zinc oxide and titanium dioxide. *Eur. Pat. Appl.*, EP 433,086 (1991).
25. H. Reiche and A. J. Bard, *J. Am. Chem. Soc.*, **101**, 3127–3128 (1979).
26. O. A. Ileperuma, W. C. B. Kiridena and W. D. D. P. Dissanayake, *J. Photochem. Photobiol. A*, **59**, 191–197 (1991).
27. R. Cai, K. Hashimoto, K. Itoh, Y. Kubota and A. Fujishima, *Bull. Chem. Soc. Japan*, **64**, 1268–1273 (1991).
28. W. J. Albery, *Acc. Chem. Res.*, **15**, 142–148 (1982).
29. B. Parkinson, *Acc. Chem. Res.*, **17**, 431–437 (1984).
30. S. Licht and D. Peramunage, *Nature*, **345**, 330–333 (1990).
31. S. Borman, *Chem. Eng. News*, 17–19 (1991).

8. Synthetic applications

There are very few synthetic procedures utilized on an industrial scale. In addition to the reaction of cyclohexane with NOCl (Section 5.3), one may note the syntheses of vitamin D (Duphar, The Netherlands), α-bromodiethyl carbonate (Asha Pharmaceutical Industries, Sweden), α-bromoethyl nitrobenzene (Ciba-Geigy, Switzerland), and 2-hydroxy-5-oxo-2.5-dihydrofuran by singlet oxygen reaction (Firmenich, Switzerland). This chapter presents many photochemical transformations, utilizing molecules which contain functionalities usually more complex than those described earlier. They are intended to illustrate the diversity of the problems attacked photochemically, and it is therefore difficult to group them under general headings. Many more interesting reactions have been excluded than included in this chapter, in which the order of presentation is quite arbitrary.

8.1 PHOTOCYCLIZATION OF 1,3,5-TRIENES, AND CYCLOREVERSION

1,3,5-Hexatrienes (**1**) undergo facile photocyclization to 1.3-cyclohexadienes (**2**), a reaction which has many applications, including the conversion of stilbenes into phenanthrenes described in the next section.

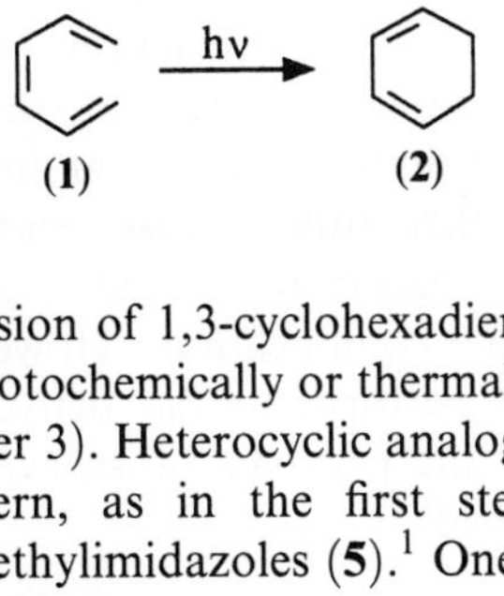

(**1**)   (**2**)

The reverse reaction, conversion of 1,3-cyclohexadienes to 1,3,5-hexatrienes, is also frequent. It can occur either photochemically or thermally, as seen earlier in the photochemistry of vitamin D (Chapter 3). Heterocyclic analogs may also undergo ring opening following the same pattern, as in the first step of the conversion of 2,3-dihydropyrazines (**3**) into N-methylimidazoles (**5**).[1] One explanation for the formation of the final product is suggested here.

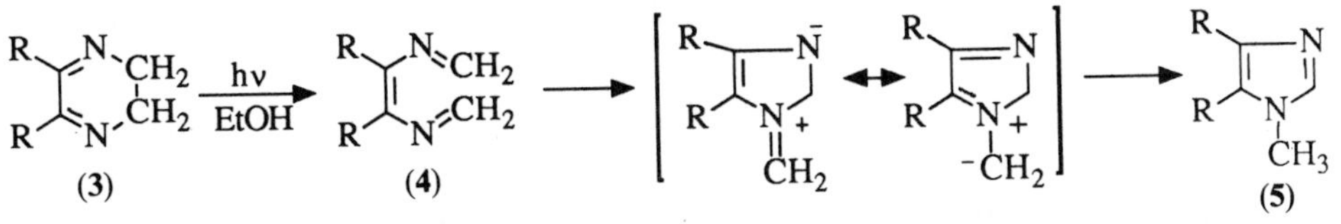

(**3**)   (**4**)   (**5**)

Occasionally, 1,3,5-hexatrienes undergo cyclization into [3.1.0]bicyclohexanes, as illustrated by the conversion of D vitamins (6) into suprasterols (7).

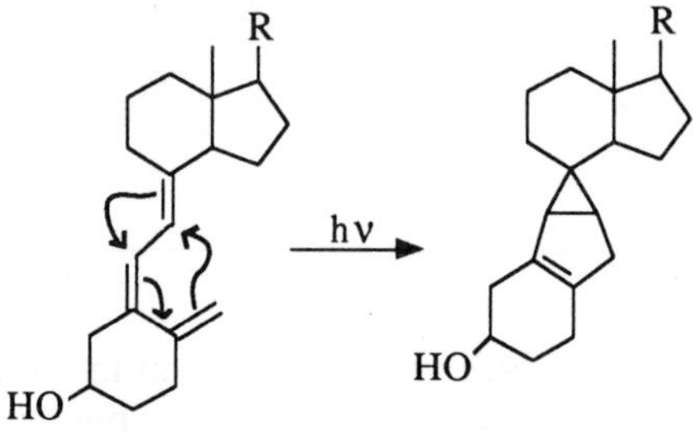

(6)   (7)

This suprasterol type of ring closure for the triene (4) initially formed in the opening of a 2,3-dihydropyrazine (see above), followed by opening of the three-membered ring and protonation–deprotonation, could also account for the final product (5).

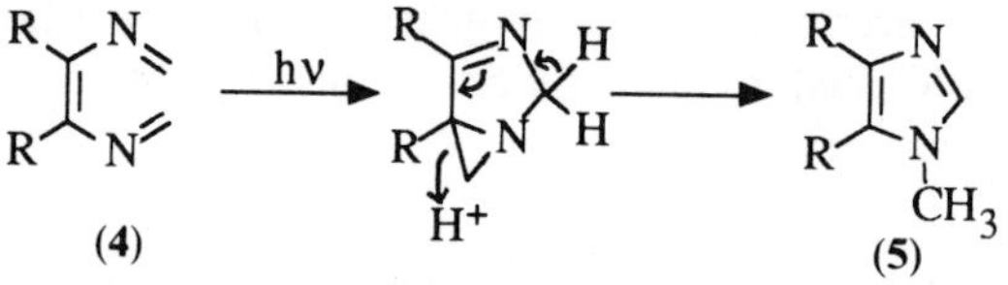

(4)   (5)

8.2 STILBENE CYCLIZATIONS AND HELICENE SYNTHESES

The initial steps of the photochemistry of stilbene (8) have raised many mechanistic questions, but the subsequent steps are important for synthetic reasons. Bimolecular reactions yielding tetraphenylcyclobutanes are classic examples of [2 + 2] cyclization reactions, as mentioned in Chapter 3. The reaction described here, however, is intramolecular; it converts the (Z) isomer into the dihydrophenanthrene (10) through a symmetry-allowed electrocyclic reaction. The initial dihydrophenanthrene product (9) can be readily oxidized to the phenanthrene (10), and it is customary to use oxygen and/or iodine in this step. The presence of O_2 and/or HI formed in the dehydrogenation step may then lead to further complications. However, the yields of photocyclization are greatly improved by performing the dehydrogenation under N_2 with a stoichiometric amount of I_2 in the presence of an excess of propylene oxide (which reacts rapidly with any HI formed).[2]

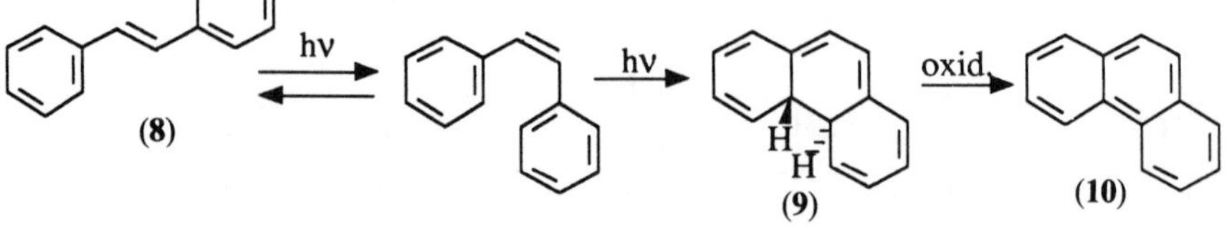

(8)   (9)   (10)

The cyclization reaction is quite general. It is a particularly useful method for making helicenes, as shown here in the case of the heptahelicene (11) (drawn in a

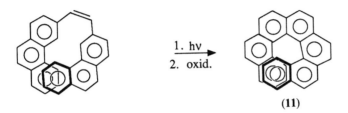

(11)

slightly distorted manner).[3] The same chemistry is involved in the conversion of [2,2]metacyclophanes (**12**) into [14]annulenes (**13**). Actually, the situation is somewhat

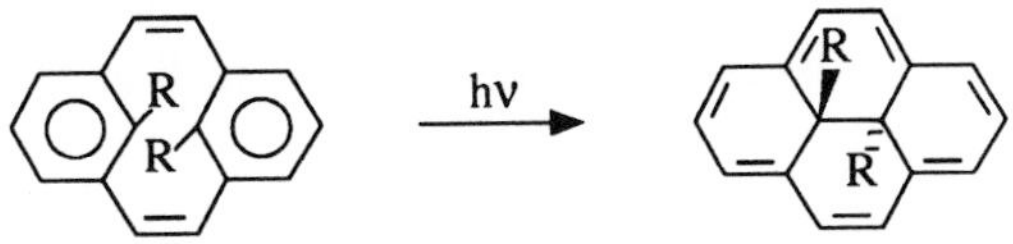

(12)   **(13)**

complicated by the fact that the cyclization may occur thermally (with different stereo-chemistry, the two R substitutents being *cis*), and the reverse reaction may occur spon-taneously. The nature of the R groups and of any additional substituents on the aromatic rings affects the course of the reactions.[4] For example, in the case of $R = CH_3$ or $R = C_2H_5$, cyclization occurs spontaneously in the dark, and it is the annulene which is photoisomerized back into the cyclophane upon exposure to visible light.[5,6]

Numerous examples of stilbene cyclization reactions have been described. Note that the cyclization may proceed even if it must take place at a position bearing a substit-uent, as shown here in the synthesis of the benzo[*c*]acridine (**14**).[7]

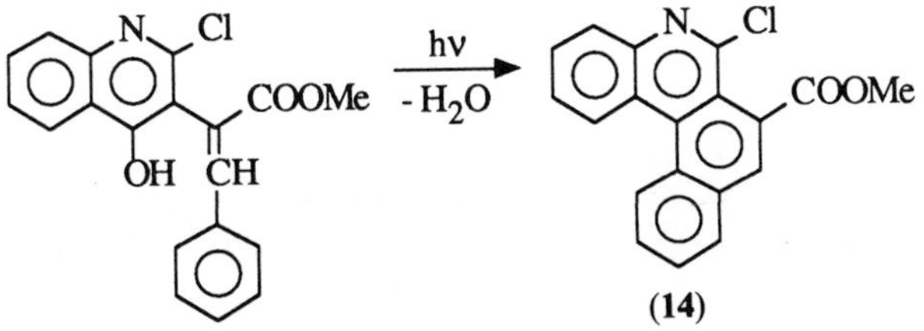

(14)

Cyclization reactions can occur in the same way with other ring systems, as shown here with the thiophene derivative (**15**).[8]

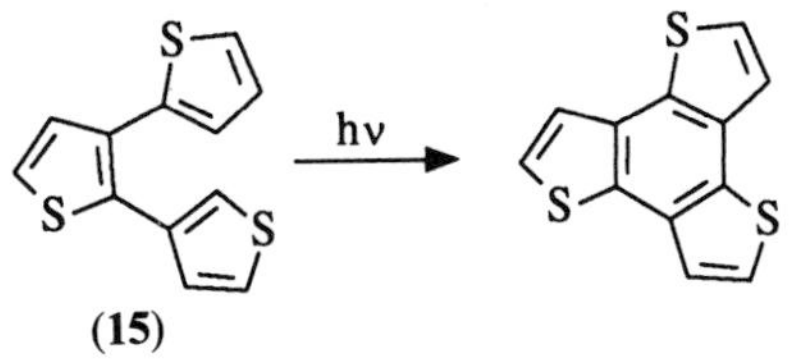

(15)

The cyclization of enamides is also interesting, because it has been a key step in the synthesis of alkaloids. The example given here is typical, leading to the ergoline alkaloid agroclavine (**17**) after reductive photocyclization of the enamide (**16**) and oxidative cleavage of the dihydrofuran ring produced.[9]

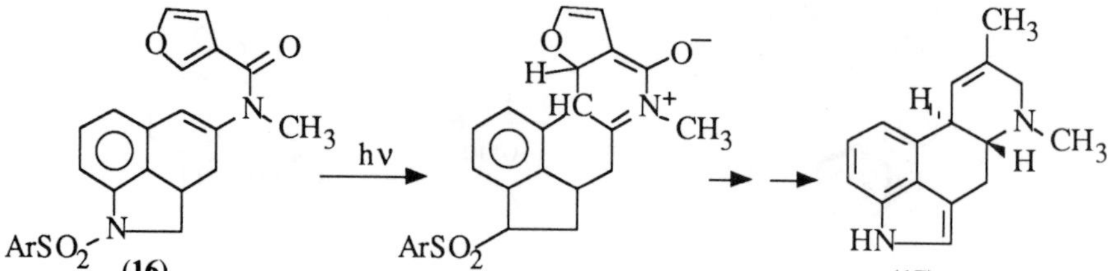

With aromatic enamides, the structure of the photoproducts depends on whether oxygen is present, and it is affected by other factors, such as the presence of base or reducing reagents.[10]

8.3 CYCLIZATION REACTIONS OF ACETYLENIC ANALOGS

Acetylenic analogs (e.g. (18), (19), and (20)) also undergo photochemical cyclization reactions and create polycyclic aromatic hydrocarbons, but they involve more complex mechanisms, in which cyclic allenes are perhaps intermediates.

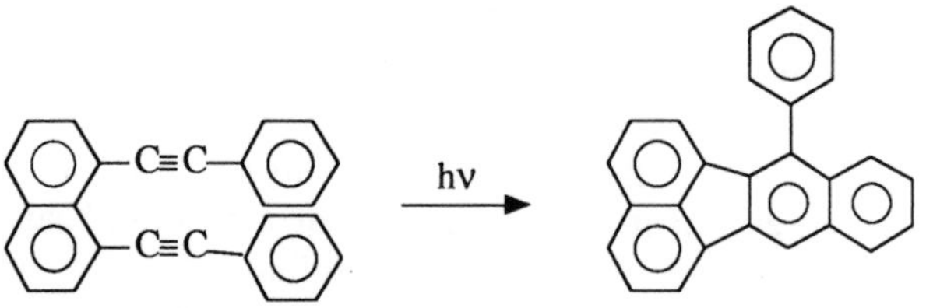

(ref. 11)

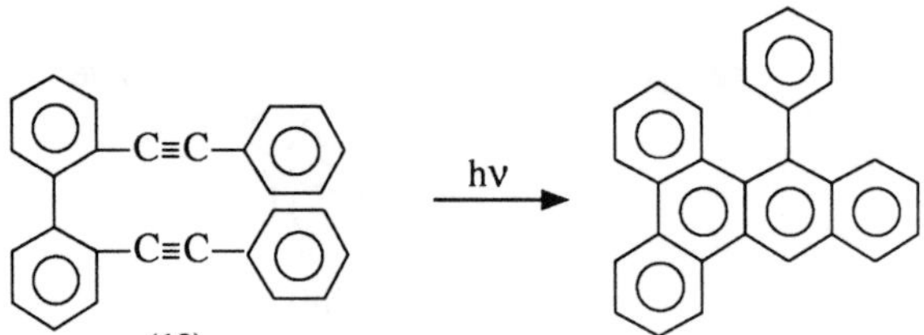

(18)

(ref. 12)

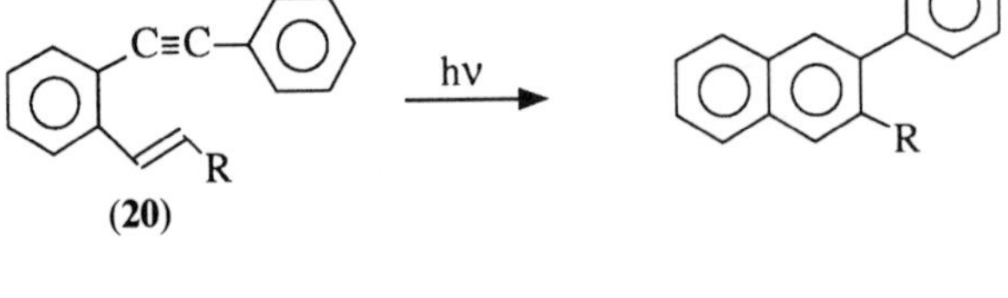

(19)

(ref. 13)

(20)

8.4. [4 + 4] CYCLOADDITION REACTIONS OF AROMATIC COMPOUNDS

The photochemical dimerization of anthracene is an old reaction. Its reversibility made it a candidate as a scheme for the utilization of solar energy. Here, it is applied to the first photochemical synthesis of a functional crown ether:[14]

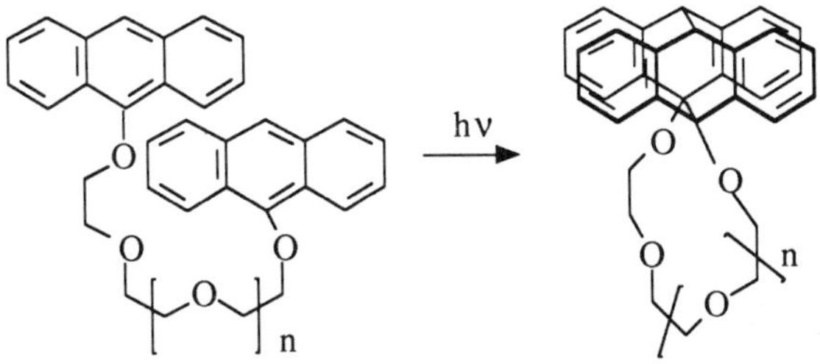

While the crown ether could be cleaved back to the starting material upon heating, its metal ion complexes were much more stable and did not revert back to the open form either thermally or photochemically. However, quantitative recovery of the open form was obtained upon treatment of the metal ion complexed crown ether with a polar solvent such as acetonitrile.

8.5. ANOTHER EXAMPLE OF [4 + 4] CYCLOADDITION: TETHERED 2-PYRIDONES

Several natural products of great interest, such as taxol, contain an eight-membered ring. A potential building block, (21), has been created by intramolecular photochemical cycloaddition, which actually produces two diastereomers.[15]

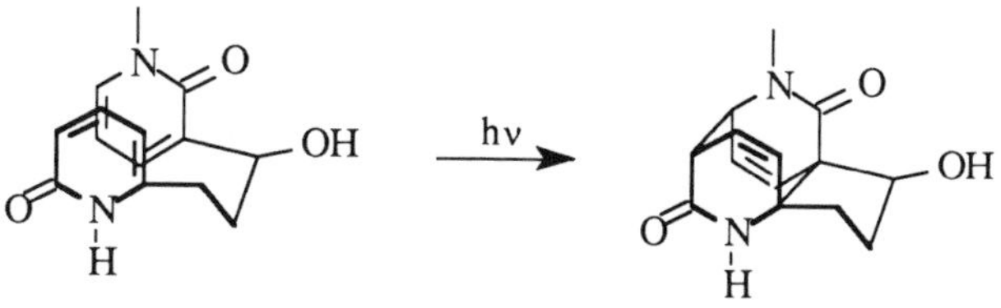

(21)

8.6. INTRAMOLECULAR DIENE–DIENE CYCLOADDITIONS

Here, an eight-membered ring is obtained by combining a photochemical [2 + 2] cycloaddition and a thermal electrocyclic process:

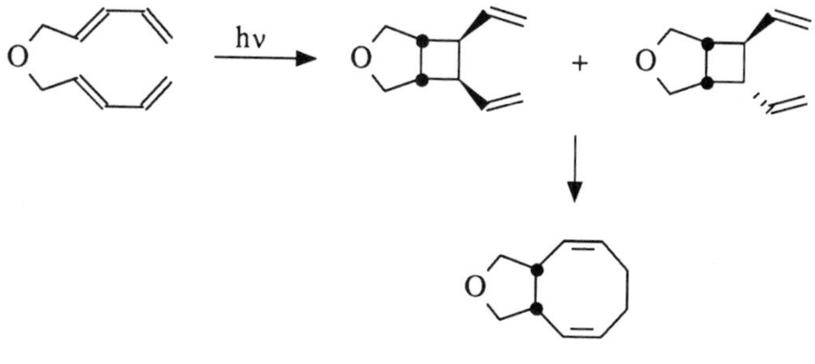

One synthesis of the triquinane coriolin (**22**) incorporated this approach.[16]

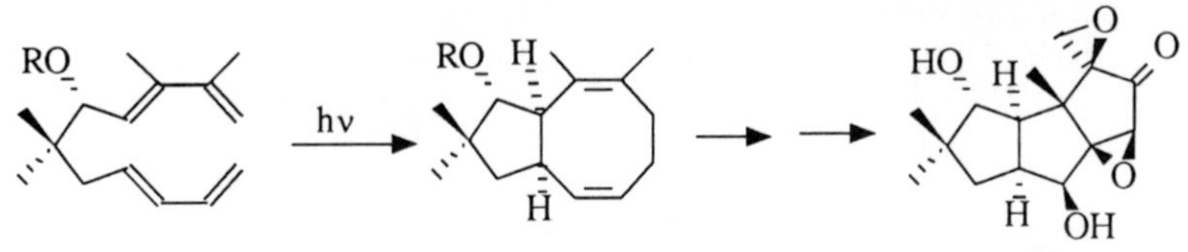

(**22**)

8.7. CYCLOADDITION REACTIONS WITH ACETYLENIC COMPOUNDS

The impressive phototoxic properties observed in some 1,3-butadiynes and 1,3,5-hexatriynes, notably phenylheptatriyne,[17] have prompted structure–activity studies with analogs. The photochemical reaction of 2,3-dimethyl-2-butene with conjugated di- and triacetylenic compounds gave interesting cycloaddition products, although the yields did not exceed 20%:[18,19]

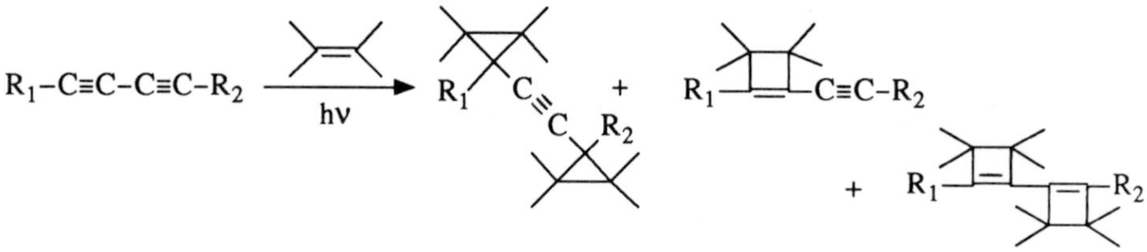

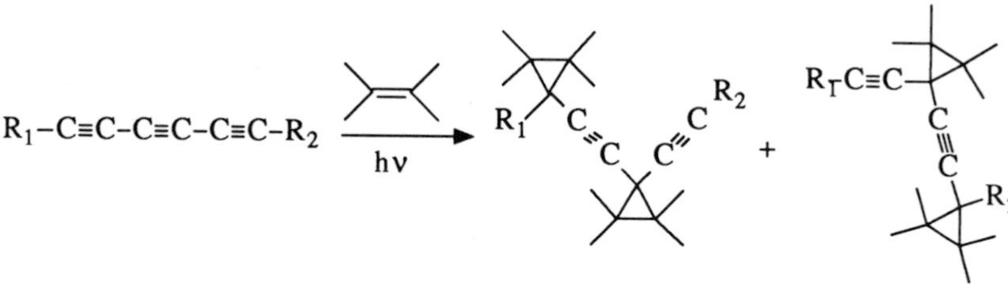

The mechanism suggested for the formation of the cyclopropane products involves the triplet state of the polyacetylene, which produces a diradical intermediate when added to one molecule of alkene. When cyclization cccurs as shown, a cyclopropane is formed, creating an intermediate which is equivalent to a carbene. Adding it to a second molecule of alkene gives the observed product.

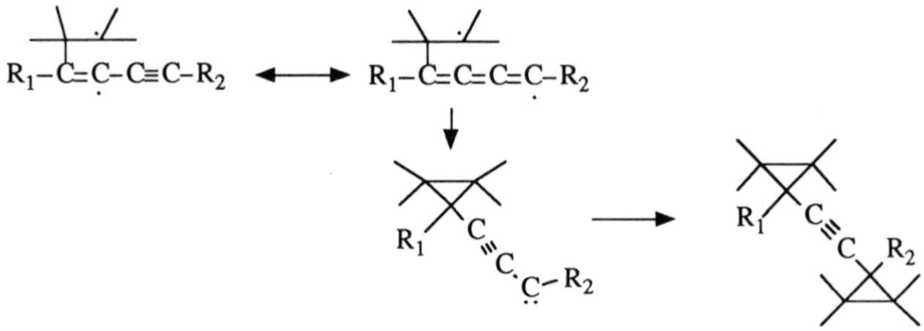

The synthesis of a three-membered ring product by addition of an alkene to an acetylene had been discovered earlier in the irradiation of tetramethylethylene with the acetylenic ketone (23) in the presence of water.[20] Here, however, a cyclopropene, (24), was produced.

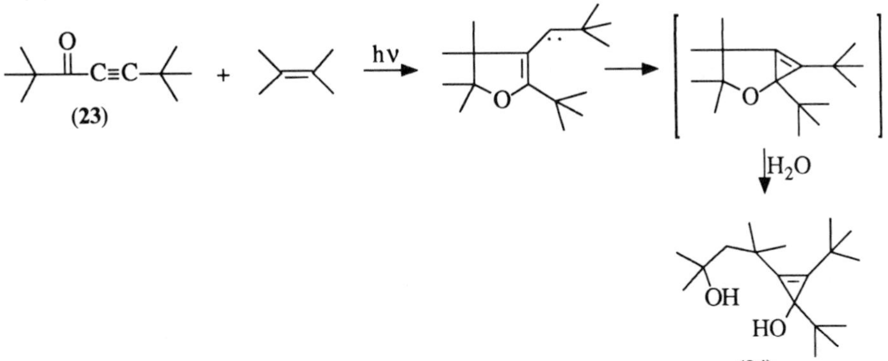

8.8. INSERTION REACTIONS OF OXACARBENES

The conversion of a carbonyl into a cyclic ether shown above brings to mind the chemistry of cyclobutanones, which undergo ring enlargement to five-membered ring oxacarbenes (Chapter 4). The carbenes can be trapped intermolecularly by reaction with water or an alcohol, but they can also be trapped intramolecularly by an alcohol, amine, or thiol. One example is shown below, but the size of the ring fused to a tetrahydrofuran may also be different, up to eight-membered, depending upon the length of the side-chain:[21]

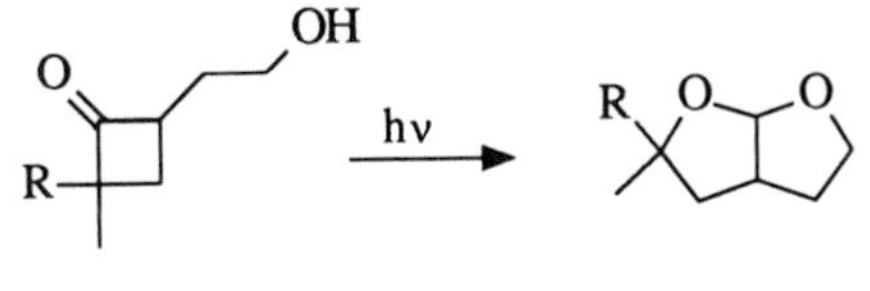

8.9. CYCLOADDITION REACTIONS OF NITRILES

The nitrile group is usually inert photochemically, allowing acetonitrile to be used as a solvent in many photochemical reactions. When the nitrile group is conjugated, however, it undergoes photocyclization reactions similar to those of acetylenes, as demonstrated in a synthesis of azocines, a class of molecules which stimulate the central nervous system, and which have hypnotic and anticonvulsive activity. The following example is interesting because photocycloaddition to the aromatic ring in phenols is seldom encountered:[22]

8.10. REACTIONS OF α, β-UNSATURATED CARBONYLS

In addition to the individual reactions of the carbonyl and alkene components des-cribed in Chapters 3 and 4, unique transformations are observed when the groups are conjugated. For instance, a hydrogen appears to be abstracted from a donor by the α-position of the carbonyl, along with alkylation at the β-position. Normally, any radical character generated in the excited carbonyl would have a resonance contri-butor at the β-position, but not at the α-position, and therefore the hydrogen abstrac-tion observed must be explained differently. A simple explanation would have been a $2\sigma + 2\pi$ process involving the alkene portion:

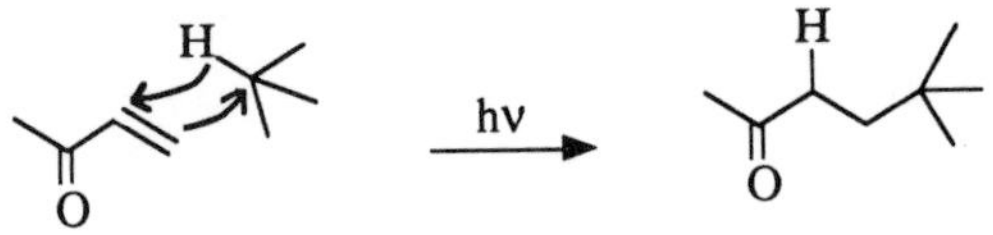

However, since the addition reaction can be sensitized and therefore does not necessa-rily occur in a concerted fashion, radical intermediates may be involved.

In a more attractive mechanism, a hydrogen is abstracted by the excited carbonyl, the radicals combine to form a carbon–carbon bond, and tautomerization produces the final product:

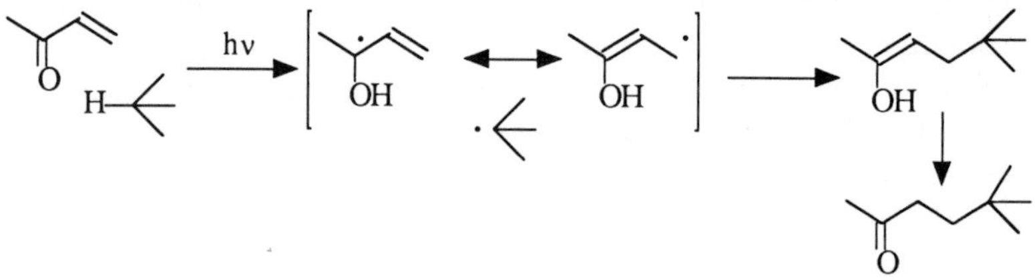

The intermolecular addition between (25) and (26) exemplifies the selectivity in the breaking of carbon–hydrogen bonds, favoring the tertiary hydrogen (as expected from the above mechanism).

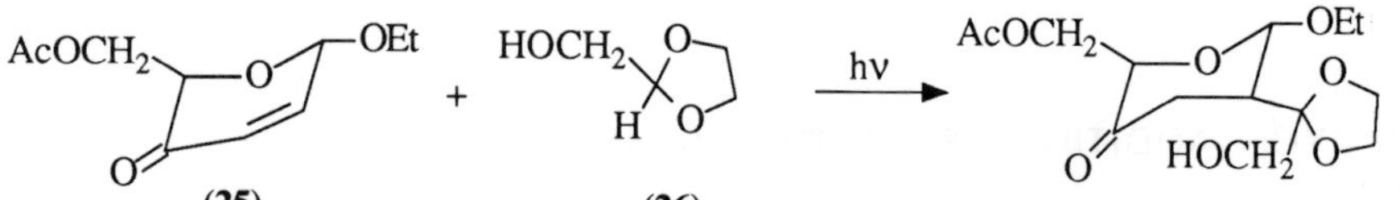

The intramolecular version of this reaction occurred quantitatively in taxinine (27) (accompanied by (E)–(Z) isomerization in the cinnamoyl side-chain).[23] However, a

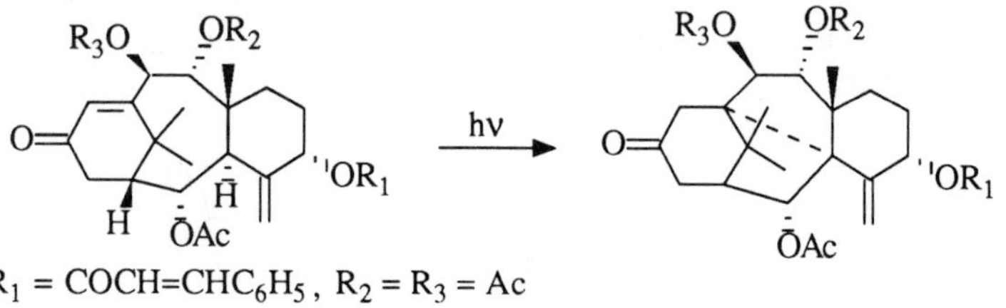

$R_1 = COCH=CHC_6H_5$, $R_2 = R_3 = Ac$

(27)

small structural change, (28), diverted the excited state toward another formal $2\pi + 2\sigma$ interaction, leading to a product, (29), containing a cyclopropane (a transformation well known in cyclohexenones). A small amount of photoproduct analogous to that shown above was also isolated.

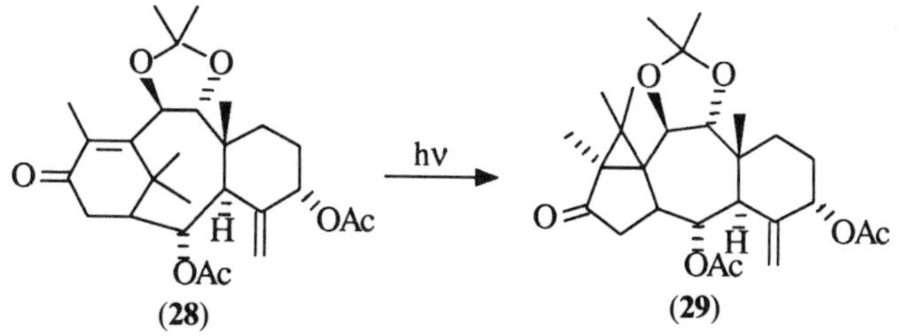

Another type of reaction of α, β-unsaturated carbonyl compounds involves intramolecular transformations:

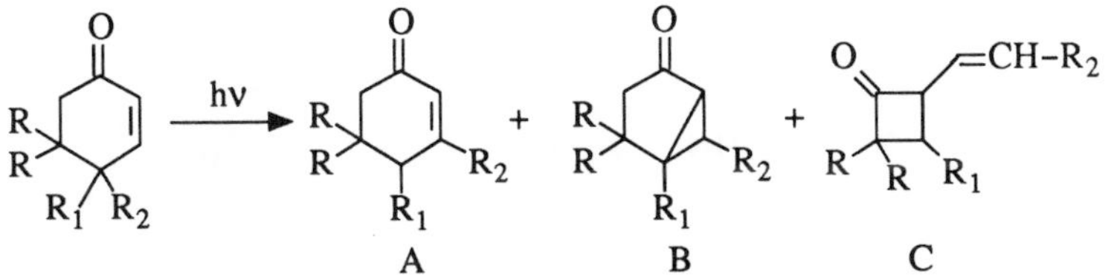

It is extremely sensitive to the nature of the substituents.[24] When $R = H$ and $R_1 = R_2 = C_6H_5$, the major product is B, in which the two phenyl groups are *trans*. Likewise, when $R = R_2 = C_6H_5$ and $R_1 = H$, the major product is B. However, when $R = R_2 = C_6H_5$ and $R_1 = H$, the major product is A, and when $R = C_6H_5$, $R_1 = H$, and $R_2 = CH_3$, the only product is C. The structure of the products is readily accounted for by $2\sigma + 2\pi$ electronic transformations in the starting material, in which the σ-electrons are either the C4–C5 bond or the bond connecting C4 to R_2.

When applied to a seven-membered ring such as (30), the same conditions led only to dimerization in which the head-to-head dimers (31) and (32) were obtained. Their stereochemistry suggests that they might have been formed by suprafacial–suprafacial cyclization of the *trans*-cycloheptenone formed upon irradiation of the (*cis*) starting material.[25] Actually, *trans*-fused cycloadducts are the usual products in reactions of cyclohexenones with alkenes, although extended irradiations have led to increasing amounts of *cis*-fused products (which may also be obtained by base treatments of the *trans*-fused isomers).[26]

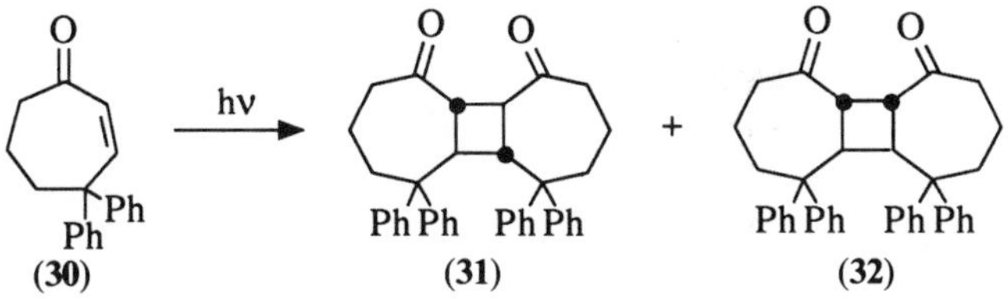

Several diterpenes possess two five-membered rings fused to a four-membered ring. Their synthesis may be approached photochemically. Thus, β-bourbonene (33) was synthesized as shown.[27] The same ring system is found in spatane diterpenes from marine organisms. It was synthesized by a similar approach, but refined in order to increase stereocontrol.[28]

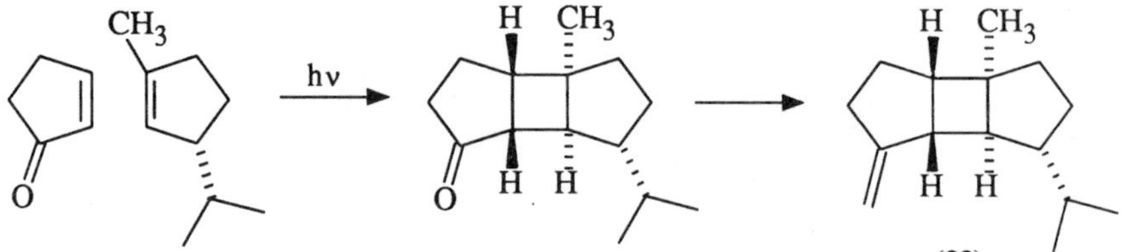

(33)

The following example combines a cycloaddition/cycloreversion reaction of an α, β-unsaturated carbonyl compound and a Mannich ring closure:

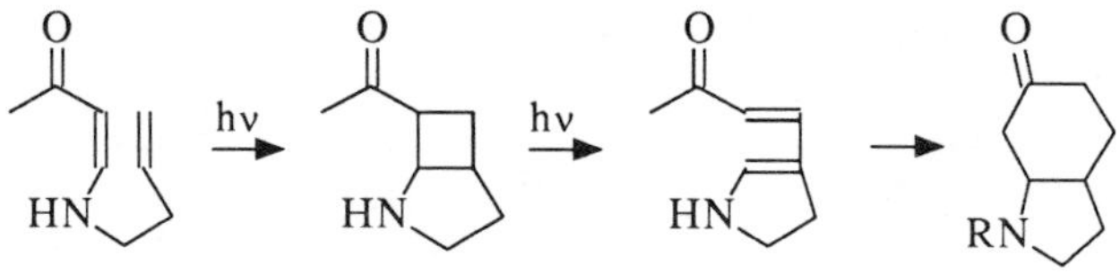

Using tryptophan in such a sequence produced, in an optically active form, a key intermediate which has been converted into vindorosin (34) by Büchi.[29]

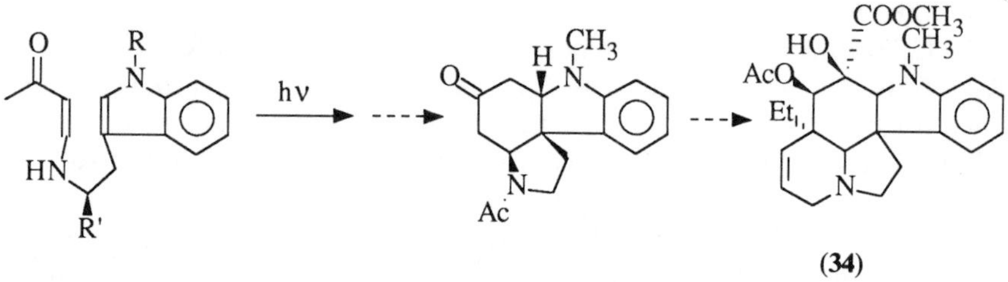

(34)

The [2 + 2] cycloaddition of an α, β-unsaturated ketone and a distant double bond allowed the synthesis of a ring system found in the ingenane diterpenes (e.g. (35)), in which a bicyclic moiety is locked in a *trans* relationship.[30] The last step shown involves base-catalyzed hydrolysis. The mixture of epimeric carboxylic acids obtained was later decarboxylated to a single ketone.

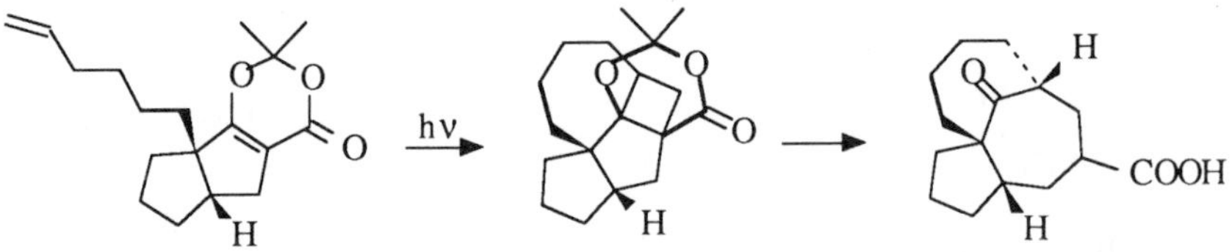

(35)

The complexity of the photochemistry of α,β-unsaturated carbonyl compounds is further illustrated by examples which reveal that the wavelength of the light used for irradiation may control the nature of the products formed:

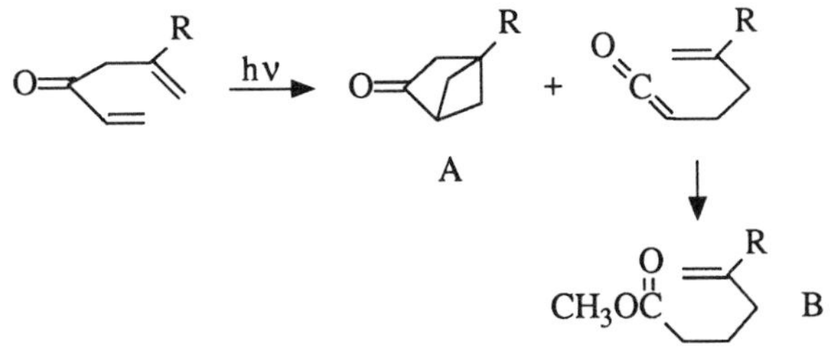

The quantum yield of formation of A(R = H) was constant (0.13) from 366 to 334 nm, but dropped to 0.046 at 313 nm. In contrast, the quantum yield of formation for B increased fourfold (from 0.00054 to 0.0023) between 366 and 334 nm, but was essentially the same at 334 and 313 nm (0.0021). In sensitized experiments, the ratio of A:B was 266:1. Only four out of the 10 related molecules which were investigated showed this wavelength dependence. In the case of the dienone with R = CH$_3$, not only were the quantum and chemical yields dependent on the wavelength, but sensitizers having the same triplet energy did not give the same results. Molecules in which the lowest triplets are π,π^* sensitized the conversion, whereas those with n,π^* did not. Finally, standard quenchers had no effect on the course of the direct conversion. Although some suggestions were made, no definite explanations for all these results have been obtained.[31]

8.11. PHOTOCHEMISTRY OF β,γ-UNSATURATED CARBONYL COMPOUNDS

These compounds also have a rich photochemistry, with 1.2 and 1.3 shifts occurring from singlet and triplet excited states, respectively. The course of the reactions may be solvent-dependent. For example, (36) undergoes initial α-cleavage to produce a ketene which is trapped by the solvent and yields (37) in methanol, or suffers a 1,3-migration of the acyl group in benzene to form (38).[32]

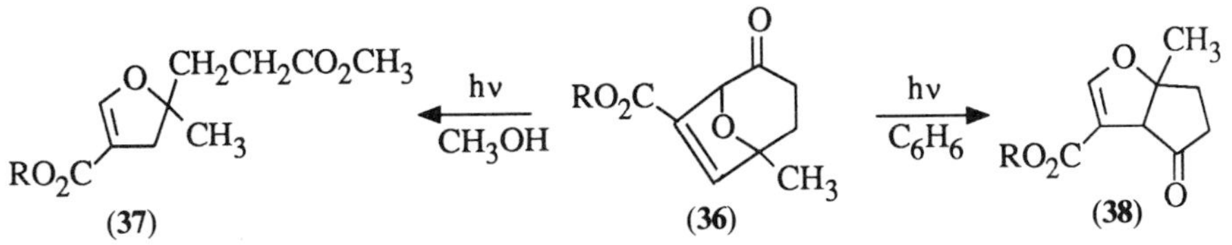

In a system as complex as that of the cembratrienone (39), different rearrangements occur competitively, producing isomers. All these were found in the same Caribbean

gorgonian organism *Eunicea calyculata*. It has been suggested that these isomeric products might be formed photochemically *in vivo*.[33]

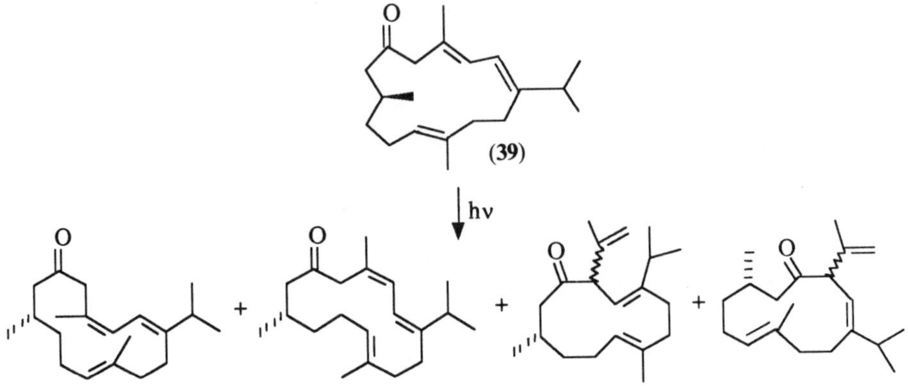

8.12. OXA-DI-π-METHANE REARRANGEMENT OF A β,γ-UNSATURATED KETONE

The total synthesis of the methyl ester of pentalenolactone P (**40**), a pentacyclic triquinane sesquiterpene, was achieved by a sequence which incorporated a di-π-methane
rearrangement of a β,γ-unsaturated ketone.[34]

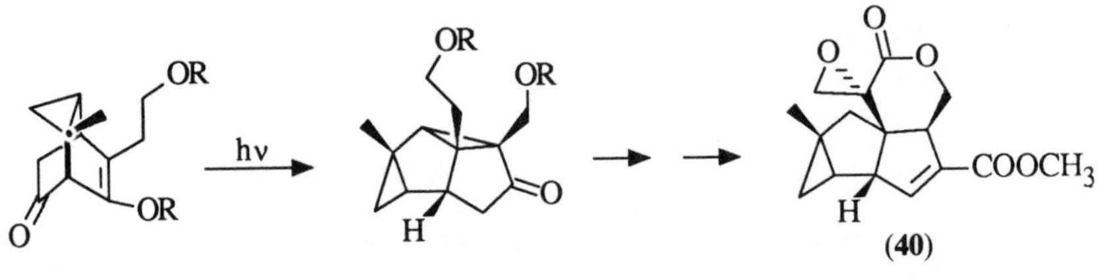

8.13. PHOTOCHEMISTRY OF γ,δ-UNSATURATED CARBONYL COMPOUNDS

These systems are similar to the dienes which undergo the di-π-methane rearrangement
(Chapter 3). An impressive example of sensitized oxa-di-π-methane rearrangement led,
after reductive cleavage of one cyclopropane bond and further elaboration, to the
synthesis of the sesquiterpene modhephene (**41**).[35]

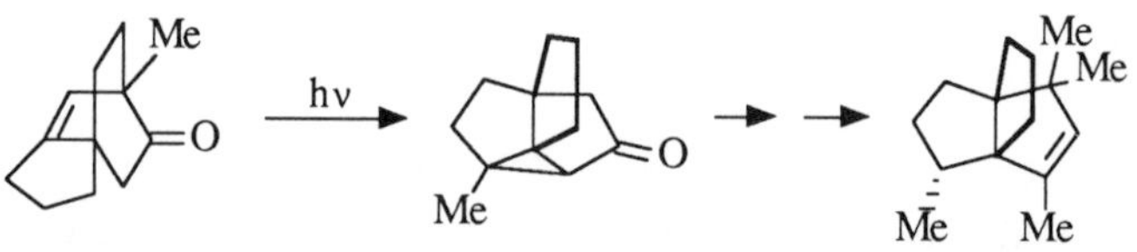

8.14. PHOTOCHEMISTRY OF δ,ϵ-UNSATURATED CARBONYL COMPOUNDS

Photoreduction by electron transfer produces a radical anion which can undergo cyclization, as do chemically or electrochemically generated species:

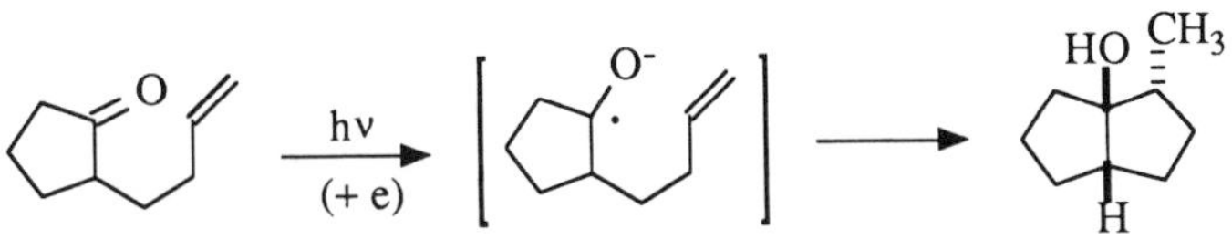

Here, the reaction was performed in hexamethylphosphoric triamide (81% yield) or in acetonitrile containing triethylamine (50% yield).[36] This model was used in a synthesis of hirsutene (43) from the acetylenic ketone (42).[37]

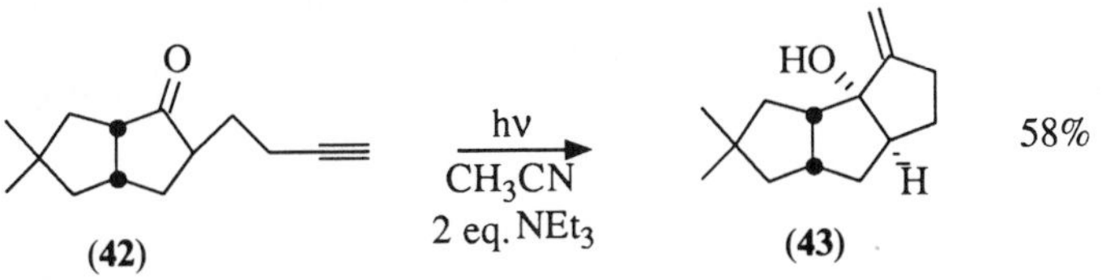

8.15. PHOTOOXIDATION REACTIONS

The antimalarial compound known as quinghaosu or artemisinin (46) is a highly oxygenated terpene derivative. It has been synthesized by photooxidation of arteannuic acid (quighao acid) (44) and from (45), an intermediate in one synthesis, which was photooxidized to (46).[38,39]

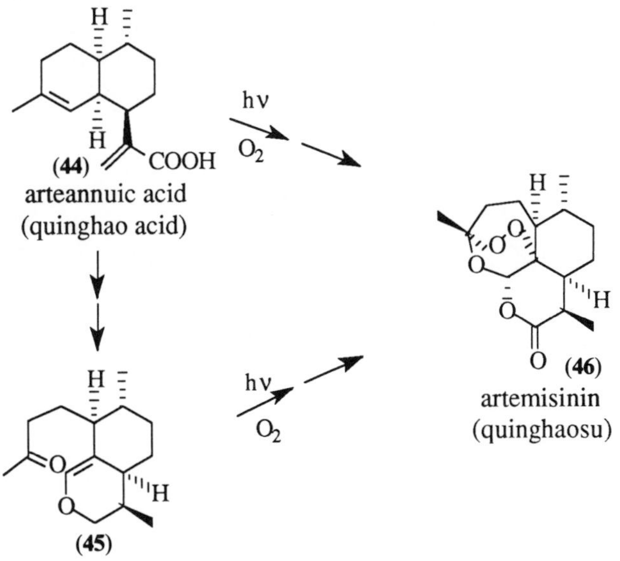

In each case the photooxidation provided allylic hydroperoxides, which were first treated with copper(II) trifluoromethanesulfonate and then with trimethylsilyl trifluoromethanesulfonate, followed by minor adjustment of the oxidation level.

The key reaction was investigated separately. It involves the initial formation of an allylic hydroperoxide, which is oxidized to the corresponding radical, undergoes cyclization, and finally opens thermally to create two carbonyl groups:[40]

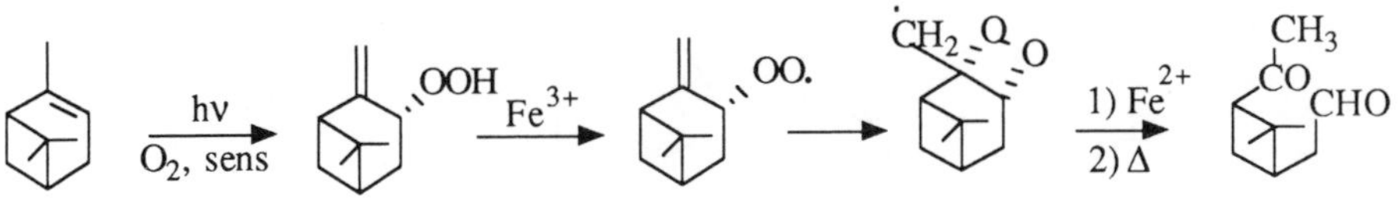

Treated separately with Fe^{3+} in the absence of O_2, the allylic hydroperoxide produces a dioxetane adjacent to a radical center. Reduction, protonation and dioxetane ring cleavage produce a dicarbonyl product (the product which would have been expected from the cycloaddition of singlet oxygen to the alkene precursor of the hydroperoxide). The above radical, following attack by molecular oxygen, redox reaction, and ring opening, leads to the product shown:

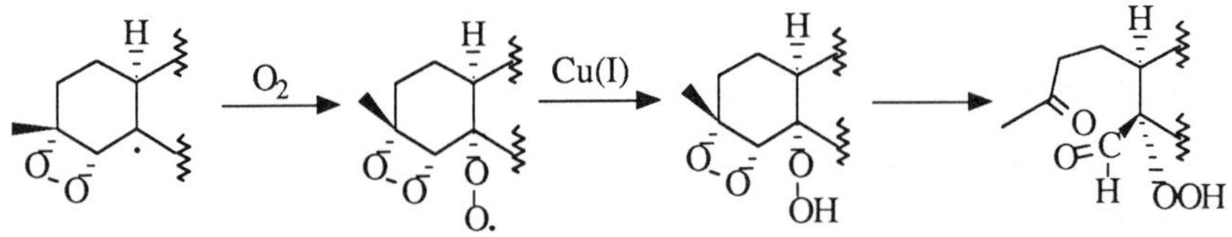

It is interesting to note that the 1,2,4-trioxane feature of artemisinin can also be generated by a Paterno–Büchi-type reaction in the presence of oxygen. That reaction normally leads to a dioxetane, the product of [2 + 2] cycloaddition of a carbonyl with an alkene, but a triplet diradical intermediate can be intercepted:

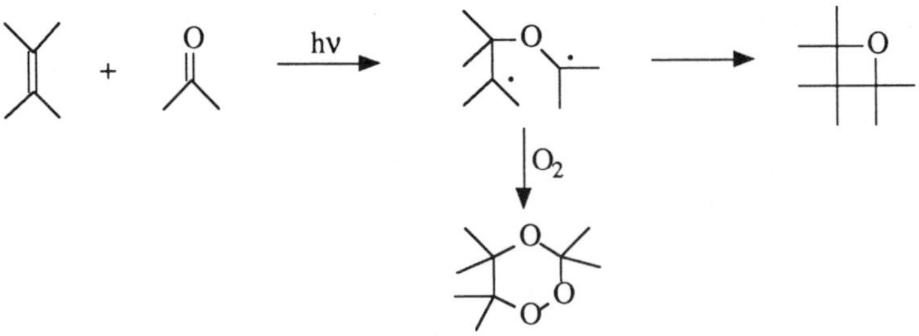

The scheme above neglects a number of potential complications, such as regioselectivity and decomposition (thermal and/or photochemical) of the trioxane.[41,42] However, it could account nicely for the conversion of arteannuic acid into artemisinin mentioned above. Note that other species, such as SO_2, have also been trapped in a similar manner.

8.16. PHOTOOXIDATION OF FURANS COUPLED WITH A DIELS–ALDER REACTION

The reaction of furans with singlet oxygen was discussed in Chapter 6. The bisfuran (47), when treated with 1 equivalent of singlet oxygen, underwent an intramolecular reaction to yield the decalin derivative (48) stereospecifically.[43]

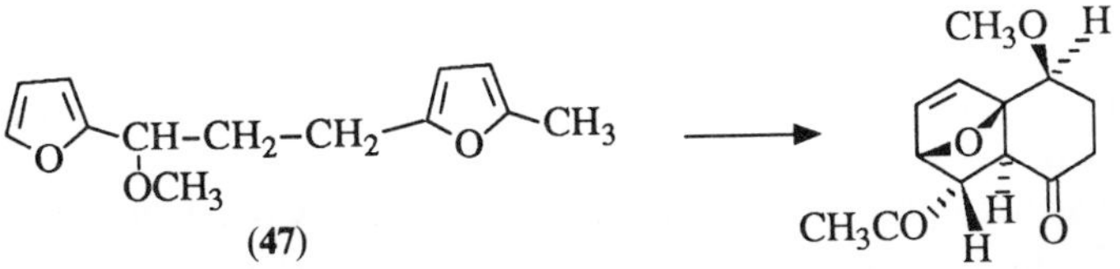

(47)

(48)

8.17. PHOTOOXIDATION OF BENZOFURANS

Benzofurans react with singlet oxygen to produce dioxetanes, which may be mutagenic:[44]

Interestingly, there is a recent patent on the use of benzofurans as fluorescent brighteners in liquid detergent compositions.[45] One can only hope that the mutagenicity results on the photoproducts will have come to the attention of the detergent manufacturers before they attempt to commercialize this class of brighteners!

Other furans have recently been found to undergo the same type of cycloaddition reaction with singlet oxygen at low temperature (Chapter 6); their mutagenicity has been ascribed to the epoxide formed upon deoxygenation of the dioxetane (Chapter 10).

8.18. PHOTOOXIDATION OF CYCLOPROPANES TO 1,2-DIOXOLANES

Tetraarylcyclopropanes are inert when irradiated alone either in the presence or absence of O_2, or under conditions where singlet oxygen is generated. When irradiated in the presence of dicyanoanthracene (DCA) and O_2, on the other hand, they produce 1,2-dioxolanes:[46,47]

An attractive explanation assumed photoinduced electron transfer from the cyclopropane to the sensitizer. The radical anion produced from DCA would transfer one electron to O_2, and $O_2^{\bar{\cdot}}$ thus formed would react with the cyclopropane radical cation, yielding the observed product. This mechanism was discounted, however, because (1) 1,8-dihydroxyanthraquinone, which also sensitized the oxygenation reaction, does not sensitize reactions proceeding via $O_2^{\overline{\cdot}}$, and (2) benzoquinone, a known quencher of $O_2^{\overline{\cdot}}$, did not inhibit the photooxidation reactions. The preferred mechanism involves photoinduced electron transfer chain reactions which provide interactions between radical cations and O_2.

8.19. ISOTOPICALLY LABELING A CARBONYL GROUP BY PHOTOOXIDATION

The sequence alkene → dioxetane → carbonyl lends itself to the synthesis of a carbonyl group possessing an isotopically labeled oxygen atom. The possibility of carrying out the conversion in non-protic media is a significant feature of this approach, which avoids the facile oxygen exchange with an aqueous solvent. Using this synthetic approach, it has been possible to obtain (**49**), the chirality of which is due solely to ^{18}O substitution.[48] The labeled diketone (**49**), even when stored in crystalline form, was very sensitive to atmospheric moisture. It lost all its label in less than 72 hours.

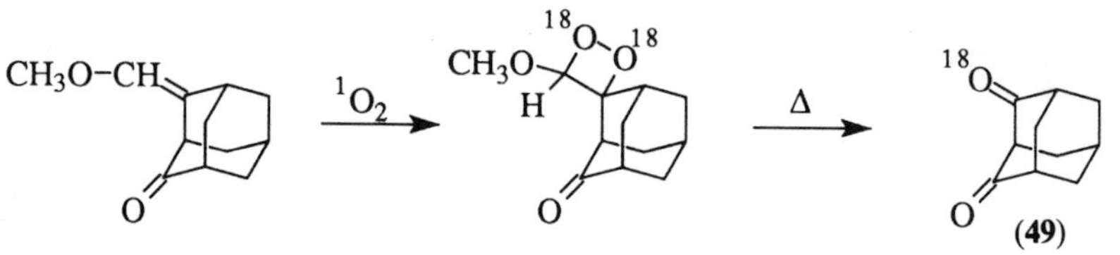

8.20. PHOTOCHEMICAL REDUCTION REACTIONS

The photochemical excitation of a carbonyl group in the presence of a hydrogen donor produces a hydroxymethyl radical intermediate, which can cyclize onto a suitably located unsaturated substituent, as illustrated:[49]

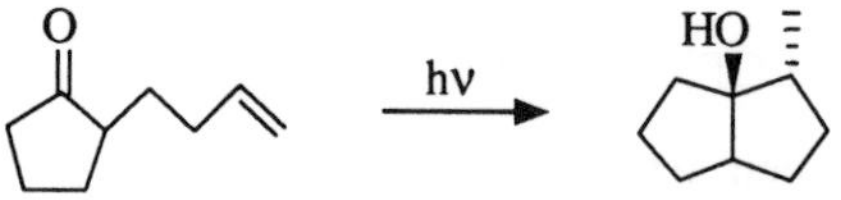

81% in HMPT
50% in Et_3N/CH_3CN

This model was used in a synthesis of hirsutene (**43**) from the acetylenic ketone (**42**).[49]

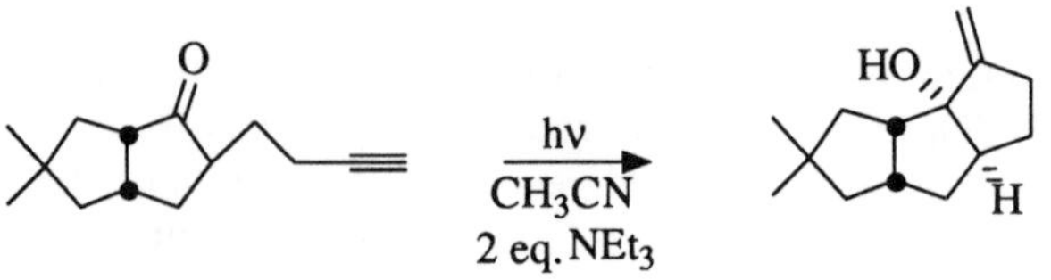

58%

8.21. ISOMERIZATION OF HETEROCYCLIC COMPOUNDS: ISOMERIZATION OF 2-PYRROLONES

A 2-pyrrolone (R = *t*-butyl) can be isomerized into its 4-isomer quantitatively:[50]

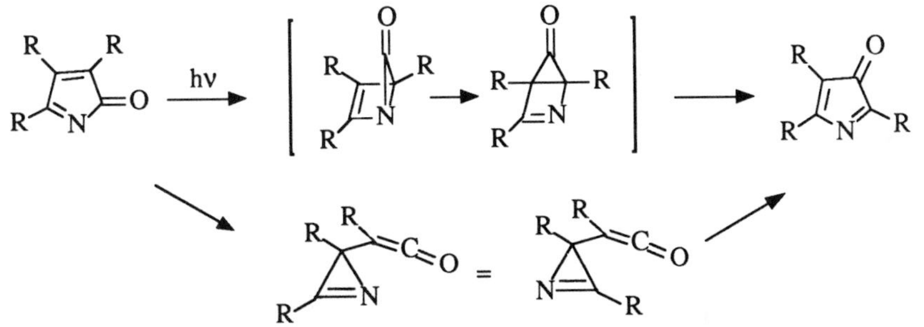

The authors suggested that a bicyclic lactam intermediate was obtained and that it iso-merized. However, isomerization to a three-membered ring followed by ring closure (on the model used for the isomerization of furans and thiophenes) would fit equally well, as illustrated in the lower part of the scheme.

8.22. ISOMERIZATION–FRAGMENTATION OF A COUMARIN

The photochemistry of coumarin has been particularly important because it allowed the very reactive and theoretically important compound cyclobutadiene (**50**) to be synthe-sized and characterized by infrared spectroscopy at 8 K.[51] Cyclobutadiene is fragmen-ted into 2 mol of acetylene upon further photolysis, and it gives a stable dimer when the temperature is raised. The dimerization is obviously a process which requires that two

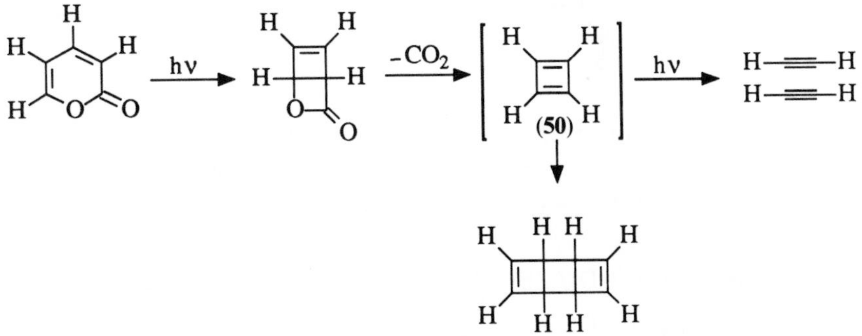

molecules be able to approach within bonding distance, and it is therefore reasonable to find that replacing all or most of the hydrogens of cyclobutadiene by bulky t-butyl groups considerably stabilizes the molecule. Another approach toward stabilizing cyclobutadiene utilizes an assembly consisting of one inert host molecule containing within its cavity only one molecule of pyrone. Cyclobutadiene photochemically synthe-sized from pyrone in the inner phase of the assembly is protected from bimolecular reactions. Its stability is such that its NMR spectrum could be recorded at 60°C. Further photolysis produced acetylene.[52] Cram's use of carcerands and hemicarcer-ands is different from work on zeolites and clathrates, since one host molecule provides one discrete molecular inner phase, and the assembly can exist in solid, solution, or gas phases. This approach invites the investigation of many other reactive molecules which are easily produced photochemically.

8.23. CYCLOADDITION REACTIONS OF HETEROCYCLIC COMPOUNDS

Cycloaddition reactions of pyrimidines are major components in the photobiology of nucleic acids: they are discussed in Chapter 9. Although dimers are particularly important, cycloaddition reactions forming other types of molecules are known. For example, cycloaddition of the uracil derivative (**51**) with benzene has been described (the different modes of cycloaddition of aromatic compounds are discussed in Chapter 3).[53]

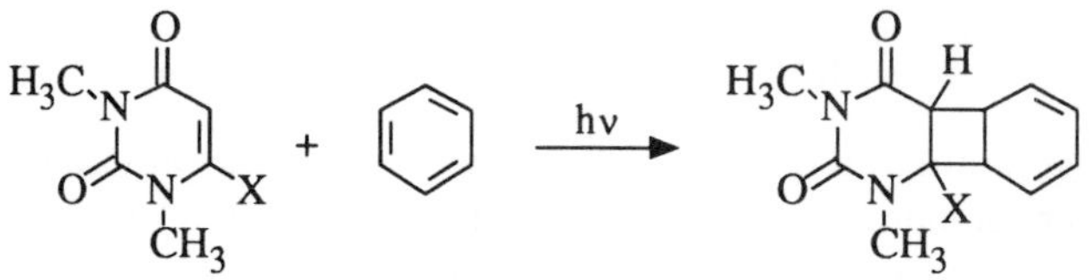

(**51**)

8.24. SYNTHESIS OF β-LACTAMS

Although the photochemical synthesis of a four-membered ring, such a β-lactam, can be approached by a $[2 + 2]$ cycloaddition reaction, it is also achieved through ring size contraction:[54]

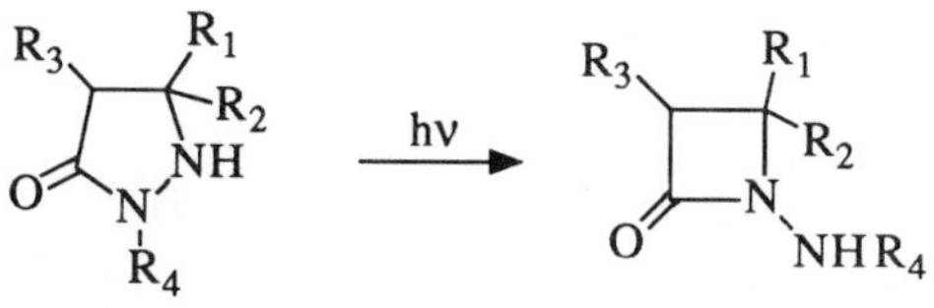

The best results were obtained when a photolabile *o*-nitrobenzyl protecting group was initially present. Although the reverse reaction could take place in the presence of trifluoroacetic acid, it was not described photochemically.

8.25. SYNTHESIS OF 1,2-DIAZEPINES BY RING ENLARGEMENT

In contrast to the above study, which showed that ring enlargement from the four- to the five-membered ring did not take place, irradiation of *N*-iminopyridinium ylides produced isomeric 1,2-diazepines in excellent yield:[55]

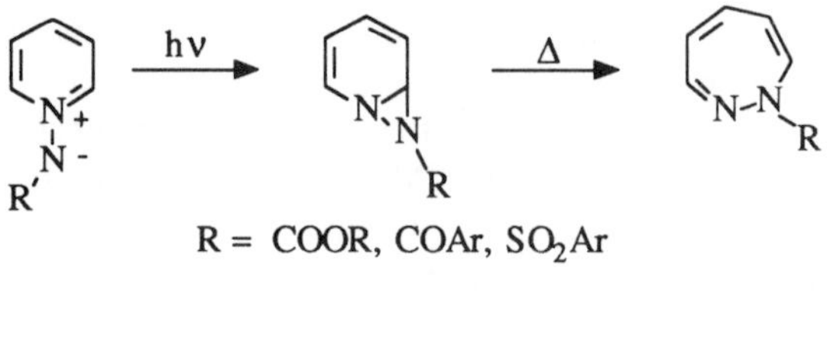

R = COOR, COAr, SO$_2$Ar

8.26. FRAGMENTATION OF SMALL HETEROCYCLIC RING COMPOUNDS

The ring opening of epoxides, aziridines, and episulfides (**52**) is commonplace in organic chemistry. Ionic processes tend to break bonds between a carbon atom and the heteroatom. Photochemically, in contrast, the carbon–carbon bond opposite to the heteroatom breaks predominantly. The process is allowed and may be concerted, as it involves $2n + 2\pi$ electrons. The initial product is therefore a carbonyl, thiocarbonyl, or imine ylide, which may then be trapped by cycloaddition with a molecule like diethyl acetylenedicarboxylate.

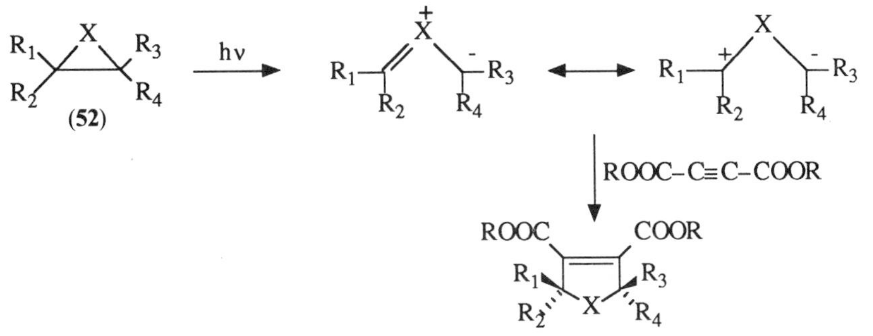

8.27. FRAGMENTATION OF OTHER HETEROCYCLIC MOLECULES

The fragmentation of a pyrimidine (e.g. (**53**)) in excellent yield is noteworthy in view of the importance of this class of molecule in biochemistry and biology.[56] A reasonable

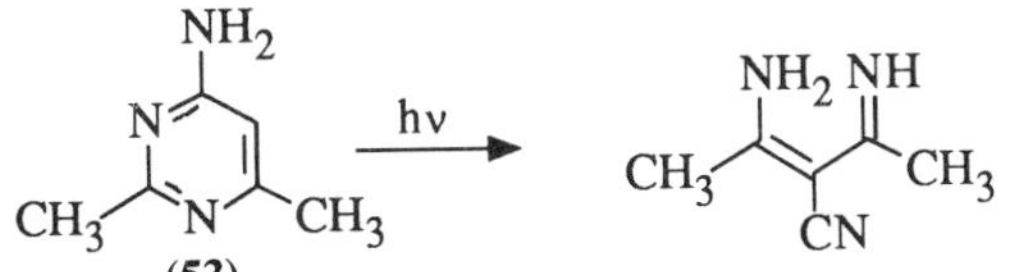

mechanism suggests isomerization to the Dewar benzene analog, 1,3-hydrogen shift, $2\sigma + 2\sigma$ electrocyclic reaction, and another 1,3-hydrogen shift (all photochemical), followed by ionic ring opening to the observed product:

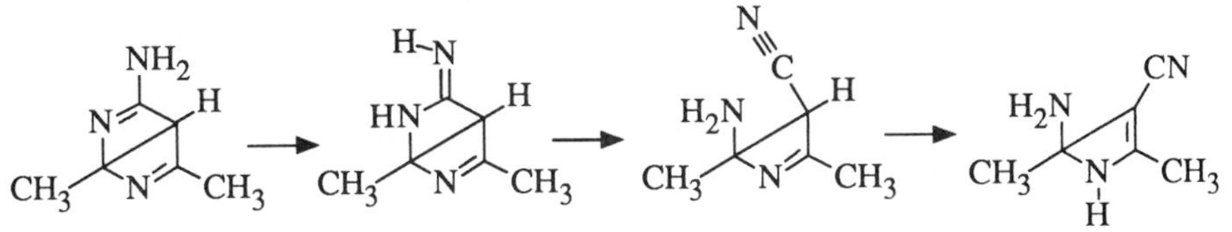

Conceivably, the ring-opening step creating the nitrile group could be an ionic reaction.

8.28. INTRAMOLECULAR REACTION OF NITROAROMATIC COMPOUNDS; A ROUTE TO THE MYTOMYCINS

The ability of a nitro group to perform an intramolecular hydrogen abstraction was first revealed in a very old rearrangement, the conversion of o-nitrobenzaldehyde (**54**) into o-nitrosobenzoic acid (**55**), in which it is the first step. This type of reaction

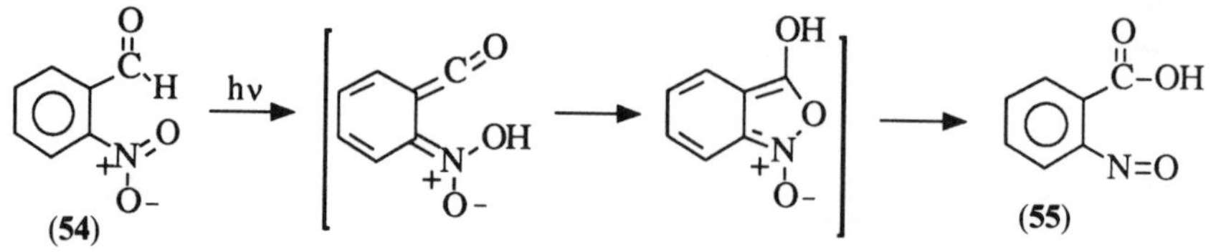

has found many important applications, for example in the design of photolabile protecting groups and in the photogeneration of bases, as discussed later. It also affords an entry into the synthesis of the mitomycins (**56**) related to mitomycin C, a clinically useful antineoplastic drug.[57]

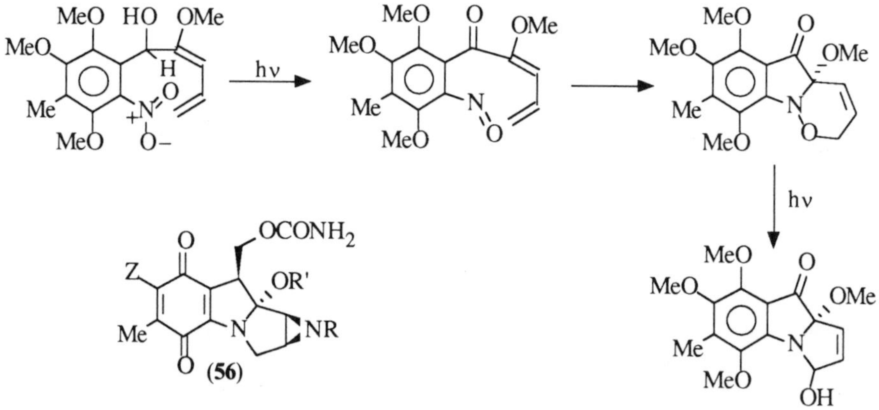

The cyclization of the nitroso compound which is initially generated is probably a thermal reaction. The mechanism of the second step is not yet known. It is logical to imagine photochemical scission of the nitrogen–oxygen bond, followed by hydrogen abstraction by the nitrogen radical, and cyclization of the aminoaldehyde produced.

8.29. DIRECTED FORMATION OF CARBON–BROMINE AND CARBON–SULFUR BONDS

The remote functionalization of steroids with a flexible side-chain bearing a carbonyl group was described in Chapter 4. A related reaction attaches a chlorine atom to the iodine of the iodobenzoate side-chain of a steroid.[58] This directs the abstraction of a hydrogen from the steroid, which subsequently becomes substituted at that position:

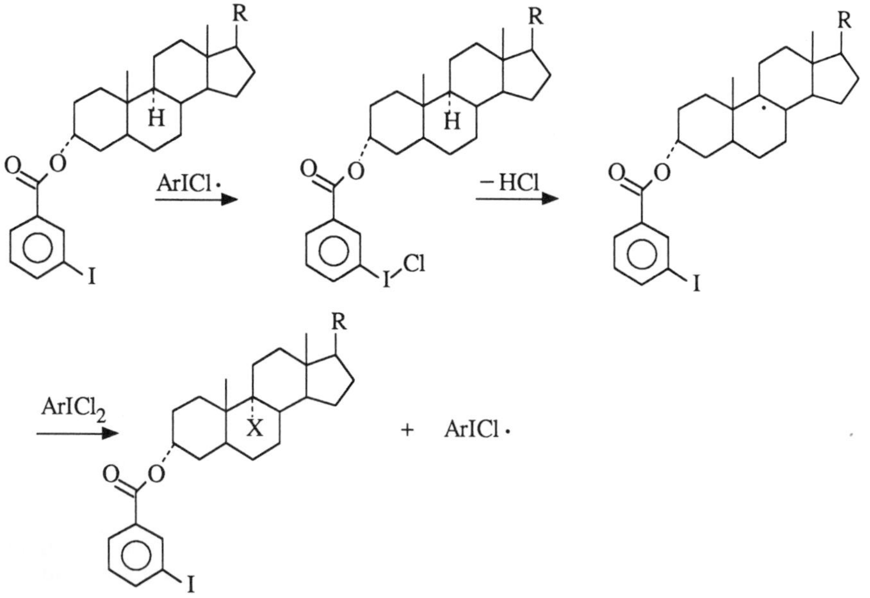

The chlorinating reagent in the example shown was $p\text{-}NO_2C_6H_4ICl_2$ which, upon photolysis, yielded over 90% of the product with R = Cl through a chain reaction. The products with R = Br and R = SCN were formed when CBr_4 or $(SCN)_2$, respectively, were present during photolysis with an aryl iodonium chloride.

8.30. PHOTOCHEMICAL SYNTHESIS OF OPTICALLY ACTIVE PRODUCTS

The achiral compound (57) crystallizes in the chiral space group $P2_12_12_1$. Upon irradiation in the solid state it undergoes a di-π-methane rearrangement and yields two regioisomers: (58) is an almost pure enantiomer, while (59) is racemic.[59]

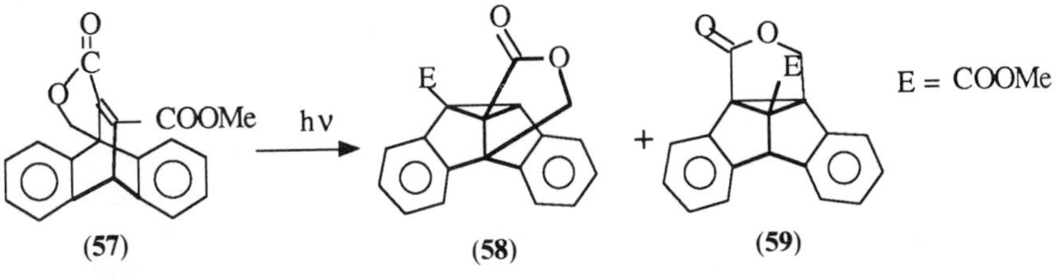

The old idea of converting racemic molecules into diastereomers by reaction with an optically active reagent, so useful for optical resolution, has been somewhat extended to photochemical syntheses. Here, crystal chirality has been used in asymmetric syntheses, as illustrated with the monoester (60), which gave crystalline salts with optically active amines. The salts were irradiated, and then treated with diazomethane to give the diester (**61**) with 14–80% enantiomeric excess.[60]

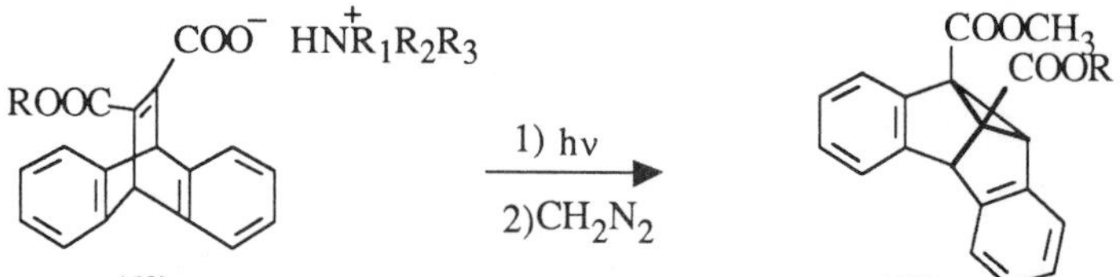

Most attempts to perform enantioselective photochemical reactions produced only modest selectivity, whether chiral sensitizers or chiral solvents were used. The most impressive result to date is shown here:

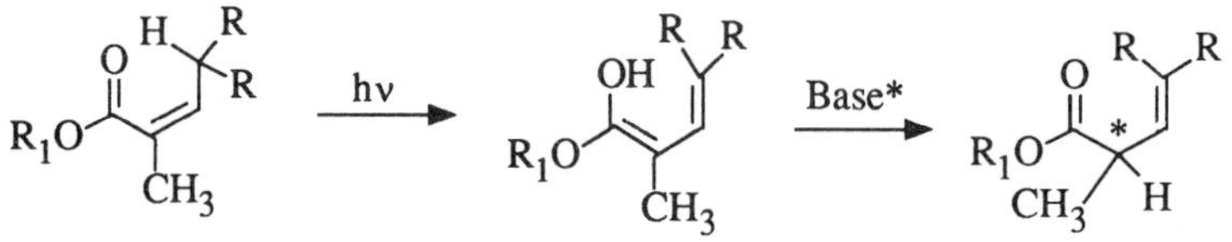

As described in Chapter 4, conjugated carbonyl compounds can be deconjugated photochemically because the conjugated enol initially formed is very acidic and tautomerizes readily in the presence of a catalytic amount of base. Since a chiral carbon is thus generated at the α-position, face selectivity in the protonation step is expected to produce enantioselectivity. The feasibility of this scheme was demonstrated by using a catalytic amount of a chiral β-aminoalcohol during photolysis: an enantiomeric excess as high as 71% was observed.[61]

Another example suggested that a $(Z)-(E)$ photoisomerization reaction performed in a chiral environment could produce appreciable enantiodifferentiation:[63]

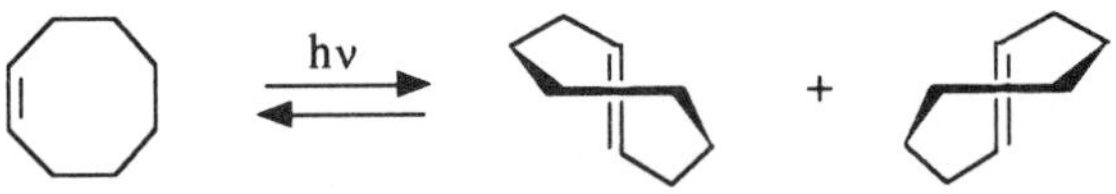

The best reported result (53% ee) was upon sensitization by a benzene hexacarboxylic ester of 1-methylheptanol at $-87°C$, but the conversion was only 6%. The results were less impressive at higher temperatures.

The effect of circularly polarized light in photochemical reactions was mentioned in Chapter 1. Preferential destruction of one antipode in a racemic mixture can occur, eventually leading to enrichment in the other antipode. The difference in reactivity between the two antipodes can be greatly enhanced by the simple process of association with a biological molecule. This was demonstrated with 1,1'-bi-2-naphthol, which was dissolved in aqueous bovine serum albumin. The $(-)$ compound was photodestroyed twice as readily as the $(+)$ antipode, leading to the observation of high enantiomeric excess with 'tolerable' photoconversion from a racemic mixture, namely 89% enantiomeric excess with 57% destruction.[63]

8.31. PHOTOREMOVABLE PROTECTING GROUPS

It is often desirable to have protecting groups which can be removed in the absence of either acidic or basic reagents. Several photolabile protecting groups have been devised. A recent review illustrated synthetic applications with peptides, carbohydrates, nucleotides, macrolides, and terpenes.[64] Selected examples are given below:

1. The *o*-nitrobenzyl group has been used in many different situations. It operates through an intramolecular redox reaction. A reasonable mechanism involves intramolecular hydrogen abstraction by one oxygen from the nitro group, cyclization, bond breaking, and proton loss:

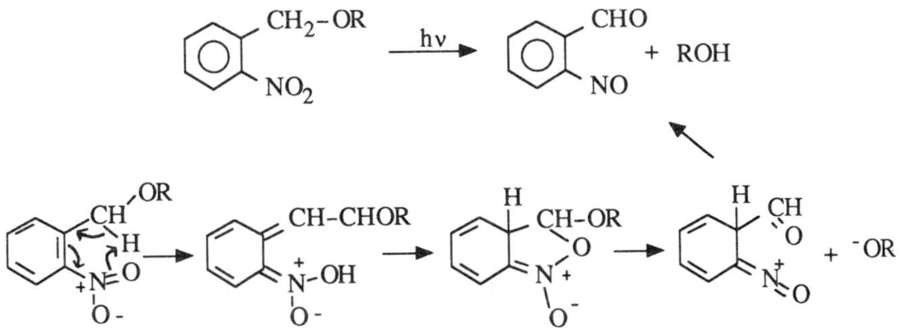

This approach has been used for the protection of other functionalities in addition to alcohols, such as amino, thiol, and carboxyl groups. It has also been used for the protection of the phosphate group in nucleotide synthesis.[65] In one interesting application, the *o*-nitrobenzyl group was attached (at its *para* position) to resins used for solid phase synthesis of polypeptides. The photochemical deprotection provided *N*-t-butoxycarbonylpeptide free acids.[66] This technique has been advocated as appropriate for the convergent solid phase synthesis of high-molecular-weight peptides.[67]

2. Tosylates are useful groups for the protection of alcohols.[68] The removal is initiated by an electron transfer from photoexcited electron-rich aromatic compounds, or it can be achieved by direct irradiation in the presence of a hydroxide base and an amine which acts as an electron donor, as shown here in a deprotection that proceeded in 95% yield without epimerization:[69]

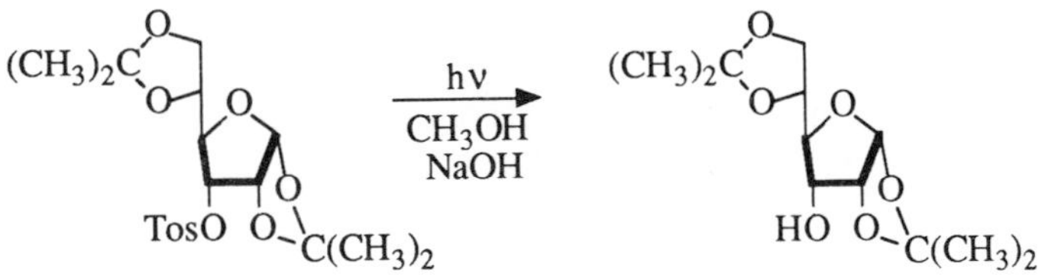

3. Benzyl groups have been used for protecting alcohols: the deprotection being achieved by photolysis in the presence of *N*-bromosuccinimide and water. A sugar

molecule possessing both benzyl and tosyl groups could be debenzylated in 95% yield in 15 minutes without loss of the tosyl groups:[70]

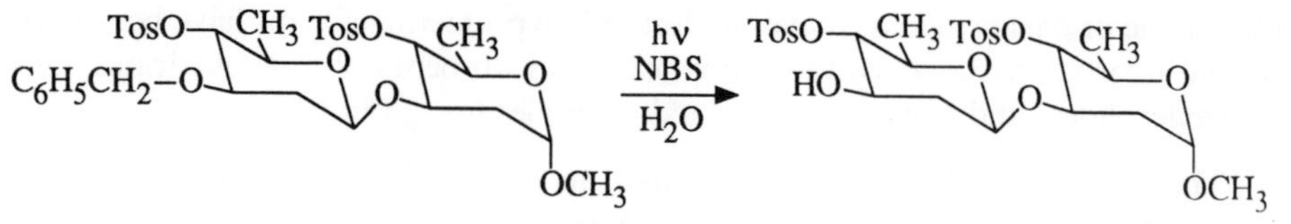

4. Protection of an amine as a formamide can be reversed by photolysis in acetonitrile:[71]

$$ArNHCHO \xrightarrow{h\nu} ArNH_2 + CO$$

This is another example of Norrish type I fission.

5. Protection of amines as benzylsulfonamides can be reversed by irradiation in the presence of 2-propanol (a good donor of hydrogen radicals):[72]

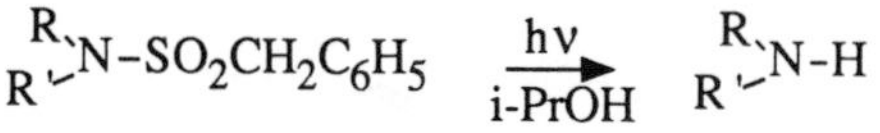

$$\underset{R}{\overset{R}{>}}N-SO_2CH_2C_6H_5 \xrightarrow[\text{i-PrOH}]{h\nu} \underset{R}{\overset{R}{>}}N-H$$

More complex aromatic sulfonamides are particularly good for the protection of the nucleic acid bases guanine, adenine, and cytosine.[73].

The detosylation of sulfonamides can also be sensitized by β-naphthoxide in methanol, a reaction with $NaBH_4$ providing the amine in the final step.[74]

6. Protection of pyridines as pyridine oxides may be reversed by photolysis in the presence of trimethyl phosphite:[75]

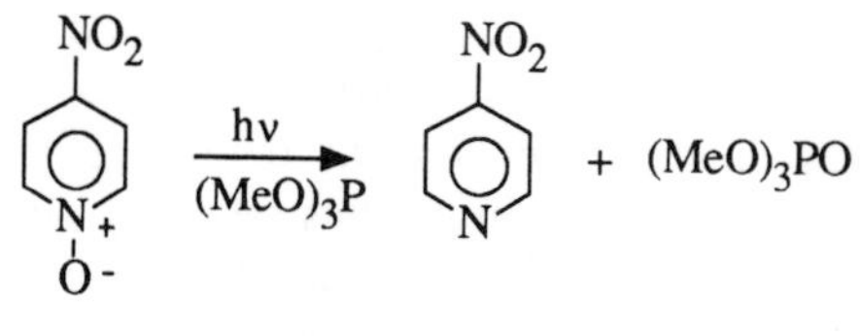

It may be noted that pyridine oxides undergo other interesting photoreactions, producing isomeric oxaziridines and leading to oxygen transfer to unsaturated molecules.

7. The deprotection of thioketals, which is often quite difficult, can be achieved under neutral conditions by benzophenone-sensitized photooxidation:[76]

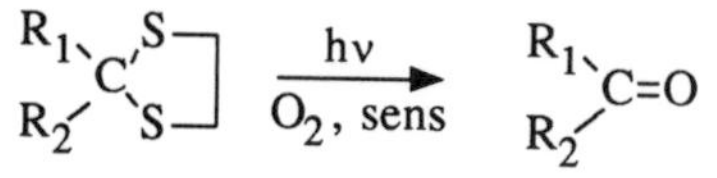

8. The conversion of oximes into carbonyls occurs in high yield upon photooxygenation in the presence of NaOH in methanol at 0°C:[77]

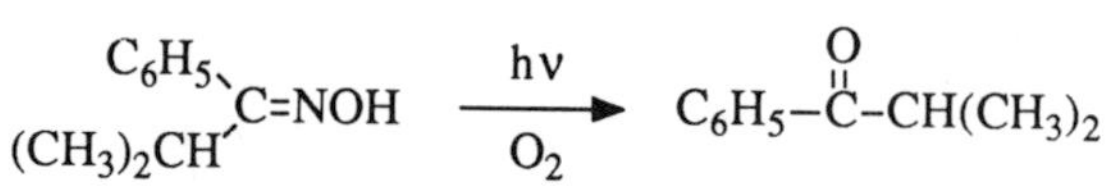

8.32. PHOTOCATALYTIC ELECTROSYNTHESIS OF A PHEROMONE

Early organic photochemistry was performed under sunlight irradiation. The ready availability of commercial mercury lamps was responsible for the explosive growth of the field, because irradiations could be reproduced in the convenience of the laboratory, regardless of time and place. Attempts to perform elaborate chemical conversions with sunlight alone are important components of a search for the utilization of solar energy. For example, a completely solar-powered photocatalytic electrosynthesis of the pheromone ($\pm$)-4-methyl-3-heptanone (**62**) has been described.[78] Solar energy was

$$CH_3CH_2Br \; + \; \underset{\underset{CH_3}{|}}{CH_2=C-COC_2H_5} \; \xrightarrow{\; 2\,\bar{e} + H^+ \;} \; \underset{\underset{CH_3}{|}}{CH_3CH_2CH_2CHCOC_2H_5}$$

(62)

used to drive this homogeneous reaction, where vitamin B_{12} operates as a photocatalyst and electrocatalyst, for biasing the cell voltage with silicon solar cells, and for solvent convection.

REFERENCES

1. P. Beak and J. L. Miesel, *J. Am. Chem. Soc.*, **89**, 2375–2384 (1967).
2. L. Liu, B. Yang, T. J. Katz and M. K. Pointdexter, *J. Org. Chem.*, **56**, 3769–3775 (1991).
3. R. H. Martin, M. Flammang-Barbieux, J. P. Cosyn and M. Gelbeke, *Tetrahedron Lett.*, 3507 (1968).
4. W. H. Laarhoven, *Org. Photochem.*, **10**, 163–308 (1989).
5. V. Boekelheide and J. B. Phillips, *J. Am. Chem. Soc.*, **89**, 1695–1704 (1967).
6. J. B. Phillips, R. J. Molyneux, E. Sturm and V. Boekelheide, *J. Am. Chem. Soc.*, **89**, 1704–1714 (1967).
7. L. Jayabalan and P. Shanmugam, *Synthesis*, 789–794 (1990).
8. N. Jayasuriya, J. Kagan, J. E. Owens, E. P. Kornak and D. M. Perrine, *J. Org. Chem.*, **54**, 4203–4205 (1989).
9. I. Ninomiya, T. Kigushi, C. Hashimoto and T. Naito, *Chem. Pharm. Bull.*, **39**, 23–30 (1991).
10. A. Couture, P. Grandclaudon and S. O. Hooijer, *J. Org. Chem.*, **56**, 4977–4980 (1991).
11. B. Bossenbroek, D. C. Sanders, H. M. Curry and H. Shechter, *J. Am. Chem. Soc.*, **91**, 371–379 (1969).
12. E. H. White and A. A. F. Sieber, *Tetrahedron Lett.*, 2713–2717 (1967).
13. P. M. op den Brouw and W. H. Laarhoven, *J. Chem. Soc. Perkin Trans. II*, 795–797 (1982).
14. J. P. Desvergne and H. Bouas-Laurent, *Isr. J. Chem.*, **18**, 220–226 (1980).
15. S. McN. Sieburth and J.-l. Chen, *J. Am. Chem. Soc.*, **113**, 8163–8164 (1991).
16. P. A. Wender and C. R. D. Correia, *J. Am. Chem. Soc.*, **109**, 2523–2525 (1987).
17. J. Kagan, *Chemosphere*, **16**, 2405–2416 (1987).
18. T. S. Lee, S. J. Lee and S. C. Shim, *J. Org. Chem.*, **55**, 4544–4549 (1990).
19. S. C. Shim and T. S. Lee, *J. Org. Chem.*, **53**, 2410–2413 (1988).
20. S. Wolff and W. C. Agosta, *J. Am. Chem. Soc.*, **106**, 2363–2367 (1984).
21. M. C. Pirrung, V. K. Chang and C. V. DeAmicis, *J. Am. Chem. Soc.*, **111**, 5824–5831 (1989).
22. N. A. Al-Jalai, *J. Photochem. Photobiol. A*, **54**, 99–104 (1990).
23. T. Kobayashi, M. Kurano, H. Sato and K. Nakanishi, *J. Am. Chem. Soc.*, **94**, 2863–2864 (1972).
24. H. E. Zimmermann and R. D. Solomon, *J. Am. Chem. Soc.*, **109**, 6276–6289 (1986).
25. R. A. Bunce, V. L. Taylor and E. M. Holt, *200th ACS National Meeting*, Washington, DC, August 26, 1990, ORGN 260.
26. D. I. Schuster, N. Kaprinidis, D. J. Wink and J. C. Dewan, *J. Org. Chem.*, **56**, 561–567 (1991), and references cited therein.
27. J. D. White and D. N. Gupta, *J. Am. Chem. Soc.*, **90**, 6171–6177 (1968).
28. R. G. Salomon, N. D. Sachinvala, S. Roy, B. Basu, S. R. Raychaudhuri, D. B. Miller and R. B. Sharma, *J. Am. Chem. Soc.*, **113**, 3085–3095 (1991).
29. J. D. Winkler, R. D. Scott and P. G. Williard, *J. Am. Chem. Soc.*, **112**, 8971–8975 (1990).
30. J. D. Winkler, K. E. Henegar and P. G. Williard, *J. Am. Chem. Soc.*, **109**, 2850–2851 (1987).
31. W. G. Dauben, J. M. Cogen, G. A. Ganzer and V. Behar, *J. Am. Chem. Soc.*, **113**, 5817–5824 (1991).
32. A. Padwa, L. Zhi and G. E. Fryxell, *J. Org. Chem.*, **56**, 1077–1083 (1991).
33. J. Shin and W. Fenical, *J. Org. Chem.*, **56**, 1227–1233 (1991).
34. H.-J. Kang, C. S. Ra and L. A. Paquette, *J. Am. Chem. Soc.*, **113**, 9384–9385 (1991).
35. G. Mehta and D. Subrahmanyam, *J. Chem. Soc. Perkin Trans. I*, 395–401 (1991).
36. D. Belotti, J. Cossy, J. P. Pete and C. Portella, *Tetrahedron Lett.*, **26**, 211–214 (1985).

37. J. Cossy, D. Belotti and J. P. Pete, *Tetrahedron*, **46**, 1859–1870 (1990).
38. R. K. Haynes and S. C. Vonwiller, *J. Chem. Soc. Chem. Commun.*, 451–453 (1990).
39. B. Ye and Y. L. Lin, *J. Chem. Soc. Chem. Commun.*, 726–727 (1990).
40. R. K. Haynes and S. C. Vonwiller, *Chem. Commun.*, 449–451 (1990).
41. W. Adam, U. Kliem, T. Mosandl, E.-M. Peters, K. Peters and H. G. v. Schnering, *J. Org. Chem.*, **53**, 4986–4992 (1988).
42. R. M. Wilson, S. W. Wunderly, T. F. Walsh, A. K. Musser, R. Outcalt, F. Geiser, S. K. Gee, W. Brabender, L. Yerino, Jr, T. T. Conrad and G. A. Tharp, *J. Am. Chem. Soc.*, **104**, 4429–4446 (1982).
43. B. L. Feringa, O. J. Gelling and L. Meesters, *Tetrahedron Lett.*, **31**, 7201–7204 (1990).
44. W. Adam, O. Albrecht, E. Feineis, I. Reuther, C. R. Saha-Moeller, P. Seufert-Baumbach and D. Wild, *Liebigs Ann. Chem.*, 33–40 (1991).
45. C. Eckhardt and K. Weber (Ciba-Geigy A.-G.), Liquid detergents containing fluorescent brighteners. *Eur. Pat. Appl. EP 394,998*, 31 October 1990.
46. K. Gollnick, X.-L. Xiao and U. Paulmann, *J. Org. Chem.*, **55**, 5945–5953 (1990).
47. K. Gollnick and U. Paulmann, *J. Org. Chem.*, **55**, 5954–5966 (1990).
48. E. W. Meijer and H. Wynberg, *J. Am. Chem. Soc.*, **104**, 1145–1146 (1982).
49. J. Cossy, D. Belotti and J. P. Pete, *Tetrahedron*, **46**, 1859–1870 (1990).
50. U. Hees, J. Schneider, O. Wagner and M. Regitz, *Synthesis*, 834–840 (1990).
51. O. L. Chapman, C. L. McIntosh and J. Pacansky, *J. Am. Chem. Soc.*, **95**, 614–617 (1973).
52. D. J. Cram, M. E. Tanner and R. Thomas, *Angew. Chem. Int. Ed. Engl.*, **30**, 1024–1027 (1991).
53. K. Seki, N. Kanazashi and K. Ohkura, *Heterocycles*, **32**, 229–230 (1991).
54. S. T. Perri, S. C. Slater, S. G. Toske and J. D. White, *J. Org. Chem.*, **55**, 6037–6047 (1990).
55. J. Streith, *Chimia*, **45**, 65–76 (1991).
56. K. L. Wierzchowski and D. Shugar, *Photochem. Photobiol.*, **2**, 377 (1963), K. L. Wierzchowski, D. Shugar and A. R. Katritsky, *J. Am. Chem. Soc.*, **85**, 827–828 (1963).
57. K. F. McClure, J. W. Benbow and S. J. Danishefsky, *J. Am. Chem. Soc.*, **113**, 8185–8186 (1991).
58. D. Wiedenfeld and R. Breslow, *J. Am. Chem. Soc.*, **113**, 8977–8978 (1991).
59. J. Chen, P. R. Pokkuluri, J. R. Scheffer and J. Trotter, *Tetrahedron Lett.*, **31**, 6803–6806 (1990).
60. A. D. Gudmundsdottir and J. R. Scheffer, *Tetrahedron Lett.*, **31**, 6807–6810 (1990).
61. O. Piva, R. Mortezaei, F. Henin, J. Muzart and J.-P. Pete, *J. Am. Chem. Soc.*, **112**, 9263–9272 (1990).
62. Y. Inoue, N. Yamasaki, T. Yokoyama and A. Tai, *J. Org. Chem.*, **57**, 1332–1345 (1992).
63. M. Zandomeneghi, *J. Am. Chem. Soc.*, **113**, 7774–7775 (1991).
64. V. N. Rajasekharan Pillai, *Org. Photochem.*, **9**, 225–323 (1987).
65. M. Rubinstein, B. Amait and A. Patchornik, *Tetrahedron Lett.*, 1445–1448 (1975).
66. D. H. Rich and S. K. Gurwara, *J. Chem. Soc. Chem. Commun.*, 610–611 (1973).
67. P. Lloyd-Williams, F. Albericio and E. Giralt, *Int. J. Peptide Protein Res.*, **37**, 58–60 (1991).
68. A. Nishida, T. Hamada and O. Yonemitsu, *J. Org. Chem.*, **53**, 3386–3387 (1988).
69. J. Masnovi, D. J. Koholic, R. J. Berki and R. J. W. Binkley, *J. Am. Chem. Soc.*, **109**, 2851–2853 (1987).
70. R. W. Binkley and D. G. Hehemann, *J. Org. Chem.*, **55**, 378–380 (1990).
71. B. K. Barnett and T. D. Roberts, *J. Chem. Soc. Chem. Commun.*, 758 (1972).
72. J. A. Pincock and A. Jurgens, *Tetrahedron Lett.*, 1029–1030 (1979).
73. G. A. Epling and M. E. Walker, *Tetrahedron Lett.*, **23**, 3843–3846 (1982).
74. J. F. Art, J. P. Kestemont and J. P. Soumillion, *Tetrahedron Lett.*, **32**, 1425–1428 (1991).
75. C. Kaneko, A. Yamamoto and M. Gomi, *Heterocycles*, **12**, 227–230 (1979).
76. T. T. Takahashi, C. Y. Nakamura and J. Y. Satoh, *J. Chem. Soc. Chem. Commun.*, 680 (1977).
77. Y. H. Kuo and T. L. Jang, *J. Chin. Chem. Soc. (Taipei)*, **38**, 203–205 (1991).
78. B. Steiger, L. Walder and R. Scheffold, *Chimia*, **40**, 93–97 (1986).

9. Photochemistry in biochemistry

9.1. LIGHT AND ENZYME ACTIVITY

Many biological processes are affected by light through reactions targeting small molecules, proteins, or nucleic acids, but often the details of the steps involving photochemistry are at best poorly understood.

In some biochemical transformations specific enzymatic steps are affected by light, because either the enzyme itself or a cofactor is photochemically modified. One example concerns photolyase, the enzyme involved in the repair of a specific kind of DNA damage, in which pyrimidine dimers are formed photochemically (Section 7.1). Considering the importance of photosynthesis in plants, one should not be surprised to find enzymatic reactions in plants which are specifically under the influence of light. One enzyme participating in the biosynthesis of chlorophyll a, for example, is protochlorophyllide reductase, a membrane-bound molecule which ensures a light-dependent hydrogen transfer from NADPH to the substrate protochlorophyllide, resulting in the reduction of one carbon–carbon bond in the molecule.[1] In this case light activates the enzyme–substrate complex.

In general, enzyme inactivation is likely to occur through a photosensitized reaction. In the environment of a cell, many endogenous molecules can serve as sensitizers, and it is difficult to obtain mechanistic details.[2] Often, experiments are conducted in which an exogenous sensitizer is introduced. Its effect depends on matters such as whether or not it crosses cellular membranes, and on where it is transported within individual cells. Almost always in these cases, light has a negative effect on enzymatic activity. Control experiments with isolated enzymes *in vitro* are necessary in support of any claims that it is a specific photosensitized reaction which is responsible for inactivation *in vivo*.

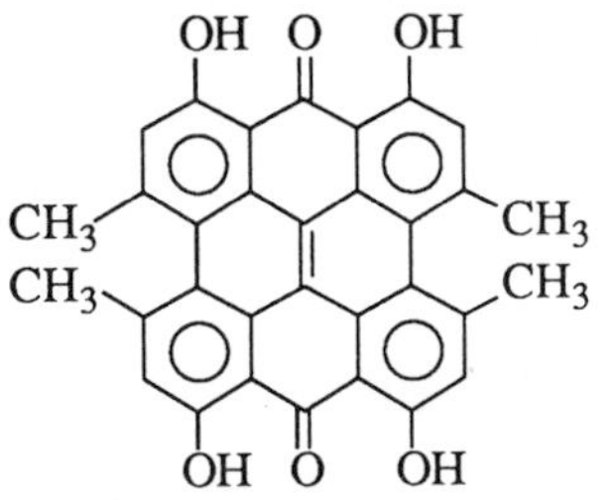

(1)

For example, hypericin (**1**) is a well-known sensitizer often found in plants of the genus *Hypericium*. It has been suggested that hypericin's role is to protect these plants from insect predators. *In vitro*, it sensitizes the formation of singlet oxygen and inhibits the enzyme succinoxidase.[3] The presumption that it is responsible for the inactivation of the same enzyme in isolated mitochrondria is strongly supported by the observation that the action spectrum for the inactivation of the enzyme *in vivo* matches the absorption spectrum of hypericin.

There are many other examples of photosensitized enzyme inactivation. The case of α-terthienyl's inactivation of acetylcholinesterase and superoxide dismutase *in vivo*, as well as several other enzymes *in vitro* is described in Section 10.16.

9.2. (E)–(Z) ISOMERIZATION AS A TOOL FOR BIOCHEMICAL CHANGES IN VITRO

Syn–anti isomerization of azobenzenes produces large geometrical changes (Chapter 12). Since enzymatic reactions usually require a delicate matching of the shapes of an enzyme and its substrate, it is reasonable to expect that imposing a geometrical change upon the enzyme will change the rate of an enzymatic reaction. The validity of the concept has been demonstrated. For example, azobenzene-4-carboxylic acid was first covalently bound via peptide bonds to lysine residues of the enzyme papain. The modified papain immobilized on alginate beads was photochromic, the azobenzene moieties being isomerized from *anti* to *syn* by irradiation with 320 nm light, and from *syn* to *anti* by treatment with visible light. Although the enzymatic activity toward a substrate was only 2.75 times greater for the enzyme carrying the *anti*- than the *syn*-azobenzene group, photoregulation of the enzymatic activity could be demonstrated by switching the irradiation back and forth between ultraviolet and visible light.[4]

The photoregulation of the enzyme α-chymotrypsin was achieved by a similar approach. The enzyme was immobilized in a cross-linked polymer containing azobenzene moieties. The photochromic polymer underwent conversion from *anti* to *syn* upon irradiation at 330–370 nm, and from *syn* to *anti* above 400 nm. The enzyme was completely inactive when entrapped in the *anti*-azobenzene polymer, but regained its activity after the polymer had been isomerized by irradiation with visible light.[5] This on-and-off switching was reversible. The change in enzymatic activity probably resulted from changes in the permeability of the polymer backbone accompanying the photochemical *syn–anti* isomerization.

In a third example, a monoclonal antibody for a peptide carrying an azobenzene group was prepared under conditions where the azobenzene was *anti*. The *anti*-hapten peptide bound strongly to the antibody, but was released upon isomerization into the *syn* form by irradiation with ultraviolet light. The process was reversible, the *anti* isomer regenerated by irradiation of the *syn*-hapten with visible light being strongly bound again by the antibody.[6]

9.3. PHOTOCHEMISTRY OF VITAMIN E

α-Tocopherol is the predominant constituent of vitamin E, a biological antioxidant of major importance. Not only does α-tocopherol (2) trap reactive peroxyl radicals, but it also scavenges singlet oxygen efficiently. The reactions with singlet oxygen have been scrutinized. The unstable hydroperoxide (3) was characterized, leading to the two products (4) and (5). In addition, spiro products (7) were isolated, with the suggestion that they were formed as indicated.[7] Interestingly, (7) did not dehydrate to (8) under

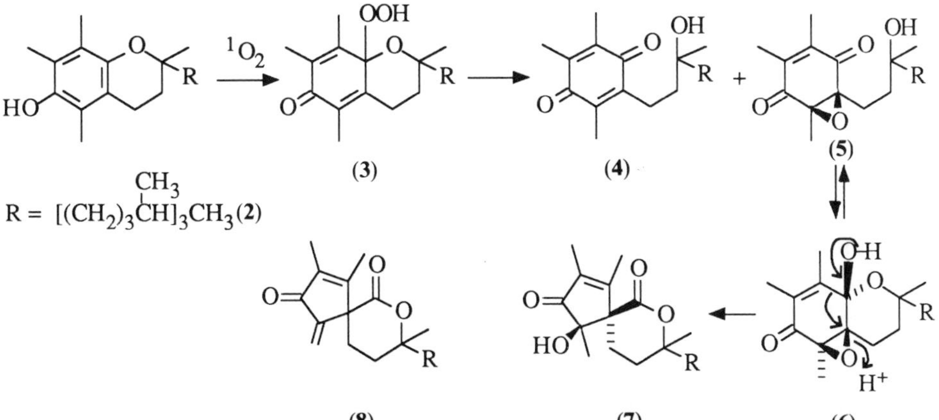

$$R = [(CH_2)_3\overset{CH_3}{\underset{|}{CH}}]_3CH_3\ (2)$$

(3)   (4)   (5)

(8)   (7)   (6)

the reaction conditions, and compounds (7) and (8) were not formed when the reaction was performed in anhydrous methanol, where the two ketals (9) and (10) were produced in addition to (4) and (5).

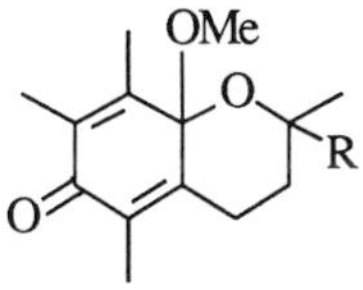

(9) 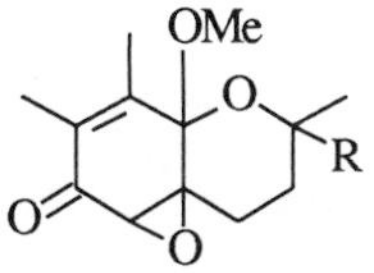   (10)

9.4. LIPID PEROXIDATION

Lipid peroxidation is a general phenomenon touching all cells in an aerobic environment. Different cell components may be affected, including membranes and soluble materials. Here we are only concerned with photoinduced reactions, although many enzymatic pathways may also lead to lipid peroxidation.

Many of the techniques of study were developed with natural membranes, particularly erythrocyte membranes, as well as synthetic ones (liposomes).

As discussed in Chapter 6, photosensitized reactions in the presence of oxygen lead to the formation of $^1O_2$ or to electron transfer processes. In the first case, alkenes are usually converted into allylic hydroperoxides. Alternatively, electron transfer reactions produce radical species (O_2^{-} and/or radicals derived from the sensitizer and from the alkene). It is easy to appreciate, therefore, that standard radical processes may lead to allylic radicals:

Whether these radicals react with molecular oxygen or with O_2^{-}, peroxidic species will eventually be formed, and their decomposition will induce the formation of other radicals. The oxygenation of these, in turn, will form new peroxidic sites, which will undergo further decomposition, and so on.

One important mode of decomposition for hydroperoxides is through a redox reaction with a metal ion such as Fe^{2+} or Cu^+:

$$ROOH + Fe^{2+} \rightarrow RO\cdot + OH^- + Fe^{3+}$$

The oxygen-centered radical formed can abstract a hydrogen and start a new oxidation reaction at a different site on the same or on a neighboring lipid.

Another species theoretically capable of inducing radical formation on a lipid is derived from $O_2^{\overline{\cdot}}$. Disproportionation of the protonated species creates O_2 and H_2O_2 (Chapter 6). Metal-catalyzed decomposition of the latter produces $\cdot OH$, a very reactive reagent for hydrogen abstraction and for many other oxidation reactions.

It is very difficult to determine with accuracy the extent to which $^1O_2$ and $O_2^{\overline{\cdot}}$ intervene in a photosensitized lipid oxidation reaction *in vivo*. One additional complication comes from the fact that peroxidic species may themselves be sources of $^1O_2$. One example mentioned in Chapter 6 is through combination–decomposition of hydroperoxyl radicals:

$$ROO\cdot + R'OO\cdot \longrightarrow ROOOOR' \longrightarrow ROOR' + O_2$$

Another example is through the disproportionation of $O_2^{\overline{\cdot}}$. Here, the formation of $^1O_2$ is very small at best.[8]

9.4.1. Diagnostic tests

Several known lipids were treated with $^1O_2$ and $O_2^{\overline{\cdot}}$ in order to establish the structure of the products and use them as diagnostic tests for oxidation mechanisms. These studies established that the detection of the 9- and 13-hydroperoxy derivatives in the oxidation of methyl linoleate (**11**) can be ascribed to radical pathways (type I) and that of the 10- and 12-hydroperoxides to $^1O_2$ reactions.

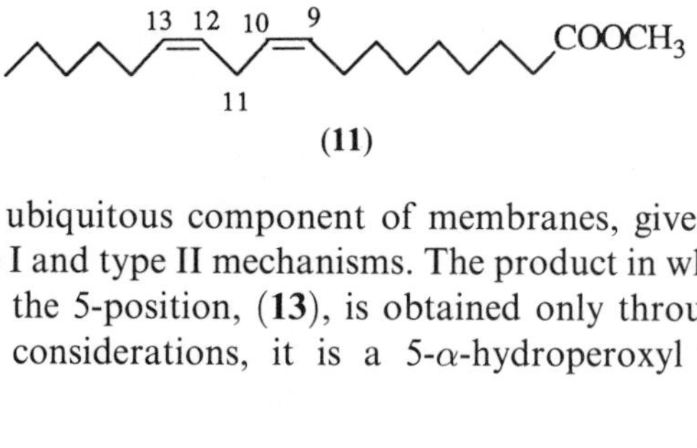

(**11**)

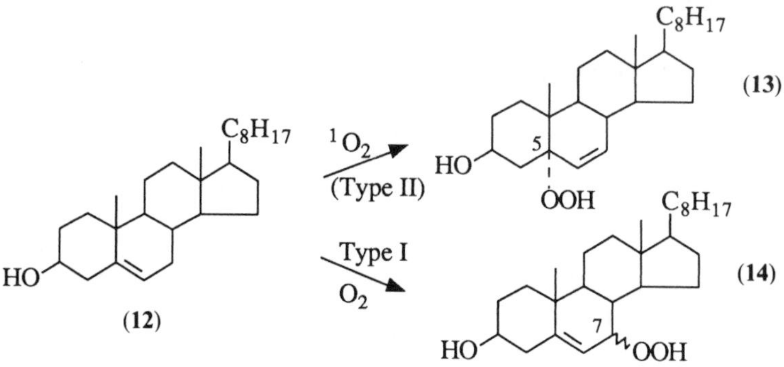

Cholesterol (**12**), a ubiquitous component of membranes, gives different oxidation products through type I and type II mechanisms. The product in which there is a hydroperoxyl substituent at the 5-position, (**13**), is obtained only through $^1O_2$ reaction. As expected from steric considerations, it is a 5-α-hydroperoxyl derivative which is obtained, where the hydroperoxyl substituent is *trans* to the angular methyl group at the adjacent position. In contrast, radical reactions yield cholesterol substituted with OOH or OH at the 7-position, (**14**). The analysis of the photooxidation products must be performed under conditions which minimize the allylic rearrangement of the 5-hydroperoxy into 7-hydroperoxy isomers.

9.4.2. Quantitative determinations

The extent of photooxidation of lipids can be evaluated either by titrating the hydroperoxide groups by iodometry, or by decomposing the oxidized lipid in the presence of thiobarbituric acid in acidic medium at 100°C. A product absorbing at 532 nm is formed, derived in large part from the condensation of malonaldehyde with thiobarbituric acid. It is difficult to get quantitative results, and there are pitfalls to be avoided.[9]

9.5. PHOTOCHEMISTRY OF AMINOACIDS

Tryptophan (**15**) is the only natural aminoacid with sufficient absorption in the ultraviolet range to react significantly in sunlight. Its photooxidation yields *N*-formylkynurenine (**16**). This photoproduct is itself a precursor to other products, particularly

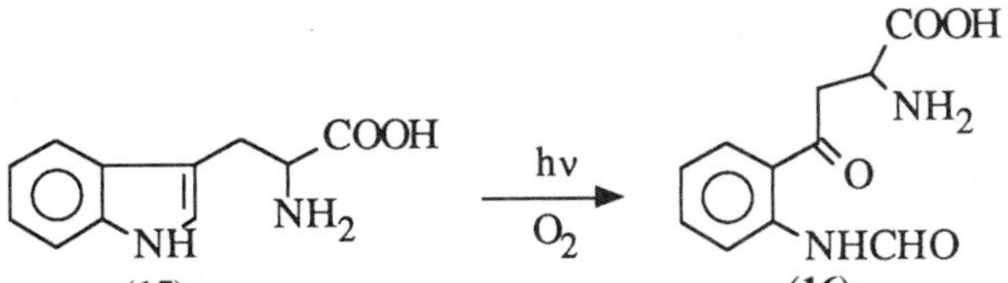

kynurenic acid (**17**), kynurenine (**18**), and 3-hydroxykynurenine (**19**). All these photoproducts absorb in the UVA region. Since these compounds are particularly prevalent

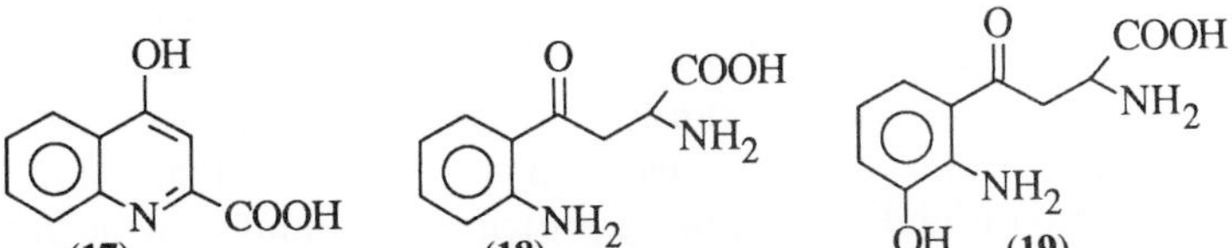

in the lens of the eye, where protein changes which occur with ageing can lead to cataract formation, the photodynamic properties of these potential sensitizers have been scrutinized.[10] Kynurenic acid was found to produce $^1O_2$ in good quantum yield, whereas kynurenine and its 3-hydroxy derivative produced neither $^1O_2$ nor $O_2^{\cdot-}$. Thus, (**18**) and (**19**) probably play a protective rather than a deleterious role in the eye.

Other photooxidation products may also be formed; one obtained in the reaction of tryptophan with $^1O_2$ is (**20**).[11]

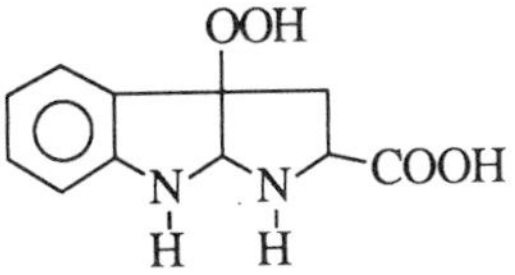

9.6. PHOTOCHEMICAL REACTIONS WITH PROTEINS

Biological photodamage to proteins is clearly visible in the characteristic skin changes in farmers, sailors, and other persons with outdoor occupations. Proteins may be involved in other photoinduced reactions, such as protein–DNA and protein–RNA cross-linking. The last two have been detected in the sensitized reactions with hemato-porphyrin derivatives used in photodynamic therapy.[12] Actually, many different types of photoreactions may take place in proteins when light is absorbed by a specific protein chromophore such as a tryptophan residue, or when there are reactions with other electronically excited molecules. In the hope that knowledge derived from small units would help to understand the behavior of the larger natural molecules, the photochemistry of dipeptides was investigated *in vitro*, revealing loss of NH_3 and CO_2 (which was pH-dependent), carbon–carbon bond cleavage, carbon–nitrogen bond cleavage, and peptide bond cleavage.[13] The importance of aromatic groups (which are the better chromophores) was proved by the structures of some of the photoproducts characterized, such as the tricyclic compound (21). Dipeptides containing phenylalanine produced extremely complex product mixtures.

GlyTrp $\xrightarrow{\text{hv}}$

(21)

Similar experiments were performed with sensitization.[14] Oxidation at nitrogen was an important reaction and, depending on the pH, singlet oxygen was quenched either physically or chemically.

It is important to note that the peptide group itself is definitely a photochemically active chromophore when light with suitably short wavelengths is available (not sunlight). The observation of electron transfer into peptide groups supports the possible involvement of proteins in long-range electron transfers.[15] Actually, long-range electron transfer has been found *in vitro* using proteins containing a heme group.[16]

As mentioned above, proteins found in the lens of the eye have been scrutinized because of their role in the formation of cataracts. Photosensitized oxidation of the eye lens proteins, the crystallins, are believed to lead to protein cross-links and high-molecular-weight aggregates. Although the aromatic aminoacids tryptophan and tyrosine are known to react with singlet oxygen, it is the photodynamic oxidation of histidine residues which is apparently required to form high-molecular-weight protein covalent aggregates. Proteins without histidine, such as melittin or bovine pancreatic trypsin inhibitor, do not form high-molecular-weight products upon reaction with singlet oxygen.[17]

With an enzymatic assay, it is easy to determine whether a specific protein has lost part or all of its activity upon irradiation, either with or without a sensitizer. For example, photosensitization with α-terthienyl led to the inactivation of acetylcholinesterase[18] and of superoxide dismutase[19] in larvae of the mosquito *Aedes aegypti*. The chemical modifications responsible for such changes in enzymatic activity are not known.

Man-made damage to the ozone layer protecting the earth from solar ultraviolet radiation has been feared to lead to increased incidence of cataracts in people who do not wear eye protection outdoors. However, there should be no such risks to those wearing either corrective eyeglasses or sunglasses (both filter ultraviolet light), or who have indoor or nocturnal occupations.

9.7. PHOTOCHEMICAL MODIFICATIONS OF NUCLEIC ACIDS

The components of nucleic acids are simple purine and pyrimidine bases, attached to a sugar, itself esterified with phosphoric acid. Ribose is used in RNA (ribonucleic acid) and deoxyribose, the corresponding sugar without the hydroxyl at the $2'$-position, is used in DNA (deoxyribonucleic acid). Familiarity with the terminology is important as one may deal with several forms: the bases themselves, their derivatives with a sugar (or nucleosides), or the phosphorylated products of these (or nucleotides). The purine bases are adenine (**22**) and guanine (**23**), the pyrimidine bases are thymine (**24**),

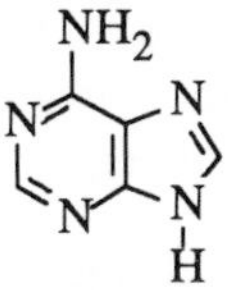

(**22**) 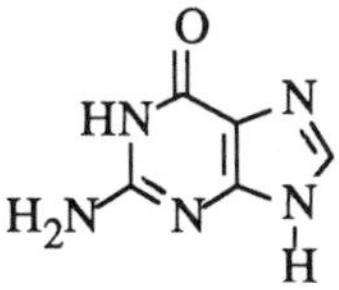   (**23**)   (**24**) 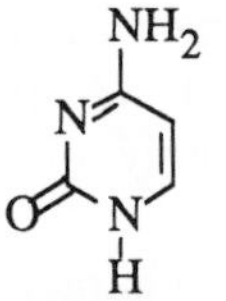   (**25**) 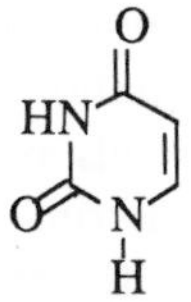   (**26**)
        Purines   Pyrimidines

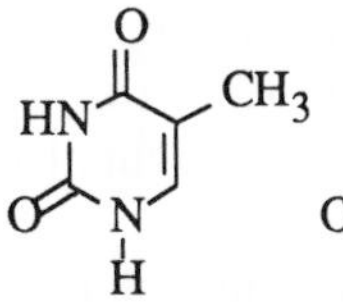

cytosine (**25**) and uracil (**26**). The corresponding nucleosides used in DNA are $2'$-deoxy-adenosine (**27**), $2'$-deoxyguanosine (**28**), $2'$-deoxycytidine (**29**), and thymidine

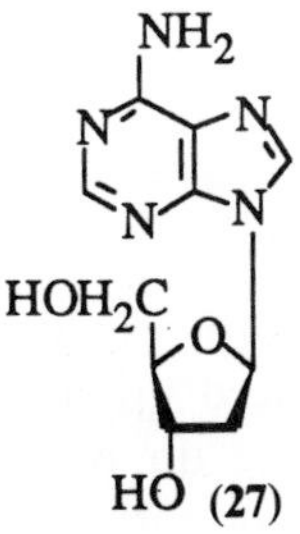

(**27**) 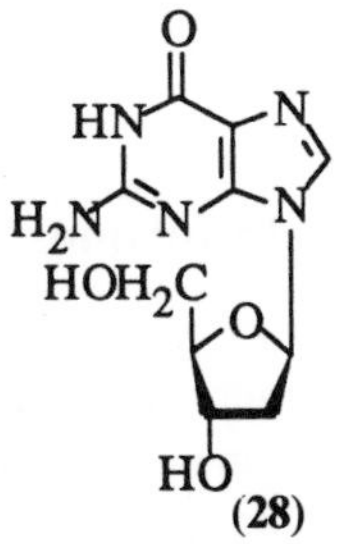   (**28**) 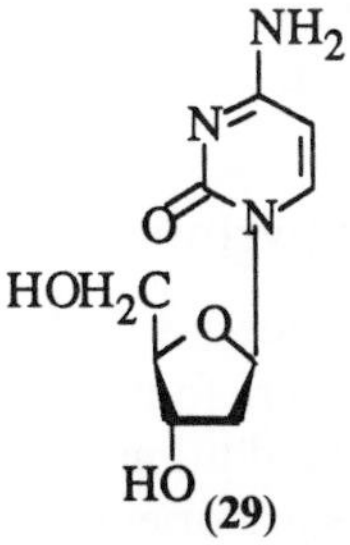   (**29**) 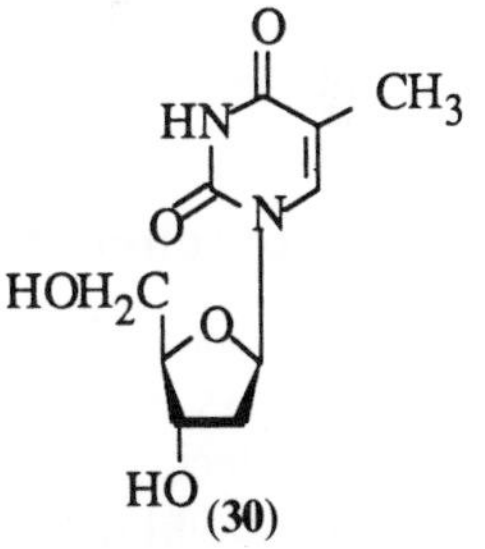   (**30**)

(**30**). The corresponding nucleosides in RNA are adenosine (**31**), guanosine (**32**), cytidine (**33**), and uridine (**34**) (the riboside corresponding to thymidine is not found in

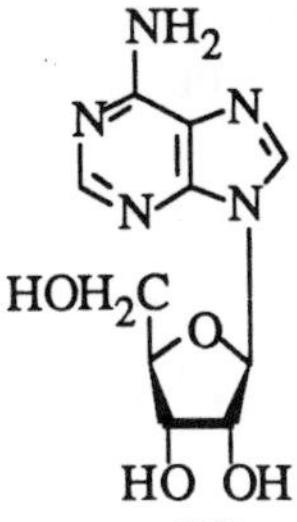

(**31**) 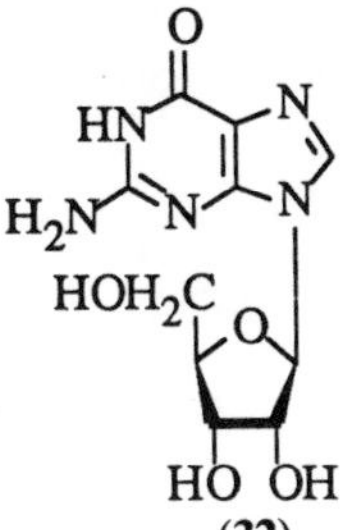   (**32**) 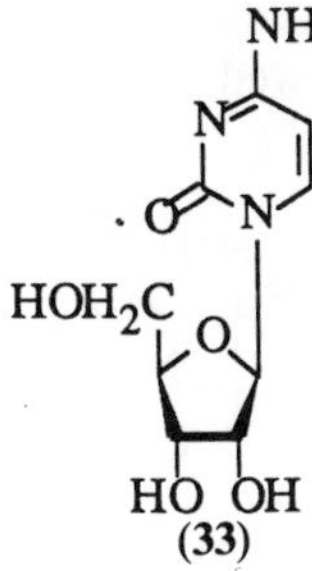   (**33**) 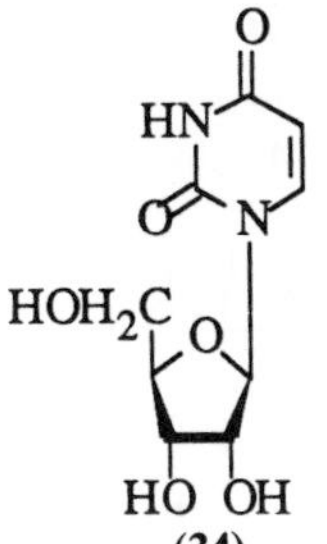   (**34**)

RNA). The names of the corresponding nucleotides must specify the hydroxyl group which has been esterified with phosphoric acid; for example, 5′-guanosine monophosphate (**35**) and 3′-guanosine monophosphate (**36**) are two very different molecules.

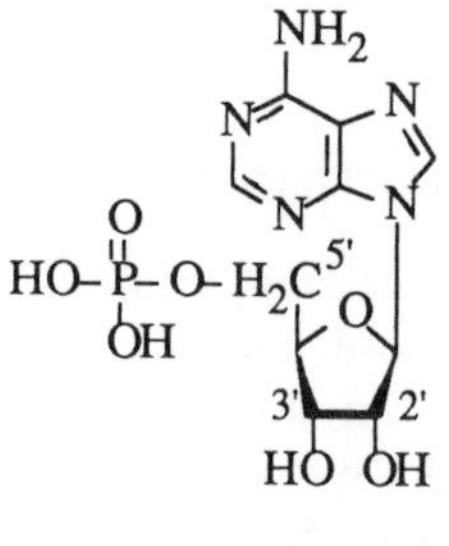

(35) 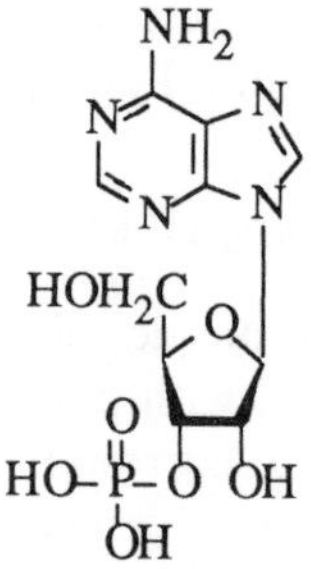   (36)

Upon irradiation, nucleic acids may either undergo structural modifications or participate in reactions with other kinds of biological molecules. An example of the latter is the formation of DNA–protein cross-links which are produced upon irradiation of human cells, probably following light absorption by endogenous sensitizers (believed to be porphyrins).[20]

The absorption characteristics of purine and pyrimidine are practically the same, whether alone, in a nucleoside or in a nucleotide: all absorb around 260 nm (this is why the purity of DNA samples is often evaluated from the ratio of the absorbance at 260 nm, which should be high, compared to that at 280 nm, which should be low, since it corresponds to the maximum absorption of proteins). All the bases absorb in the ultraviolet beyond the range readily available on earth from sunlight. Thus, they cannot undergo much direct photochemistry in nature, though sensitized reactions are possible. Much is known, however, about the direct photochemistry of nucleic acids and their components in the laboratory.

9.7.1. Photochemistry of pyrimidines: cyclobutanes

Unlike purines, pyrimidines undergo cycloaddition reactions and form cyclobutane derivatives. As mentioned earlier, four diastereomers may be produced upon suprafacial–suprafacial [2 + 2] cycloaddition of an alkene to a pyrimidine double bond, depending upon how the rings are oriented and whether they are *syn* or *anti*. Irradiating a mixture of two different pyrimidines could therefore give 12 diastereomeric photoproducts, four dimers from each of the pyrimidines, and four mixed products.

The structures of the dimers (**37**)–(**40**) result from the more likely suprafacial–suprafacial approach of two cytosine molecules. The antarafacial–antarafacial mode of cycloaddition, although allowed by the rules of orbital control, is much less likely and has been ignored.

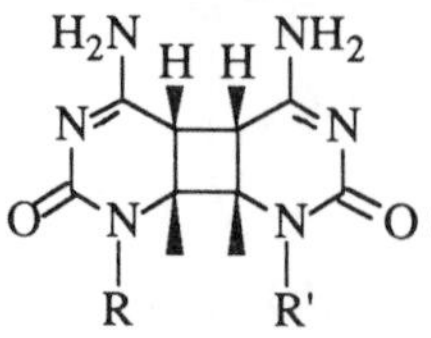

(37) 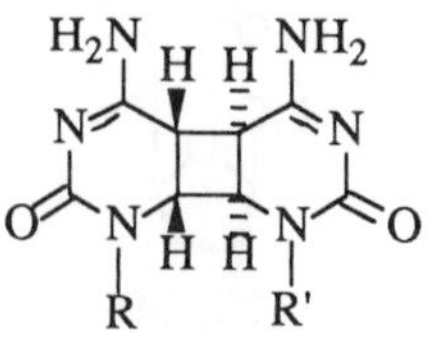   (38) 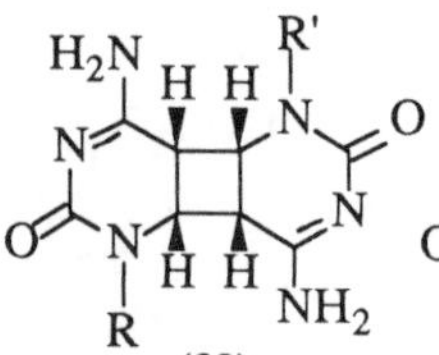   (39) 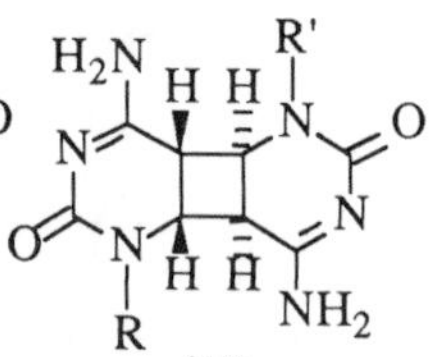   (40)

Constraints enforced by the structure of a natural molecule of nucleic acid limit the number of possible adducts in the cycloaddition of any specific pair of bases, but extensive possibilities for pairing exist, since the conformation of the molecule determines what neighbors are present near each base.

A shorthand notation for a cyclobutane dimer between two bases, B_1 and B_2, is $B_1 \diamond B_2$.

In principle, [2 + 2] cycloaddition is a reversible reaction. However, once aromaticity has been lost, adducts no longer absorb light at the same wavelengths as the starting materials. When direct photolysis is involved, light of shorter wavelength is required to reverse the cycloaddition (although not necessarily in the presence of sensitizers). Photochemical cyclobutane formation, leading to DNA dimers, can often be reversed upon irradiation with light of shorter wavelength, at least *in vitro*. The situation may be quite different in biological systems, where mutations associated with the formation of cyclodimers may be due to more than one chemical step. A short-wavelength ultraviolet light treatment may thus reactivate a biological system impaired by an initial irradiation, but it may be an altered system which is regenerated. For example, mutations in ultraviolet-irradiated phage S13 have been attributed to a deamination of cytosine in cyclobutane dimers.[21]

In vivo, there are biological repair mechanisms which cleave pyrimidine photodimers created by ultraviolet irradiation. The enzyme DNA photolyase performs this task, and requires irradiation with visible light. The first step in a proposed mechanism for its action involves an electron transfer from the dimer to an electronically excited quinone.[22]

Much work has been done with the irradiation of individual nucleic acid bases and analogs in different environments (for example ice versus water solution). Other photochemical [2 + 2] cyclization reactions of pyrimidines are known; that with benzene was reported in Chapter 8.

9.7.2. Addition reactions of pyrimidines

As mentioned in Chapter 7, photoinduced electron transfer reactions logically account for the addition of protic species (such as H_2O or C_2H_5OH) to pyrimidines, as shown here in the hydration of cytosine (**25**). Thermal treatment of the photohydration product

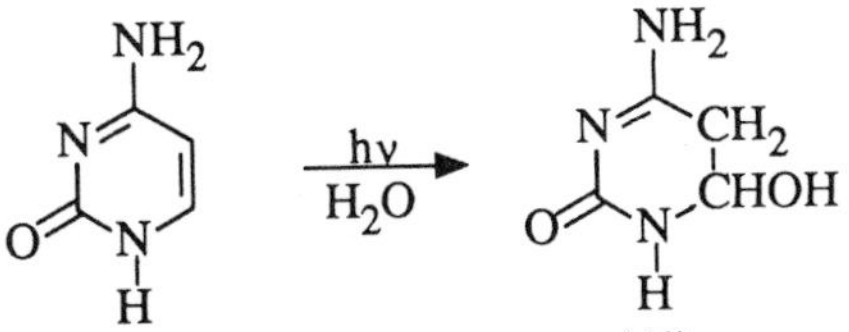

(**25**)   (**41**)

(**41**) can reverse the reaction. Similar behavior has been observed with pyrimidine nucleosides, nucleotides, and polynucleotides. One must note that the hydration products from cytosine derivatives are less stable than uracil itself toward deamination and conversion into uracyl derivatives.

Other types of dimeric structures, such as (42) and (43) from thymine (24), are known to be formed photochemically.

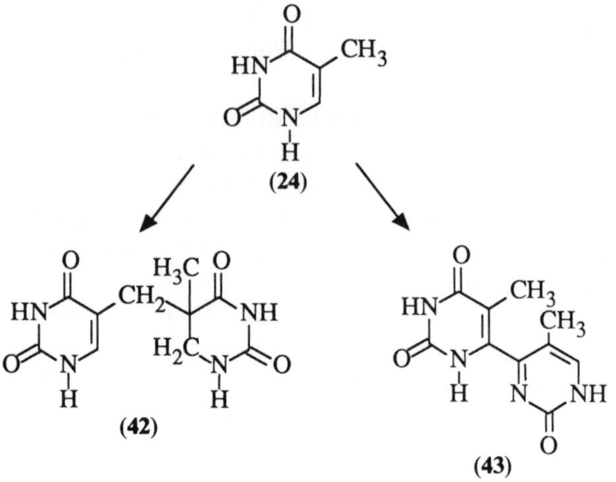

(24)

(42)

(43)

It may be possible to study modifications of nucleic acid bases with fluorescence techniques, which are extremely sensitive. For example, irradiation of the ribonucleoside derivative (44) in aqueous solution induces a transformation to (45), which has an intense green fluorescence.[23]

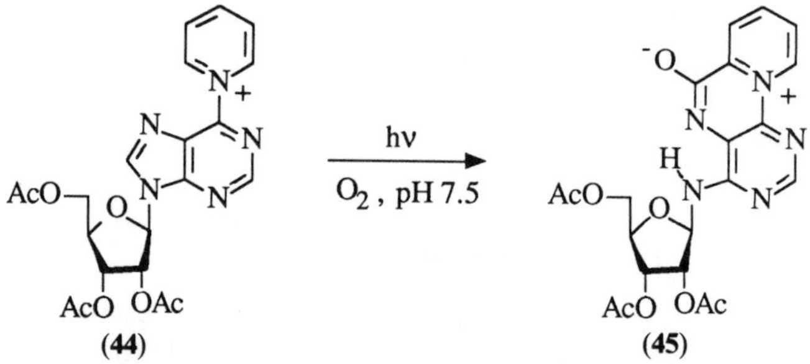

(44)

(45)

The following reaction sequence, described in Chapter 4, involves the photochemical formation of an oxetane which is then opened with the assistance of the adjacent amide nitrogen:

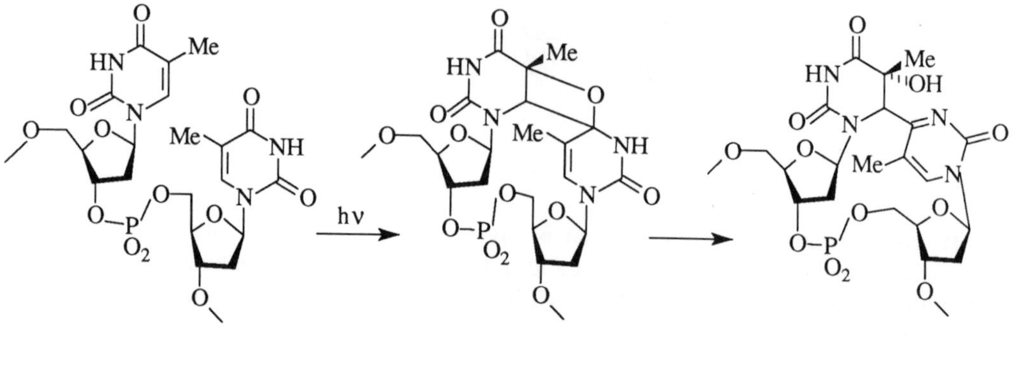

Further irradiating such a product with short-wavelength ultraviolet light may yield the Dewar valence isomer of the pyrimidone, as shown in the case of a $2'$-deoxycytidylyl-$(3'-5')$thymidine (**46**).[24]

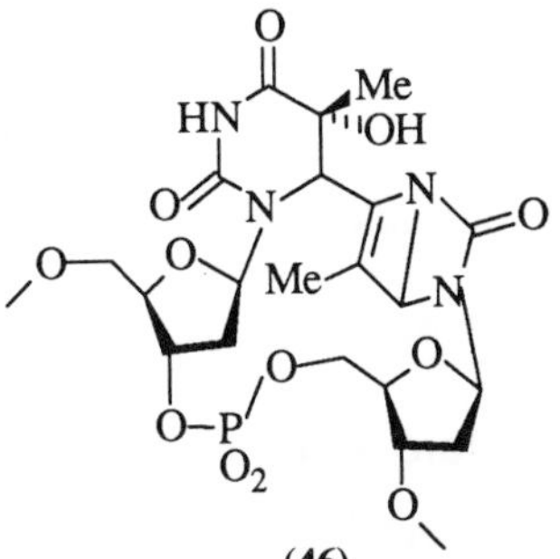

(**46**)

There has been an understandable tendency to study the photochemistry of each class of biological molecules independently. However, cells contain a large variety of molecules, allowing for interactions which may have important biochemical consequences. One example is the cross-linking of proteins to DNA in living systems, perhaps a major cause of their damage from ultraviolet light in biological systems.[25] Chemical studies have also revealed a photoinduced exchange reaction when thymidine (**30**) is irradiated in the presence of amines.[26] This transformation must

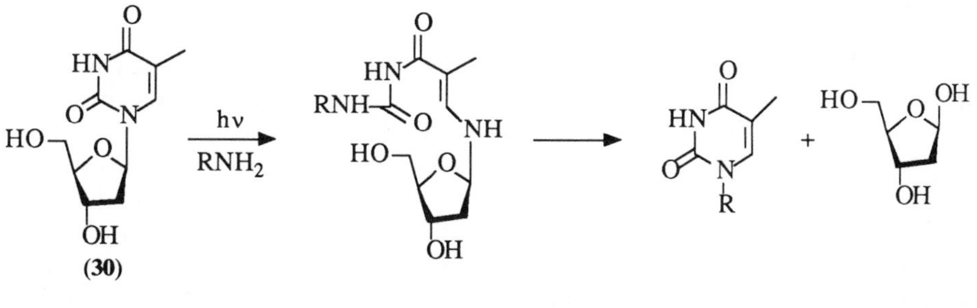

(**30**)

go through an intermediate, (**47**), capable of undergoing either elimination or hydrolysis. This type of reaction is valuable for cleaving DNA at specific sites, especially at

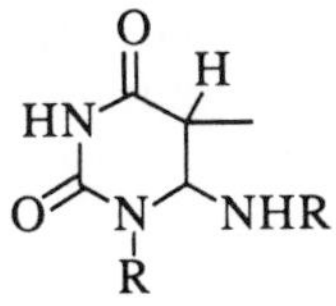

(**47**)

thymidine sites when the base spermine is used. It could also account for the cross-linking of DNA with proteins, particularly histones.

Finally, another type of photoinduced fragmentation of pyrimidines was reported in Chapter 8:

This reaction has no known biochemical importance.

9.8. PHOTOCHEMICAL CLEAVAGE OF DNA

9.8.1. Strand breaks in nucleic acids by direct irradiation

Irradiation of DNA produces strand breaks with an efficiency which depends on the wavelength and intensity of the light. Since several different naturally occurring molecules may act as sensitizers when cells are irradiated, the mechanism(s) of DNA cleavage *in vivo* cannot be easily elucidated. Reactive oxygen species are often believed to mediate the cytotoxicity, and H_2O_2 has been strongly implicated in the inactivation of bacterial cells. However, comparison of the effects of ultraviolet light and H_2O_2 in mammalian cells indicates that different inactivation mechanisms operate.[27]

Although DNA can be cleaved photochemically under a variety of conditions, it is appropriate to distinguish laser-induced cleavage reactions from effects due to lower-intensity light, because the use of lasers in medicine is rapidly gaining importance.

9.8.2. DNA cleavage with lasers

Upon irradiation of DNA with powerful lasers, the excited state which is initially formed may absorb a second photon. The energy of the resulting excited molecule may be higher than the ionization energy for the DNA bases. In aqueous solution, such treatments may produce breaks in the DNA either in a single strand or in complementary double strands.[28] For example, direct irradiation of DNA in plasmids pBR322 and pL32 with a 248 nm laser cleaves phosphodiester bonds attached to the 3'-position of deoxyguanosine residues.[29] The mechanism proposed is a two-photon excitation leading to photoionization and formation of a base radical cation. The guanine base has the lowest oxidation potential, and is thus most likely to give up an electron; the resulting radical cation is an intermediate in strand cleavage.

The formation of single- and double-strand breaks has been analyzed in double-stranded DNA upon nanosecond laser-induced photoionization,[30] and there is a review on DNA strand break induced by laser light.[31] The laser power can have a marked effect on the photosensitized phosphodiester bond cleavage of DNA deposited as a thin layer on a glass plate. Perhaps contrary to intuition, less cleavage occurs at higher laser powers, the energy absorbed being dissipated through vaporization of the sample (this does not apply to solutions).[32]

9.8.3. DNA cleavage: direct and photosensitized reactions

DNA can be cleaved photochemically either *in vitro* or *in vivo* with non-laser ultraviolet light sources. *In vivo*, for example, dose-dependent single-strand breaks were observed in the DNA of *Proteus mirabilis* after ultraviolet irradiation, but the breaks were rapidly repaired in the cells.[33]

Here again, it is difficult to analyze mechanistically the photosensitized damage to DNA in cells because many endogenous components can act as sensitizers. For example, NADH photosensitizes single-strand break formation in double-stranded plasmid DNA upon near-ultraviolet irradiation *in vitro*.[34]

Many photosensitizers have been shown to cleave DNA, some *in vivo*, but most *in vitro*. For example, sensitizers such as $Co(dip)_3^{3+}$,[35] and gilvocarcin V[36] (51) lead to DNA cleavage in mammalian cells (the latter also produces DNA-to-protein cross-links). *In vitro*, methylene blue (48), rose bengal (49), proflavine complexes,[37] α-terthienyl,[38] porphyrin derivatives, daunomycin (50),[39] gilvocarcins (51),[40] psoralens,[41] polycyclic aromatic hydrocarbons such as 7,12-dimethylbenz[*a*]anthracene[42] and 1-nitropyrene,[43] uranyl salts,[44] cobalt and ruthenium complexes,[45,46] are among the molecules known to cleave DNA.

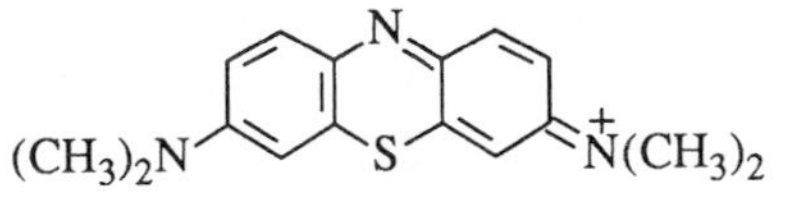

(48)

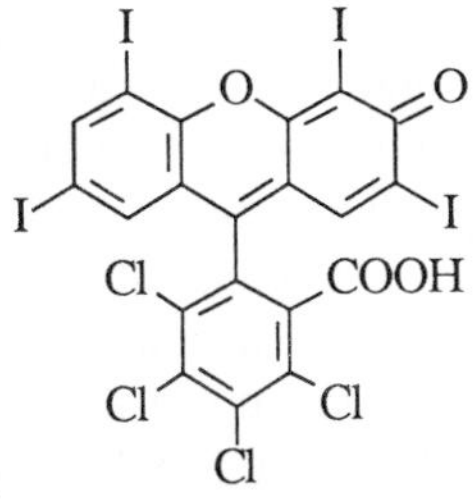

(49)

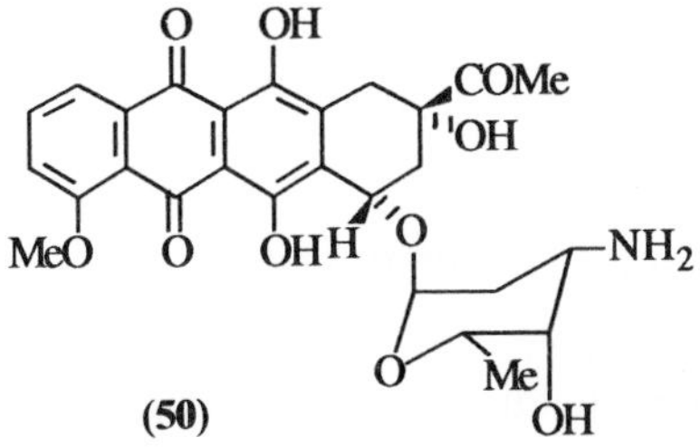

(50)

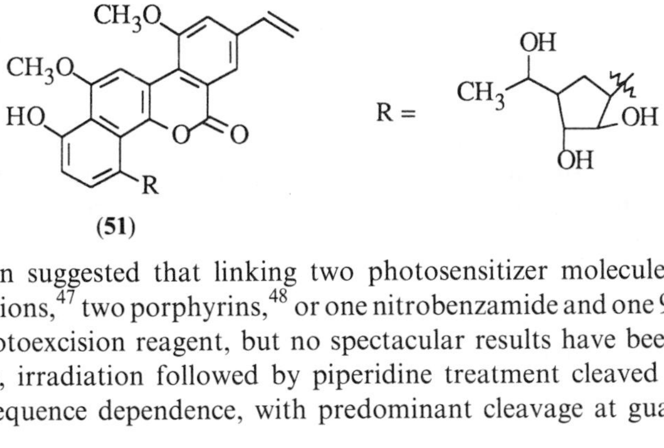

(51)

It has been suggested that linking two photosensitizer molecules (e.g. two diazapyrenium cations,[47] two porphyrins,[48] or one nitrobenzamide and one 9-aminoacridine[49]) creates a photoexcision reagent, but no spectacular results have been obtained. In the last example, irradiation followed by piperidine treatment cleaved supercoiled DNA with some sequence dependence, with predominant cleavage at guanine and thymine residues. However, it is not indispensable to have two sensitizer moieties in order to obtain double-strand cleavage, as will be demonstrated in Section 9.8.4.2.

9.8.4. Mechanism of DNA cleavage

The mechanism of strand scission of DNA in photosensitized reactions is not always known. Two types of DNA cleavage reactions have been observed. In one, at least one strand is cleaved directly. In the other, base-labile sites are created photochemically, demonstrated by the cleavage which occurs when the photolyzed DNA is treated with a base, either inorganic (such as NaOH) or organic (such as piperidine). As detailed in Chapter 6, mechanistic studies on photodynamic reactions generally use standard reagents for exploring whether singlet oxygen is produced. For example, the formation of singlet oxygen is frequently detected through quenching by NaN_3 or observation of enhanced reaction in D_2O. However, this approach does not yield reliable conclusions, as demonstrated when the bacteriophages λ_{vir} and T4D were inactivated with ultraviolet light in the presence of the drug chlorpromazine (**52**). DNA strand breaks could be detected,[50] but a higher inactivation rate was observed under nitrogen than under air bubbling. The inactivation was repressed by NaN_3 and enhanced in D_2O, both under aerobic and under anaerobic conditions. Some pathways other than singlet oxygen must therefore be involved, and neither NaN_3 nor D_2O can be used with confidence in this type of mechanistic study. Actually, mounting evidence indicates that oxygen is not always required in DNA cleavage, which frequently takes place more readily under anaerobic conditions (at least *in vitro*).

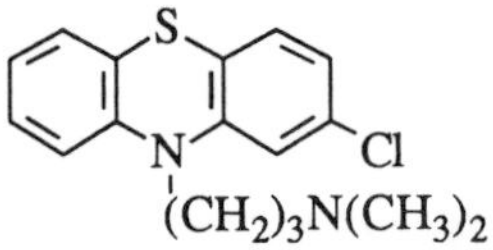

(**52**)

9.8.4.1. Site specificity in the cleavage of DNA

Many sensitizers, such as acridine, ethidium bromide, certain cationic porphyrins, methylene blue, or the dimethyldiazaperopyrenium dication (**53**)[51] cleave DNA

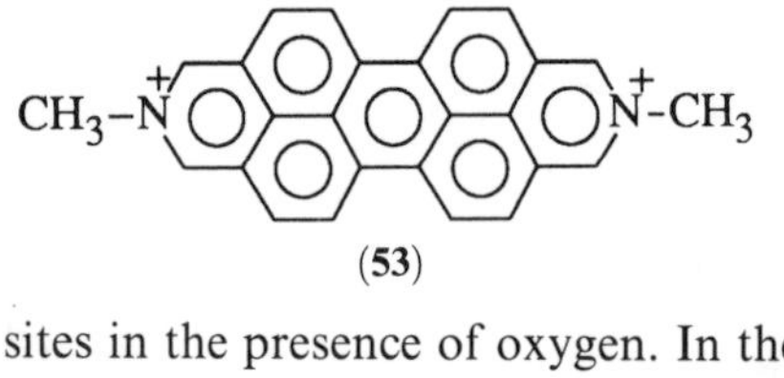

(**53**)

preferentially at guanine sites in the presence of oxygen. In the case of methylene blue, irradiation of DNA in the presence of oxygen formed 8-hydroxyguanine (**54**) in 2–4%

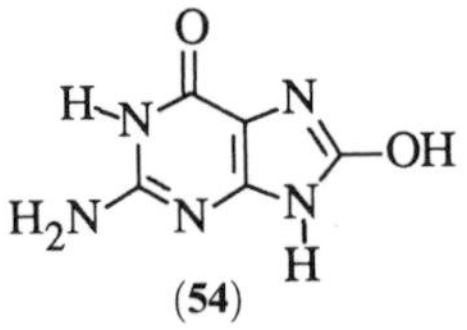

(**54**)

of the guanine residues present.[52] Irradiation with methylene blue, followed by piperidine treatment, cleaved DNA at all the guanine sites, and this reaction could be used for
sequencing DNA.[53] The sites of cleavage depend on the sensitizers used. For example,
the antitumor antibiotics esperamicin and neocarzinostatin sensitize DNA cleavage
with some preference for thymine and adenine,[54] while certain phenothiazine drugs
appear to cleave independent of sequence.[55] Some metal bleomycins, either excited
directly or sensitized by trisbipyridylruthenium(II) $(Ru(bpy)_3^{2+})$, cleave DNA with a
preference for pyrimidines adjacent to a guanine residue.[55,56] Alkylamines photosensitize the cleavage of DNA at thymine sites.[57] Some derivatives of aminoacridine show a
structure-dependent preference for different bases.[58] For example, (55) cleaves at A–G,
whereas the cleavage induced by (56) is practically sequence-independent.

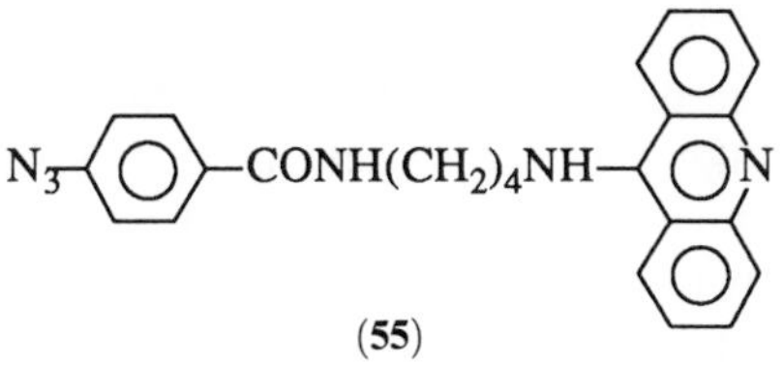

(55) 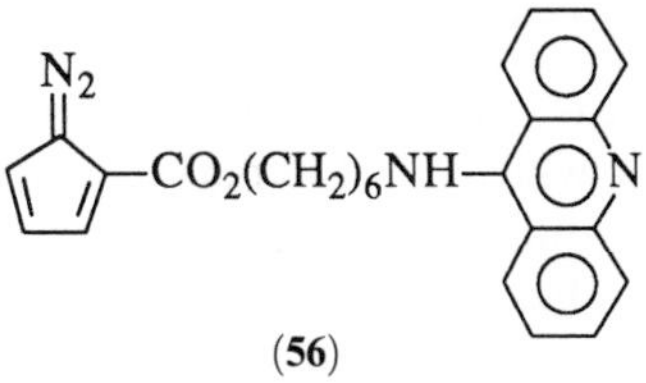   (56)

9.8.4.2. Sequence-targeted cleavage of DNA

The challenge of cleaving DNA at a specific site may be approached by attaching a
sensitizer to a molecule capable of recognizing a segment of DNA. For example, attaching a reagent to a sequence complementary to a portion of single-stranded DNA
can lead to the recognition of the targeted sequence. Irradiation must therefore be
more likely to produce damage near the end of the selected sequence. The same approach has been used successfully to cleave double-stranded DNA, through the formation of a triple-stranded complex.[59] In the example shown here (see (57)), an ellipticine
derivative was attached at the end of an 11-nucleotide fragment. The compound was then

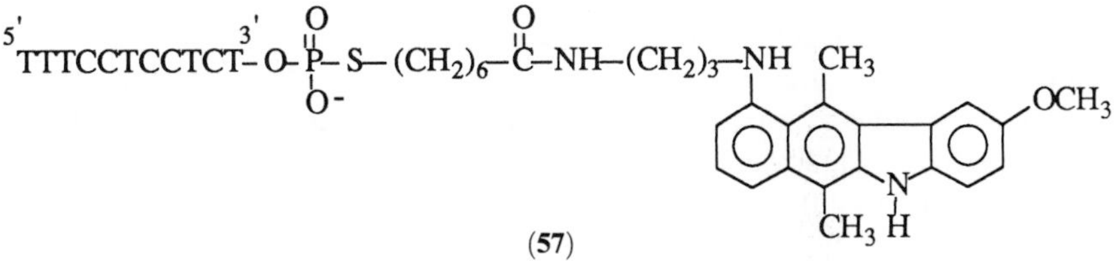

(57)

presented to a double-stranded DNA fragment with 32 base pairs, which contained a
sequence complementary to the sequence included in the photosensitizer. Upon irradiation of the mixture, cleavage of the polynucleotide occurred with almost perfect selectivity at sites adjacent to the targeted complementary sequence (and on the opposite
strand too). It was suggested that narrow localization of the cleavage sites argued
against oxidative cleavage by diffusible species, such as ·OH, and for direct reaction
of the photoexcited sensitizer with the nearest neighboring nucleotides. Although the
detailed mechanism remains to be elucidated, the experiment proved that it was
possible to design synthetic endonucleases which are sequence-specific and which can be
activated by light.[60]

9.8.4.3. Photocleavage of DNA according to topology

In addition to targeting specific bases or sequences of bases for cleavage by sensitized processes, it is also possible to selectively target regions of DNA on the basis of their topology. Polypyridyl metal complexes, because they are sensitive to stereochemistry, have been suggested as selective photocleaving agents for Z-DNA regions, A-DNA regions, and cruciforms.[61] For example, Λ-tris(4,7-diphenyl-1,10-phenanthroline)-cobalt(III) binds to left-handed Z-DNA and photosensitizes its cleavage.[62] It induces cleavage at four specific sites in pBR322 DNA.[63] It also targets a family of conformationally distinct sites along SV40 DNA, allowing site-specific cleavage along the strand.[64]

9.8.4.4. Photodynamic reactions with DNA and nucleotides

In most cases where photosensitized damage of DNA has been observed, neither the extent to which photodynamic reactions take place nor mechanistic data have been firmly established, especially when the damage is observed *in vivo*. Though valuable, model studies *in vitro* must be interpreted with care. For example, the observation of enhanced damage in D_2O and decreased damage in the presence of sodium azide, which usually indicates singlet oxygen participation, cannot be trusted in DNA photochemistry, as mentioned earlier.

Flash photolysis studies, focusing on the fate of the triplet excited state of a sensitizer in the presence of oxygen and either nucleotides or DNA, provide direct evidence for the initial stages of the photochemistry. Sensitization reactions by rose bengal, for example, produced both singlet oxygen and the superoxide ion.[65] Analyzing the reactivity of these two species and of the electronically excited sensitizer with DNA and the four nucleotides proved that only guanine residues showed any appreciable reaction at neutral pH and that the rate of oxidation increased at higher pH values. The photocleavage of DNA by methylene blue is very similar. Known to occur in aerated solutions, the DNA damage consisted of breakages at the sugar phosphate bond $5'$ to guanine residues (after treatment with alkali), corresponding to Maxam and Gilbert G-specific reactions.[66]

The exact molecular mechanism may be controlled by conformational effects, which determine the extent of interaction between the sensitizer and the specific sites in which a base, such as guanine, is situated. This interaction depends, in turn, on factors such as the ionic strength or pH, as demonstrated with the small duplex $d(CG)_6$, where the simple change from 0.1 to 4.0 M in the medium increased fourfold the rate of oxidation at guanine sites.[67]

9.8.4.5. The chemistry of photosensitized cleavage reactions

In many cases, the biologically important phenomena associated with the photodynamic reactions of DNA depend on cleaving the phosphodiester backbone of the molecule. This was the important result of the photooxidation reactions sensitized by rose bengal or methylene blue mentioned earlier, which allowed sequencing of a DNA through specific cleavage at guanine sites.

Of course, other photoinduced reactions which do not result in backbone cleavage may occur either in competition with, or prior to, the cleavage reactions. Although photooxidation reactions of DNA often take place at guanine sites, and guanine has been shown to yield a crystalline polymer upon irradiation in the presence of oxygen,[68] the photooxidation of guanine, guanosine, or guanosine phosphates has not been as extensively studied as might have been expected.

The photodynamic cleavage reactions of DNA do not all necessarily follow the same mechanism. Cleavage could occur from reaction in the deoxyribose moiety at the oxygen-bearing 1'-, 3'-, 4'- or 5'-positions. The main reaction considered in these transformations is hydroxylation (which may result from initial hydroperoxidation, or reaction with ·OH of an initially formed radical). A simple ionic mechanism leads to cleavage of the polymer as shown in the scheme below:

1. 1'-Hydroxylation:

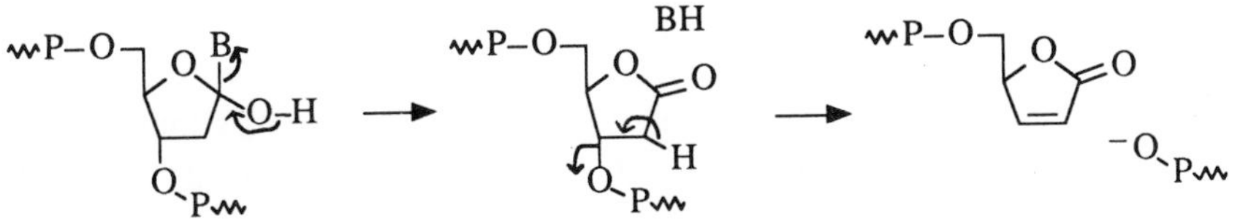

2. 3'-Hydroxylation:

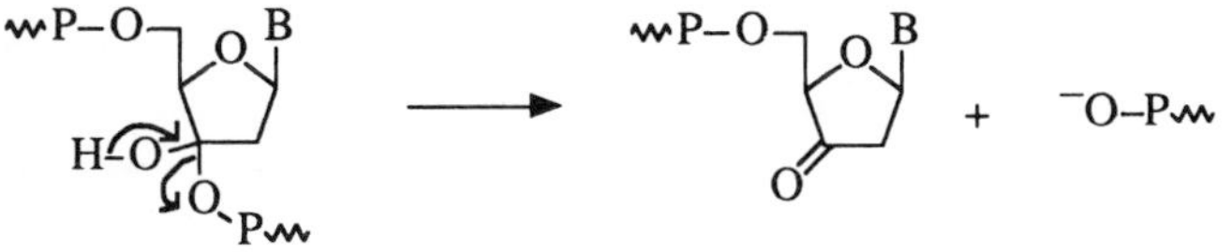

3. 4'-Hydroxylation:

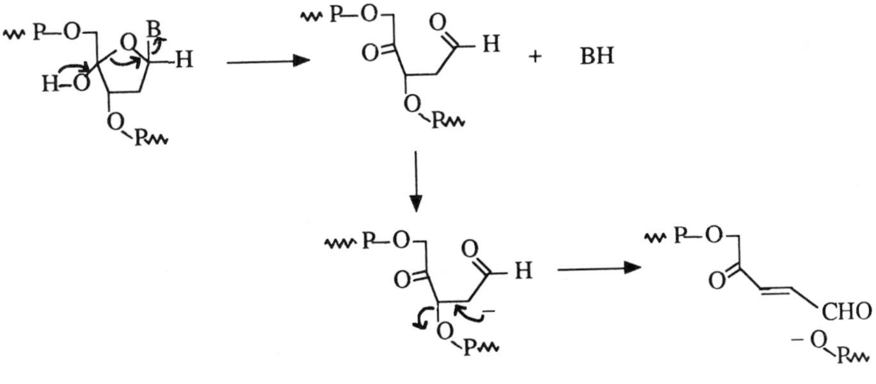

The intervention of an enolate ion (as shown) is not essential, as an enol would also lead to a similar elimination reaction.

4. 5'-Hydroxylation:

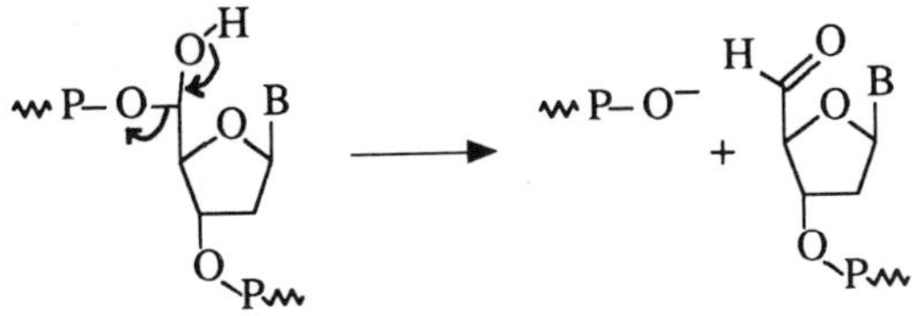

Actually, the mechanism of photodynamic cleavage of DNA is very poorly known at best. Cleavage through reaction at the 4'-position has been the most frequently considered, particularly in relation to mechanistic studies with the bleomycin antibiotics. An alternative mechanism which has been proposed involves the decomposition of a hydroperoxide as shown.[69]

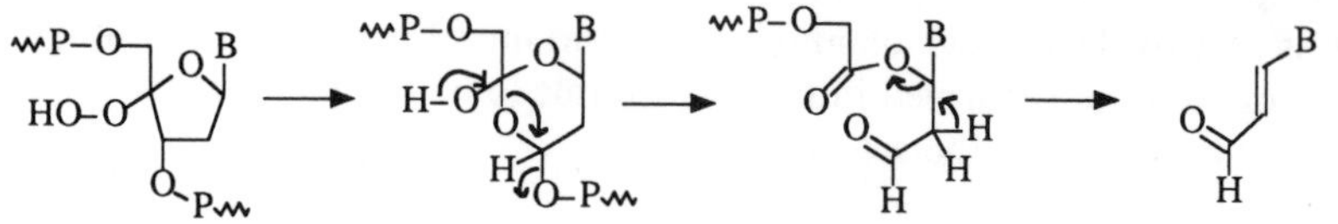

9.8.4.6. Non-photodynamic sensitized cleavage of DNA

The mechanism of oxygen-independent photosensitized cleavage reactions of irradiated DNA is usually not known. Redox reactions may be involved, as suggested in the case of nitro compounds:[70]

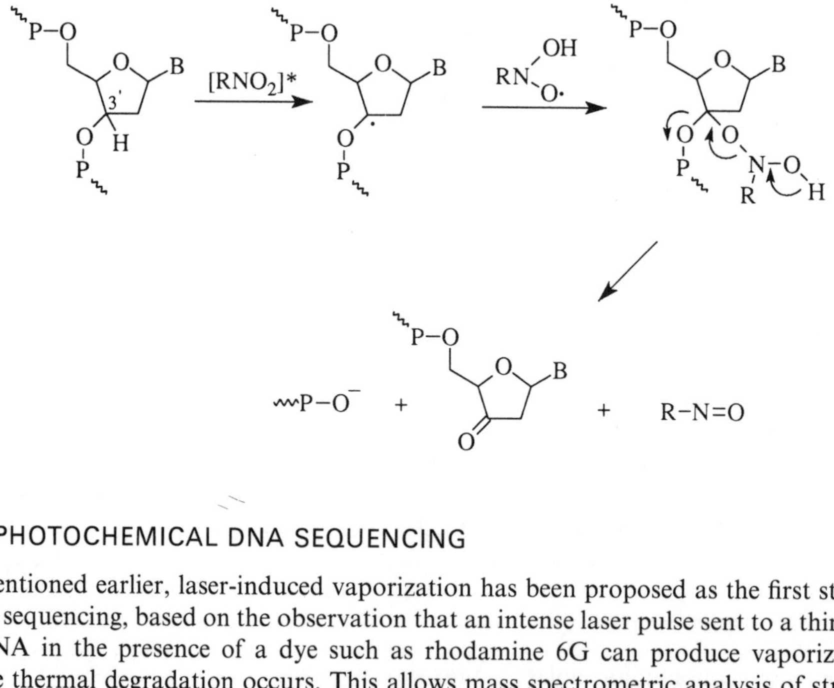

9.9. PHOTOCHEMICAL DNA SEQUENCING

As mentioned earlier, laser-induced vaporization has been proposed as the first step in DNA sequencing, based on the observation that an intense laser pulse sent to a thin film of DNA in the presence of a dye such as rhodamine 6G can produce vaporization before thermal degradation occurs. This allows mass spectrometric analysis of strands of DNA containing as many as 1000 bases.[71] Higher laser powers increase vaporization and decrease decomposition among the vaporized molecules; increased production of smaller fragments is observed at lower powers. The same observation has been made with dATP, indicating photochemical activation of the P–O–P bond at intermediate and low vaporization powers. The mechanism suggested was either a two-photon excitation of the weak 274 nm absorption band of the phosphate, or conversion of the photon energy into thermal energy before vaporization occurs, resulting in phosphodiester bond cleavage.

Sequencing DNA is usually achieved through selective cleavage reactions of a radio-actively labeled sample. Techniques based on labeling with chemiluminescent substrates may be equally valuable.[72] A laser-based technique for sequencing large DNA fragments (40 kilobases or higher) at a rate of 100–1000 bases per second relies on the fluorescent labeling of the bases in a single fragment of DNA, which is then attached to a support. The individual labeled bases are detected as an endonuclease cleaves them from the DNA fragment in a flowing sample stream.[73]

9.10. DETECTION OF PHOTOADDUCTS, AND MAPPING OF DNA

DNA adducts produced *in vivo* have been mapped with single nucleotide resolution. After ultraviolet treatment of the cells, DNA was isolated and cleaved as (6–4) adducts with piperidine. Gene-specific fragments were then amplified by use of a ligation-mediated polymerase chain reaction (LMPCR).[74]

9.11. DNA FOOTPRINTING

Interactions between proteins and DNA are important in all aspects of molecular biology. Footprinting is a technique which allows the sites of contact between DNA and proteins to be determined; this can be done with photochemical techniques, even *in vivo*. Normally, irradiation of DNA induces covalent bond formation between bases (such as the pyrimidine–pyrimidine dimers described in Section 9.7.1). When DNA interacts with a protein at the time it is photolyzed, the pattern of covalent bond formation is altered. After purification and end-labeling of the DNA with ^{32}P, sequencing reveals the sites which were protected from photochemical damage or made more sensitive to it through contact with the protein. This technique was successfully used for probing contacts between the *lac* repressor and the *lac* operator in *Escherichia coli*.[75]

9.12. PHOTOAFFINITY LABELING

The principle of photoaffinity labeling is marvelously simple.[76] Invented by Westheimer, it is based on the fact that biochemical reactions may include a step in which a substrate interacts with its receptor (for example at the active site of an enzyme). If a chemical reaction forms a covalent bond between the substrate and an aminoacid residue of the receptor, locating the modified aminoacid on the primary structure can lead to the identification of one of the components of the receptor. Many of the aminoacid components of the receptor can be revealed when the chemical reaction performed on the substrate is not specific. Traditionally, the key step in these labeling experiments has been the photochemical decomposition of a suitable precursor, which forms either a carbene or a nitrene. This approach is logical because (1) electronic excitation is very fast, and (2) carbenes and nitrenes generally react rapidly, often with little selectivity. The three-dimensional shape of the active enzyme–substrate complex can be mapped by locating all the points of attachment of the label to the protein.

Examples of photoaffinity labeling were provided in Chapter 5 (Section 5.9.5). It is important to stress that the chemical yield of an actual photolabeling step (formation of the covalent bond between the reagent and an enzyme site) does not need to be high, as the reagent is usually radioactive and radioactivity measurements are very sensitive.

Molecules containing a phenyl group substituted with an azide have been used frequently, despite the drawback of the nitrene isomerizing into a seven-membered ring compound (Chapter 5).

Numerous substrates substituted with a *p*-azidophenyl have been successfully used for photolabeling. For example, dextranesucrase from *Streptococcus sanguis* was photolabeled with *p*-azidophenyl α-D-glucopyranoside.[77] The human serum vitamin D binding protein was labeled with an *N*-(4-azido-2-nitrophenyl)aminophenyl ether of 25-hydroxyvitamin D$_3$.[78] Other labeling compounds which lose nitrogen upon irradiation have been used. For example, two transmembrane proteins responsible for the transport of anions and glucose in human erythrocytes were studied with diazofluorene (**58**).[79]

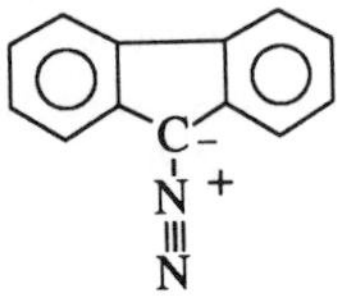

(58)

Sound work with photoaffinity labeling requires understanding the photochemical behavior of the substrate itself, which may undergo unforeseen chemistry. For example, short-wavelength light (254 nm) causes rapid loss of opiate binding activity in morphine and related opium alkaloids. Longer-wavelength light (366 nm) was found to induce an interesting rearrangement, illustrated with codeinone and some of its derivatives (**59**). The initial step may be viewed as an eight-electron electrocyclic process, followed by ring opening of the cyclopropane upon nucleophilic attack by the solvent and protonation of an adjacent carbonyl group.[80]

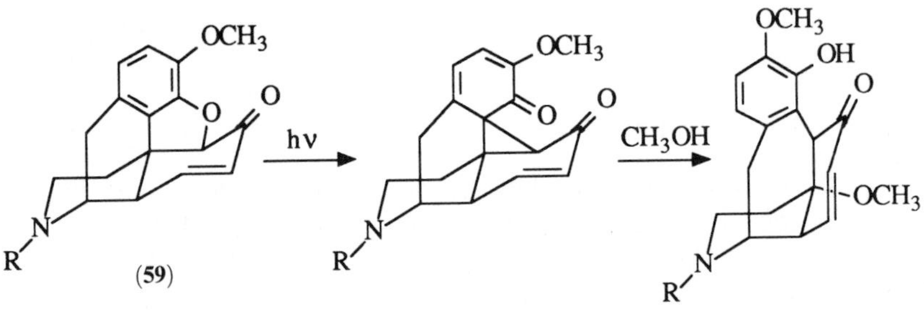

(59)

The iodinated azide (**60**) illustrates the importance of carefully investigating the photochemistry of a prospective labeling agent, as iodine is lost here before nitrogen. Such a precursor labeled with ^{125}I would therefore be rather ineffective as an affinity labeling compound, and better precursors should either carry ^{125}I at a less reactive position or else (perhaps preferably in view of the facile photocleavage of carbon–iodine bonds) bear a different radiolabel.[81]

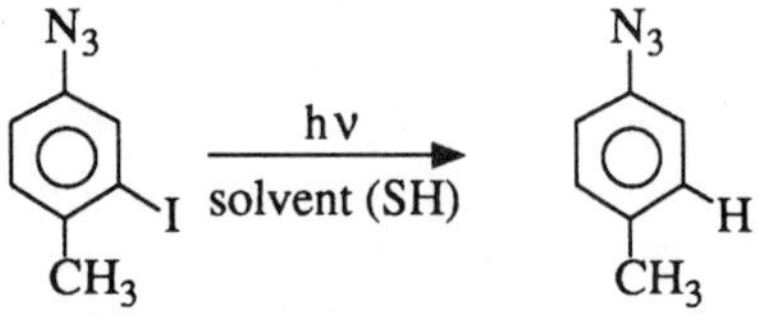

(60)

An example of photoaffinity labeling with a diazirine derivative was provided in Section 5.9.4. Other substrates besides azido and other nitrogen-containing substrates are used in photoaffinity labeling. For example, photoaffinity labeling of the 1,25-dihydroxyvitamin D$_3$ receptor (Chapter 3) has been achieved very simply with the substrate itself, ditritiated at the positions 26 and 27.[82] Additional examples of application of photoaffinity labeling as well as mechanistic considerations have been provided by Schuster et al.[83]

9.13. PHOTOCHEMICAL REACTIONS OF PSORALENS: A MORE DETAILED ANALYSIS

Psoralens have gained enormous importance in molecular biology and in medicine, as a component in the PUVA therapy of psoriasis and other diseases. Some aspects of these applications will be reviewed in the following chapters. The photochemistry of psoralens has been investigated extensively, particularly in relation to biochemical modifications, with the main focus on the following points: [2 + 2] cyclization reactions with the furan moiety; [2 + 2] cyclization with the pyrone moiety; participation in photodynamic reactions, creating either $^1O_2$ or O_2^-; reactions with such reactive oxygen products. In order to cover the reactions more thoroughly, additional processes should also be analyzed with respect to psoralens, namely [4 + 2] cyclization reactions with the pyrone moiety, formation of dihydropsoralen derivatives, and electron transfer or energy transfer reactions with molecules other than O_2.

9.13.1. [2 + 2] cyclization reactions

The photochemistry of psoralens with unsaturated molecules has been investigated extensively. Simple inspection of the psoralen ring system immediately suggests the complexity of these reactions. An unsymmetrically substituted alkene can lead to two regioisomers at each site for cyclization. When adducts have formed at both sites, the resulting cyclobutanes are either *syn* or *anti* with respect to the psoralen ring system. Neglecting enantiomers, the total of possible racemic products is already impressive: eight monoadducts (Fig. 9.1) and 16 di-adducts. This total increases if the alkene undergoes *cis–trans* equilibration prior to cycloaddition or if an optically active substrate is involved, as in DNA derivatives. A surprisingly small number of these structures have been identified in the treatment of psoralens with DNA or its components (particularly pyrimidines). For example, only two main cycloaddition products were

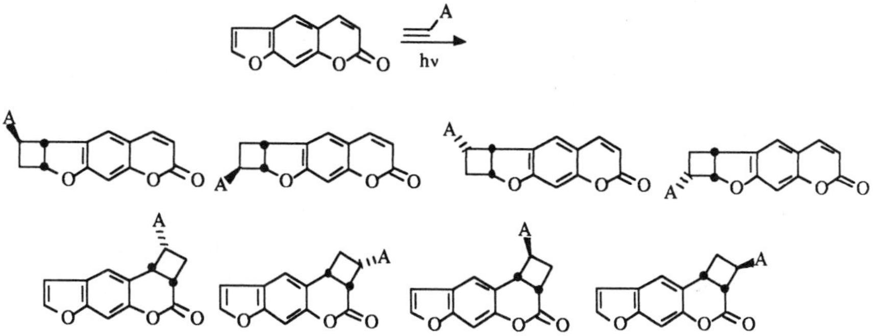

Fig. 9.1 Structure of the monoaddition products of psoralen with an alkene (CH$_2$=CHA).

observed in the photochemical treatment of DNA with a pyridopsoralen, the most promising class of monofunctional psoralens for the treatment of psoriasis.[84]

Monoadducts or interstrand cross-links may be formed in the photoinduced cyclization of a psoralen with DNA. In general, cyclobutane rings can be cleaved by photolysis with shorter wavelength light. However, a cyclobutane ring formed on the furan side can also be cleaved by treatment with base. Although the chemical details are not known, this cleavage does not result in breaking the DNA backbone.[85]

9.13.2. [4 + 2] cycloaddition reactions

It is important to note that in addition to [2 + 2] cycloadditions to alkenes, some coumarins undergo [4 + 2] cycloaddition:[86]

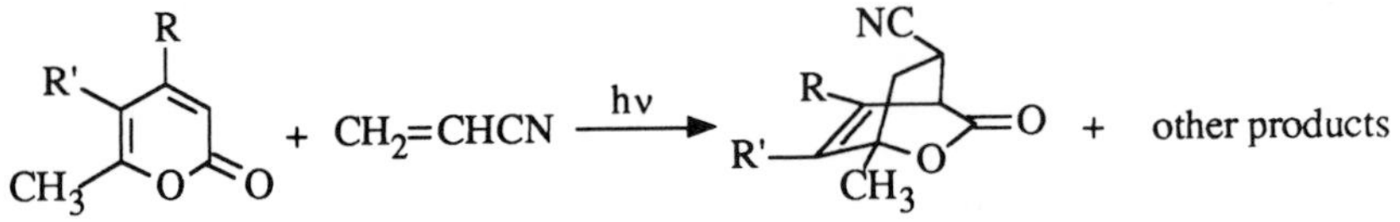

This process has not yet been identified in the photochemistry of psoralens.

9.13.3. Production of $^1O_2$ and O_2^-

Like most organic molecules which absorb light, psoralens can transfer energy to oxygen and produce $^1O_2$. A good correlation between generation of $^1O_2$ and skin photosensitization was found with some psoralens, although it does not necessarily prove a direct causality for the biological effect,[87] particularly since other psoralens showed no such correlation.[88] Most psoralens also generate O_2^- when irradiated in aqueous solution, but the biological consequences are not clear (a more critical discussion is presented in ref. 89).

9.13.4. Photooxidation of psoralens

Simple furans have often been used as $^1O_2$ traps, and it is not surprising to find that psoralens themselves react in the same manner.[90] A reasonable path to the product (62) observed with 8-methoxypsoralen (8-MOP) (61) is illustrated. The new biologically active photoproduct (63) from 8-MOP has been recently detected, and synthesized chemically.[91]

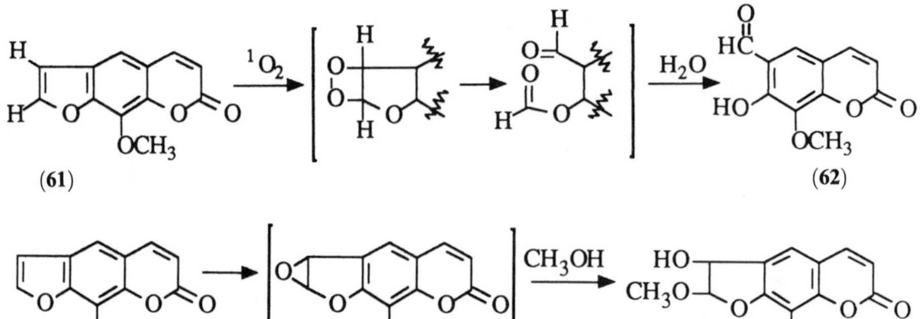

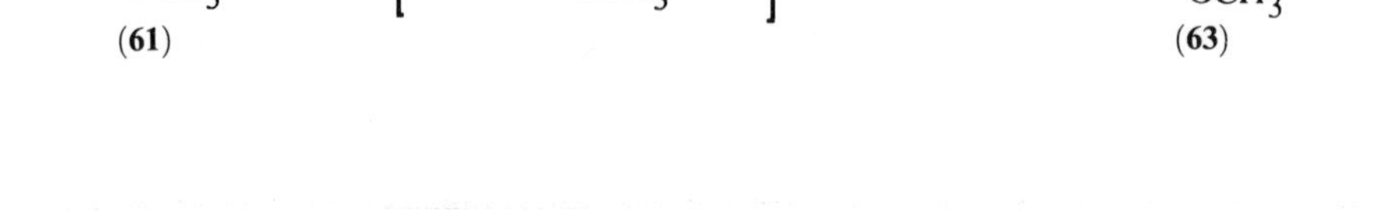

Considering the fact that the procedure involved treatment of 8-MOP with H_2O_2 and NaOCl, an epoxide was the likely precursor to the isolated product, and was probably also formed in low yield in the reaction of $^1O_2$ with the same substrate (see Chapter 6). The observation that the photoproduct was easily conjugated with proteins also favors the epoxide structure.

8-Methoxypsoralen was found to undergo photooxygenation of the furan ring under electron transfer conditions (see Chapter 7). The structure shown was obtained with the sensitizers 9,10-dicyanoanthracene or tetrachlorobenzoquinone (chloranil), the substituent R being characteristic of the solvent used, OCH_3 in methanol, $OCOCH_3$ in acetic acid.[92]

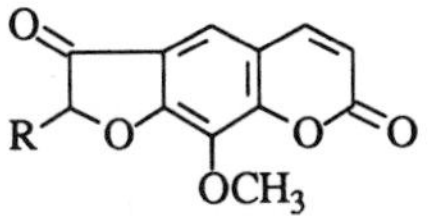

9.13.5. Photosensitization reactions

Pyridopsoralens have photobiological properties that set them apart from other types of psoralens. In particular, they do not induce erythema on the skin of patients treated for psoriasis by PUVA therapy, and they are strictly monofunctional (producing cyclo-addition reactions only on the furan side).

The three pyridopsoralens (**64**), (**65**), and (**66**) were used to sensitize the photolysis of thymine. Six diastereomers of cyclobutadithymidine (**67**) were obtained, as well as other diastereomeric dimers (**68**).[93]

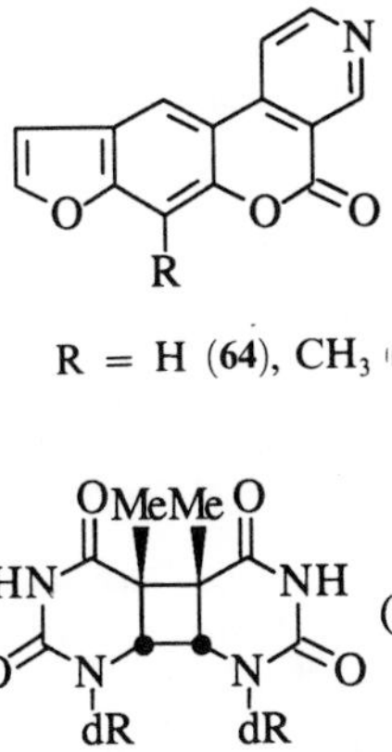

R = H (**64**), CH$_3$ (**65**)

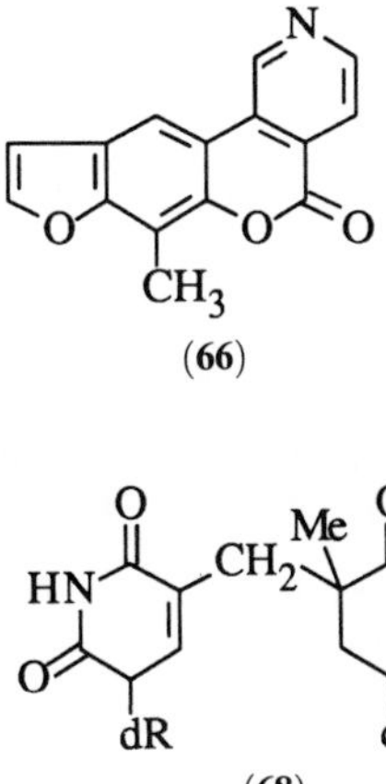

(**66**)

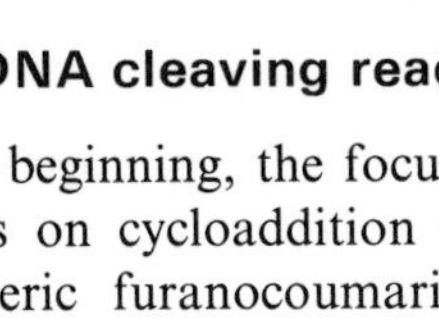

(and isomers)

(**67**)

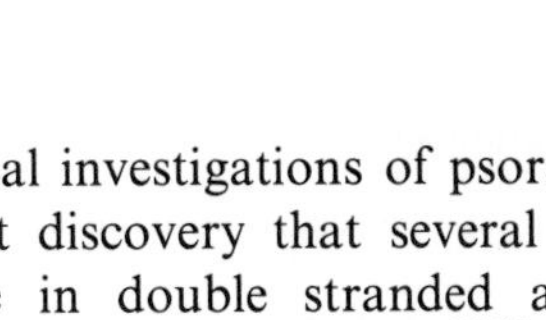

(**68**)

9.13.6. DNA cleaving reactions

From the beginning, the focus of the photochemical investigations of psoralens with DNA was on cycloaddition reactions. The recent discovery that several psoralens and isomeric furanocoumarins induced cleavage in double stranded and single stranded pBR322 and M13 circular DNA came therefore as a surprise.[94,95]

164   ORGANIC PHOTOCHEMISTRY: PRINCIPLES AND APPLICATIONS

REFERENCES

1. T. P. Begley and H. Young, *J. Am. Chem. Soc.*, **111**, 3095–3096 (1989).
2. D. H. Hug and J. K. Hunter, *J. Photochem. Photobiol. B*, **10**, 3–22 (1991).
3. C. Thomas, R. S. MacGill, G. C. Miller and R. S. Pardini, *Photochem. Photobiol.*, **55**, 47–53 (1992).
4. I. Willner, S. Rubin and A. Riklin, *J. Am. Chem. Soc.*, **113**, 3321–3325 (1991).
5. I. Willner, S. Rubin and T. Zor, *J. Am. Chem. Soc.*, **113**, 4013–4014 (1991).
6. M. Harada, M. Sisido, J. Hirose and M. Nakanishi, *FEBS Lett.*, **286**, 6–8 (1991).
7. M. d'Ischia, C. Costantini and G. Prota, *J. Am. Chem. Soc.*, **113**, 8353–8356 (1991).
8. T. Nagano and I. Fridovich, *Photochem. Photobiol.*, **41**, 33–37 (1985).
9. A. W. Girotti, *Photochem. Photobiol.*, **51**, 497–509 (1990).
10. C. M. Krishna, S. Uppuluri, P. Riesz, J. S. Zigler, Jr and D. Balasubramanian, *Photochem. Photobiol.*, **54**, 51–58 (1991).
11. S. G. Bertolotti, N. A. Garcia and G. A. Arguello, *J. Photochem. Photobiol. B*, **10**, 57–70 (1991).
12. J. P. J. Boegheim, G. C. M. Warmerdam, T. M. A. R. Dubbleman and J. van Seveninck, In G. Jori and C. Perria (eds) *Photodynamic Therapy of Tumors and Other Diseases*, pp. 133–136. Libreria Progetto, Padova (1985).
13. R. R. Hill, J. D. Coyle, D. Birch, E. Dawe, G. E. Jeffs, D. Randall, I. Stec and T. M. Stevenson, *J. Am. Chem. Soc.*, **113**, 1805–1817 (1991).
14. S. G. Bertolotti, N. A. Garcia and G. A. Arguello, *J. Photochem. Photobiol., B*, **10**, 57–70 (1991).
15. R. R. Hill, J. D. Coyle, D. Birch, E. Dawe, G. E. Jeffs, D. Randall, I. Stec and T. M. Stevenson, *J. Am. Chem. Soc.*, **113**, 1805–1817 (1991).
16. H. B. Gray, *Aldrichim. Acta*, **23**, 87–93 (1990).
17. D. Balasubramanian, X. Du and J. S. Zigler, Jr, *Photochem. Photobiol.*, **52**, 761–768 (1990).
18. J. Kagan, M. Hasson and F. Grynspan, *Biochem. Biophys. Acta*, **802**, 442–447 (1984).
19. M. Nivsarkar, G. P. Kumar, M. Laloraya and M. M. Laloraya, *Arch. Insect Biochem. Phys.*, **16**, 249–255 (1991).
20. J. G. Peak, M. J. Peak, R. S. Sikorski and C. A. Jones, *Photochem. Photobiol.*, **412**, 295–302 (1985).
21. I. Tessman and M. A. Kennedy, *Mol. Gen. Genet.*, **227**, 144–148 (1991).
22. D. Burdi and T. P. Begley, *J. Am. Chem. Soc.*, **113**, 7768–7770 (1991).
23. B. Skalski, R. P. Steer and R. E. Verrall, *J. Am. Chem. Soc.*, **113**, 1756–1762 (1991).
24. T. Douki, L. Voituriez and J. Cadet, *Photochem. Photobiol.*, **53**, 293–297 (1991).
25. K. C. Smith (ed.), *Aging, Carcinogenesis and Radiation Biology: The Role of Nucleic Acid Addition Reactions*. Plenum Press, New York (1976).
26. I. Saito and T. Matsuura, *Acc. Chem. Res.*, **18**, 134–141 (1985).
27. J. G. Peak, B. Pilas, E. J. Dudek and M. J. Peak, *Photochem. Photobiol.*, **54**, 197–203 (1991).
28. T. W. Masnyk and K. W. Minton, *Photochem. Photobiol.*, **54**, 99–107 (1991).
29. D. T. Croke, W. Blau, C. OhUigin, J. M. Kelly and D. J. McConnell, *Photochem. Photobiol.*, **47**, 527–536 (1988).
30. E. Bothe, H. Görner, J. Optiz, D. Schulte-Frohlinde, A. Siddiqui and M. Wala, *Photochem. Photobiol.*, **52**, 949–959 (1990).
31. D. Schulte-Frohlinde, M. G. Simic and H. Görner, *Photochem. Photobiol.*, **52**, 1137–1151 (1990).
32. R. J. Levis and L. J. Romano, *J. Am. Chem. Soc.*, **113**, 7802–7803 (1991).
33. K. Stoerl and C. Mund, *Stud. Biophys.*, **61**, 231–243 (1977).
34. T. G. Burchuladze, E. G. Sideris and G. Y. Fraikin, *Biofizika*, **35**, 722–725 (1990).
35. L. B. Chapnick, L. A. Chasin, A. L. Raphael and J. K. Barton, *Mutation Research*, **201**, 17–26 (1988).
36. M. J. Peak, J. G. Peak, C. M. Blaumueller and R. K. Elespuru, *Chem.-Biol. Interactions*, **67**, 267–274 (1988).
37. J. Piette, M. Lopez, C. M. Calberg-Bacq and A. Van de Vorst, *Int. J. Radiat. Biol.*, **44**, 427–433 (1981).
38. T. P. Wang, J. Kagan, R. W. Tuveson and G. R. Wang, *Photochem. Photobiol.*, **53**, 463–466 (1991).
39. P. J. Gray, D. R. Phillips and A. G. Wedd, *Photochem. Photobiol.*, **36**, 49–57 (1982).
40. T. T. Wei, K. M. Byrne, D. Warnick-Pickle and M. Greenstein, *J. Antibiot.*, **35**, 545–548 (1982).
41. J. Kagan, X. Chen, T. P. Wang and P. Forlot, *Photochem. Photobiol.*, **56**, 185–193 (1992).
42. H. Utsumi, H. Kitani, C. M. Chang-Liu and M. M. Elkind, *Carcinogenesis (London)*, **8**, 1439–1444 (1987).
43. J. Kagan, T. P. Wang, A. Benight, R. W. Tuveson, G.-R. Wang and P. P. Fu, *Chemosphere*, **20**, 453–466 (1990).
44. P. E. Nielsen, C. Jeppensen and O. Buchardt, *FEBS Lett.*, **235**, 122–124 (1988).
45. J. K. Barton and A. L. Raphael, *J. Am. Chem. Soc.*, **106**, 2466 (1984).
46. M. B. Fleisher, K. C. Waterman, N. J. Turro and J. K. Barton, *Inorg. Chem.*, **25**, 3549–3551 (1986).
47. A. J. Blacker, J. Jazwinski, J.-M. Lehn and F. X. Wilhelm, *J. Chem. Soc., Chem. Commun.*, 1035–1036 (1986).
48. B. E. Bolwer, L. S. Hollis and S. J. Lippard, *J. Am. Chem. Soc.*, **106**, 6102–6104 (1984).
49. P. E. Nielsen, C. Jeppesen, M. Egholm and O. Buchardt, *Biochemistry*, **27**, 6338–6343 (1988).

50. H. Fujita, A. Endo and K. Suzuki, *Photochem. Photobiol*, **33**, 215–222 (1981).
51. A. Slama-Schwok, J. Jazwinski, A. Bere, T. Montenay-Garestier, M. Rougee, C. Helene and J. M. Lehn, *Biochemistry*, **28**, 3227–3234 (1989).
52. J. E. Schneider, S. Price, L. Maidt, J. M. C. Gutteridge and R. A. Floyd, *Nucleic Acids Res.*, **18**, 631–635 (1990).
53. T. Friedmann and D. M. Brown, *Nucleic Acids Res.*, **5**, 615–622 (1978).
54. Y. Uesawa, J. Kuwahara and Y. Sugiura, *Biochem. Biophys. Res. Commun.*, **164**, 903–911 (1989).
55. J. Piette, M.-P. Merville-Louis and J. Decuyper, *Photochem. Photobiol.*, **44**, 793–802 (1986).
56. R. Subramanian and C. F. Meares, *J. Am. Chem. Soc.*, **108**, 6427–6429 (1986).
57. I. Saito, H. Sugiyama and T. Matsuura, *J. Am. Chem. Soc.*, **105**, 956–962 (1983).
58. P. E. Nielsen, C. Jeppesen, M. Egholm and O. Buchardt, *Nucleic Acids Res.*, **16**, 3877–3888 (1988).
59. H. E. Moser and P. B. Dervan, *Science*, **238**, 645–650 (1987).
60. L. Perrouault, U. Asseline, C. Rivalle, N. T. Thuong, E. Bisagni, C. Giovannangelli, T. Le Doan and C. Helene, *Nature*, **344**, 358–360 (1990).
61. J. K. Barton, *Pure Appl. Chem.*, **61**, 563–564 (1989).
62. J. K. Barton and A. L. Raphael, *J. Am. Chem. Soc.*, **106**, 2466–2468 (1984).
63. J. K. Barton and A. L. Raphael, *Proc. Natl Acad. Sci. USA*, **82**, 6460–6464 (1985).
64. B. C. Müller, A. L. Raphael and J. K. Barton, *Proc. Natl Acad. Sci. USA*, **84**, 1764–1768 (1987).
65. P. C. C. Lee and M. A. J. Rogers, *Photochem. Photobiol.*, **45**, 79–86 (1987).
66. T. Friedman and D. M. Brown, *Nucleic Acids Res.*, **5**, 615–622 (1978).
67. S. E. Rokita, S. Prusiewicz and L. Romero-Fredes, *J. Am. Chem. Soc.*, **112**, 3616–3621 (1990).
68. C. Santhosh and P. C. Mishra, *J. Photochem. Photobiol. A*, **61**, 281–284 (1991).
69. G. H. McGall, J. Stubbe and J. W. Kozarich, *J. Org. Chem.*, **56**, 48–55 (1991).
70. P. E. Nielsen, C. Jeppesen, M. Egholm and O. Buchardt, *Biochemistry*, **27**, 6338–6343 (1988).
71. R. J. Levis and L. J. Romano, *J. Am. Chem. Soc.*, **113**, 7802–7803 (1991).
72. C. Martin, L. Bresnick, R. R. Juo, J. C. Voyta and I. Bronstein, *BioTechniques*, **11**, 110–113 (1991).
73. L. M. David, F. R. Fairfield, C. A. Harger, J. H. Jett, R. A. Keller, J. H. Hahn, L. A. Krakowski, B. L. Marrone, J. C. Martin *et al.*, *Genet. Anal.: Tech. Appl.*, **8**, 1–7 (1991).
74. G. P. Pfeifer, R. Drouin, A. D. Riggs and G. P. Holmquist, *PNAS*, **88**, 1374–1378 (1991).
75. M. M. Becker and J. C. Wang, *Nature*, **309**, 682–687 (1984).
76. V. Chowdhry and F. H. Westheimer, *Ann. Rev. Biochem.*, **48**, 293–325 (1979).
77. S. F. Kobs and R. M. Mayer, *Carbohydr. Res.*, **211**, 317–326 (1991).
78. R. Ray, R. Bouillon, H. Van Baelen and M. F. Holick, *Biochemistry*, **30**, 7638–7642 (1991).
79. A. K. Lala, H. F. Batliwala and S. Bhat, *Pure Appl. Chem.*, **62**, 1453–1456 (1990).
80. A. G. Schultz, N. J. Green, S. Archer and F. S. Tham, *J. Am. Chem. Soc.*, **113**, 6280–6281 (1991).
81. E. Leyva, D. H. S. Chang, M. S. Platz, D. S. Watt, P. J. Crocker and K. Kawada, *Photochem. Photobiol.*, **54**, 329–333 (1991).
82. T. A. Brown and H. F. DeLuca, *Biochim. Biophys. Acta*, **1073**, 324–328 (1991).
83. D. I. Schuster, W. C. Probst, G. K. Ehrlich and G. Singh, *Photochem. Photobiol.*, **49**, 785–804 (1989).
84. A. Moysan, A. Viari, P. Vigny, L. Voituriez, J. Cadet, E. Moustacchi and E. Sage, *Biochemistry*, **30**, 7080–7088 (1991).
85. A. T. Yeung, W. J. Dinehart and B. K. Jones, *Biochemistry*, **27**, 6332–6338 (1988).
86. K. Somekawa, T. Shimo, H. Yoshimura and T. Suishu, *Bull. Chem. Soc. Jpn.*, **63**, 3456–3461 (1990).
87. N. J. De Mol and G. M. J. Beijersbergen van Henegouwen, *Photochem. Photobiol.*, **33**, 815–819 (1981).
88. D. Vedaldi, F. Dall'Acqua, A. Gennaro and G. Rodighiero, *Z. Naturforsch.*, **38c**, 866–869 (1983).
89. W. R. Midden, In F. P. Gasparro (ed.) *Psoralen DNA Photobiology*, vol. 2, pp. 2–50. CRC Press, Boca Raton (1988).
90. H. H. Wasserman and D. R. Berdahl, *Photochem. Photobiol.*, **35**, 565–567 (1982). M. K. Logani, W. A. Austin, B. Shah, and R. E. Davies, *Photochem. Photobiol.*, **35**, 569–573 (1982).
91. N. Mizuno, K. Esaki, J. Sakakibara, N. Murakami and S. Nagai, *Photochem. Photobiol.*, **54**, 697–701 (1991).
92. J. Xu, Y. Song and Y. Shang, *J. Chem. Soc., Chem. Commun.*, 1621–1622 (1991).
93. A. Moysan, A. Viari, P. Vigny, L. Voituriez, J. Cadet, E. Moustacchi and E. Sage, *Biochemistry*, **30**, 7080–7088 (1991).
94. J. Kagan, X. Chen, T. P. Wang and P. Forlot, *Photochem. Photobiol.*, **56**, 185–193 (1992).
95. X. Chen and J. Kagan, unpublished results.

10. Photochemistry and biology

10.1. GENERALITIES

Organisms and cellular components undergo a variety of transformations in the presence of radiation capable of interacting with native chromophores. However, it is often difficult to identify the chromophore(s) involved in a given biological change. Usually, the shorter the wavelength, the larger the potential number of molecules which can be electronically excited. However, there is no direct correlation between the initial photochemical transformation of a molecule within a cell and the severity of the biological damage inflicted. Much depends on factors such as the type and location of damage within the cell and, particularly, whether the cell has a repair mechanism for this damage. Thus, formation of singlet oxygen upon photosensitization with visible light may be much more lethal than DNA damage from short-wavelength ultraviolet light if the latter is repaired by an endogenous enzyme system. Of course, the photochemical behavior of the major individual constituents of cells can be investigated separately, but important interactions between different kinds of molecules will be absent, and photosensitized effects due to metabolites present in minute quantities in cells are hard to identify and study separately.

Cells exposed to light may undergo drastic chemical modifications of structural proteins, lipids, nucleic acids, etc., but irradiation may also allow enzyme systems to be turned on or off *in vivo*. It is possible to design systems which act similarly *in vitro*. For example, it was recently shown that irradiation in the presence of a thiol and a titanium(III) salt activated the enzyme 2-(methylthio)ethanesulfonic acid reductase from methanobacterium.[1]

10.2. PHOTOTOXICITY

Phototoxic reactions are characterized by the observation that organisms, or selected cells, are more severely damaged upon simultaneous treatment with a chemical and light than with either the chemical or light alone. In general, this is evidenced by reduced or arrested growth of the biological target, or even by its death. Phototoxicity may also be expressed in the opposite manner, with *increased* rate of growth, as shown recently with yeasts.[2]

All the photobiological transformations in nature are the result of photochemical reactions induced by sunlight, particularly its ultraviolet component. Since the ultraviolet radiation reaching the earth depends on the extent to which it has been filtered by stratospheric ozone, the reports of the depletion of the ozone layer by environmental pollution have generated serious concern. It has been estimated that a 1% decrease in ozone produces a 1.6% increase in annual carcinogenic ultraviolet, leading to an increase of 2.7% in non-melanoma skin cancer. Geographical latitude makes little difference.[3]

Ozone reduction in the stratosphere is also expected to decrease immunity,[4] because irradiation with UVB is known to suppress the cell-mediated immune response. The field of photoimmunology is complex; urocanic acid (**1**), by isomerizing from *trans* to *cis*, has been suggested as the skin photoreceptor which initiates immune suppression.[5]

(**1**)

10.3. PHOTODYNAMIC REACTIONS

In Chapter 6 we learned that photobiologists describe light-dependent reactions involving oxygen as *photodynamic reactions* and further divide them into type I reactions, which have a free radical component because of an electron transfer step, and type II, which produce $^1O_2$.

While this definition clearly characterizes the formation of $^1O_2$ as resulting from type II photodynamic reactions, it is ambiguous with respect to the formation of $O_2^{\overline{\cdot}}$, which could occur by electron transfer to $^1O_2$ initially formed in a type II reaction. Almost invariably, the transfer of an electron to O_2, which must occur in the formation of $O_2^{\overline{\cdot}}$, is taken as proof of a type I photodynamic process. The terminology was recently questioned, however, and it has been suggested that only the *primary* step involving the electronically excited sensitizer molecule should be considered when classifying a photodynamic reaction.[6] In that case, type II reactions are all those directly involving oxygen and the electronically excited sensitizer, and type I reactions are those in which radicals or radical ions are produced by hydrogen atom or electron transfer from the electronically excited sensitizer. The definition makes good sense from a theoretical viewpoint but is not likely to be used much by photobiologists dealing with $O_2^{\overline{\cdot}}$ formation. According to the latest definition, $O_2^{\overline{\cdot}}$ formed by electron transfer from a radical or radical anion belongs to a type I mechanism (as in the past), but to a type II mechanism if the electron is transferred directly from the electronically excited sensitizer. In most complex biological systems, however, the only information usually available is that $O_2^{\overline{\cdot}}$ is formed (for example because added superoxide dismutase suppresses an observed transformation), and it is impossible to distinguish the mechanistic options experimentally. Should physical photochemists insist on mechanism-based terminology, photobiologists will be wise to omit any mention of types I and II in relation to $O_2^{\overline{\cdot}}$ formation.

10.4. HEMOLYSIS OF RED BLOOD CELLS

Erythrocytes have been widely used as a test system in photobiology because they hemolyze when chemical damage leads to physical damage of the cell membrane. This hemolysis is readily detected experimentally. If need be, the ghosts (membrane material) can be separated from the medium by centrifugation, and it is even possible to place the ghosts in an appropriate medium and reseal them. The physical properties of intact and hemolyzed erythrocytes are very different. For example, a suspension of intact erythrocytes diffracts light at 610 nm, and the extent of hemolysis in a photosensitized reaction can therefore be measured from the decreased absorbance at that wavelength as the reaction proceeds (complete hemolysis, needed in order to determine the 100% point in a kinetic measurement, may be obtained by addition of a detergent). Although erythrocyte membranes are a complex assembly of protein and lipid components, no other cytoplasmic membranes and no nuclear DNA are present in the cells. Many of the studies reported in the literature have been performed with human erythrocytes, but the potential dangers associated with the presence of HIV-1 or other viruses in blood samples from unknown donors suggest that it might be safer to use blood samples from readily available animals.

In most of the studies with erythrocytes, the rate of hemolysis was used as a measure of a sensitizer's activity. Usually, singlet oxygen is the major agent formed during sensitized reactions which induce hemolysis. Even in the absence of a sensitizer in solution, hemolysis can be induced by singlet oxygen generated photochemically in the gas phase.[7] Since the protein and lipid components of the erythrocyte membranes can be analyzed separately, it is possible to probe whether any photosensitized damage has been inflicted upon each of these components.[8] The use of standard diagnostic techniques described in Chapter 6 (quenching by NaN_3, enhancement in D_2O) often produces definitive results with hydrophilic sensitizers, for example with the dyes rose bengal or merocyanin 540,[9] but usually not with highly hydrophobic sensitizers.

One should be aware that the sensitivity of erythrocytes to photosensitized hemolysis depends on their age, the greatest variations being observed in the first 2 days after the blood has been drawn. The most reproducible results are obtained with older erythrocyte samples, for example from blood bank samples which are no longer appropriate for transfusion.[10]

10.5. CIRCADIAN RHYTHMS

The photoperiod controls many biological events in mammals, mediated by the release of the pineal hormone melatonin (2). This topic is beyond the scope of this book, but it is worth pointing out that it affects human biochemistry too. In the isolation of Antarctica, the prolonged absence of sunlight during winter facilitates the study of circadian rhythms. A small group of human subjects was found to have melatonin and other biochemical markers very well synchronized up to the last sunset sleep. During the 126 days of winter, when there was no sunlight, synchronization was lost (even though the subjects had knowledge of time) but it returned when the sun reappeared.[11]

$$CH_3O \quad \text{—} \quad CH_2CH_2NHCOCH_3$$

N
|
H

(2)

Furanocoumarins such as 8-methoxypsoralens are discussed in several places in this book because of their practical importance as photosensitizers. It is interesting to note that they have biological activity in the dark too, including the ability to increase the serum melatonin level in humans.[12] This property has been suggested to be valuable for overcoming the effect of jet lag, but no one has yet shown that the treatment is effective and does not cause skin photosensitization in travelers.

10.6. *CIS–TRANS* ((Z)–(E)) ISOMERIZATION OF ALKENES IN NATURE

As described in Chapter 3, alkenes readily undergo (E)–(Z) isomerization in the presence of light. The isomerization of urocanic acid and its importance in immunology are mentioned in Sections 10.1 and 10.7. The (E)–(Z) isomerization of the double bond present between positions 11 and 12 in retinal is a critical step in the phenomenon of vision. Isomerization of the double bond between positions 13 and 14, on the other hand, is essential in controlling the phototactic and photophobic responses of flagellated microorganisms.[13] Other important application of (E)–(Z) isomerization of alkenes include medical applications, to be discussed in Chapter 11. The treatment of hyperbilirubia in infants, for example, is based on such a photochemical isomerization. In this case, one diastereomer is metabolized faster and has physiological properties quite different from the others. The position of the photoequilibrium can therefore be shifted until the concentration of the toxic diastereomer is brought down to an acceptable level.

Usually, a photoequilibrium is established upon irradiation of one alkene, but any further chemical reactions of one of the diastereomers obviously shift the equilibrium. A large number of coumarins have been isolated from plants, and their biosynthesis includes a photoisomerization, from (E) to (Z), which allows lactonization to proceed. Here, the end of the biosynthesis of the dihydroxycoumarin (**3**) from caffeic acid is shown.

(**3**)

An interesting example of biological activity due to (E)–(Z) isomerization where there is no change in the isomer concentration is kryptocyanin (**4**), useful in the photosensitized treatment of tumors. Usually, drugs used for such treatments are powerful photodynamic compounds, and it is the $^1O_2$ (and/or perhaps $O_2^{\cdot-}$) which eventually kills the cells in which the sensitizer is electronically excited. Photoexcited kryptocyanin, however, does not populate its triplet state appreciably and, therefore, does not produce $^1O_2$. The only efficient photoprocess involves internal conversion from the first singlet excited state to the ground state. This non-radiative process, which involves dissipation of 165 kJ/mol, has been postulated to simply transfer heat to the surroundings, in an amount sufficient to disrupt the cell.[14]

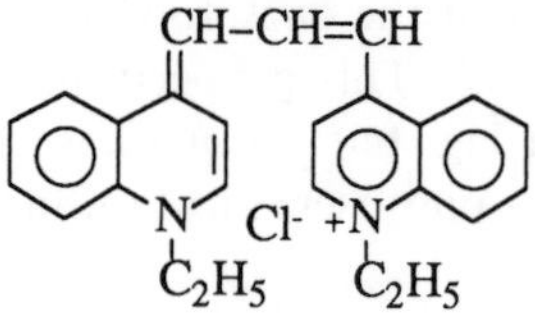
CH-CH=CH
N
Cl- +N
C2H5
C2H5

(4)

10.7. UROCANIC ACID AND THE IMMUNE SYSTEM

The isomerization of urocanic acid (**1**) requires UVB irradiation, which is independently known to suppress cell-mediated immunity; *cis*-urocanic acid administered to mice has been shown to induce the same effects. Both urocanic acid and histamine are enzymatically derived from histidine. Both histamine and *trans*-urocanic acid induce $3', 5'$-cyclic monophosphate *in vitro*, in skin fibroblast cell cultures, but *cis*-urocanic acid does not. However, it inhibits the adenosine $3', 5'$-cyclic monophosphate-inducing effect of both its *trans* isomer and histamine, by uncoupling the receptor from its signaling system.[15]

10.8. [2+2] CYCLOADDITION REACTIONS OF ALKENES IN PLANTS

Several cyclobutane derivatives have been isolated from plants, along with the alkenes known to be their precursors in photochemical reactions. Although the co-occurrence of precursors and potential photoproducts does not necessarily prove that there is a light-dependent step in the biosynthesis, the implication is very strong. For example, truxillic acid (**7**) and truxinic acid (**6**), known photodimers of cinnamic acid (**5**), are also natural products (a number of diastereomers are known). The structure of truxillic acid was actually published in 1889, one year before that of cinnamic acid. Likewise, cordopatine (**8**) and isocordopatine (**9**), found in *Echynops* plants (family Asteraceae) coexist with a photochemical precursor.[16]

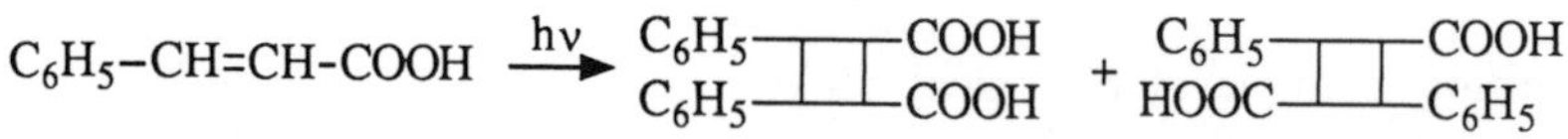
C6H5–CH=CH-COOH
hν
C6H5
C6H5
COOH
COOH
+
C6H5
HOOC
COOH
C6H5

(5) (6) (7)

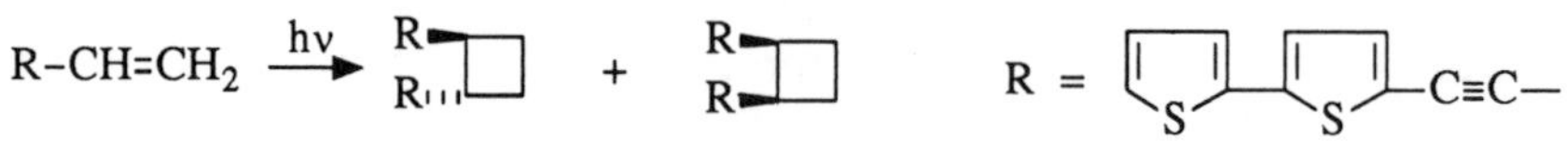
R-CH=CH2
hν
R
R
+
R
R
R =
S
S
C≡C–

(8) (9)

10.9. PSORALENS AND THEIR BIOLOGICAL ACTIVITY IN NATURE

Psoralens are phototoxic against many organisms, ranging from microorganisms to insects, fish, and mammalian species. The presence of phototoxic psoralens in many plants has interesting ecological consequences since it may determine how these plants interact with other organisms. For example, cattle and sheep suffer from photosensitization when they feed on *Ammi* plants.[17] Likewise, *Pituranthos triradiatus* is a plant which is rarely grazed. In a test where leaves were offered to starved hyraxes*, photosensitization in sunlight occurred quickly, and 80% of the animals died within a day. Young shoots, which contained very little psoralens, had a greater appeal to the animals.[19]

Of course, a plant rich in furocoumarins does not necessarily spell doom for a feeding organism, if this organism contains the proper enzymes for rapid detoxification or if it stays away from ultraviolet light. Some insects do have such enzymes. For example, the larvae of *Phytomyza sphondylii* (Diptera: Agromyzidae) feed on leaves of *Heracleum lanatum*, a plant from the family Apiaceae (Umbelliferae) which is rich in furanocoumarins. These larvae rapidly inactivate 8-methoxypsoralen.[20] The larvae of other insects may feed at night or roll themselves into plant leaves which filter harmful radiations.

10.10. PHOTOINSECTICIDES

Photoinsecticides have been mentioned in discussing the phototoxicity of psoralens and α-terthienyl. Many dyes have insecticide activity, particularly against larvae,[21] but only one is approved in the USA by the Environmental Protection Agency. It is erythrosin B (**10**), a xanthene dye also used as a food additive, which is now used for controlling houseflies in poultry facilities.

(10)

A newer approach to the design of insecticides has great similarities with the photodynamic herbicides described below. Treatment of insect larvae with δ-aminolevulinic acid and 2,2′-dipyridyl leads to the accumulation of divinyl protoporphyrin IX, which kills larvae in the presence of light by a photodynamic process and which is also toxic in the dark. Alternatively, insect larvae can be killed after direct treatment by a protoporphyrin.[22] Some of the expected advantage os these 'porphyric insecticides' include their biodegradability and the presumption that insects will not easily develop resistance to them. The fact that the same compounds have both insecticide and herbicide activity suggests that difficulties may arise in designing practical applications.

---

*A reader who is not an expert in zoology might be interested to learn that a hyrax is a mammal with peculiar teeth: the molars resemble those of the rhinoceros and the incisors those of rodents. The lorax described in Dr. Seuss' famous book of the same name has been rumored to be a close relative of the hyrax.[18]

10.11. PHOTOSENSITIZING PORPHYRINS AS HERBICIDES

Several porphyrin intermediates of heme or chlorophyll biosynthesis are powerful photosensitizers capable of generating singlet oxygen. Interference with the enzymatic mechanisms which control these syntheses in plants allows the accumulation of porphyrins which may act as herbicides in the presence of light. The concept behind the development of tetrapyrrole-dependent photodynamic herbicides involves use of a porphyrin biosynthetic precursor (δ-aminolevulinic acid) combined with one of several chemicals (called modulators), which alter the pattern of tetrapyrrole accumulation. Different classes of modulators have been recognized.[23] In the case of 2,2′-dipyridyl, a heme and chlorophyll pathway inhibitor, the accumulation of porphyrin has a strong phototoxic effect. Exogenous porphyrins are much less active.[24]

The herbicidal effect of other organic molecules may be based on different mechanisms. For example, 1,10-phenanthroline is a potent photosensitizing molecule, producing oxygen-dependent redox reactions of complexed metal ions leading to the production of the superoxide anion. It is also a potent photodynamic herbicide modulator which induces the accumulation of large amounts of tetrapyrroles, whether or not δ-aminolevulinic acid is also added.[25]

Another approach in the design of herbicides consists of weakening the natural defense mechanism with which plants protect themselves from photodynamic damage. Thus, preventing plants from synthesizing the usual carotenoid pigments which are powerful quenchers of singlet oxygen will increase the lethality of this agent normally produced by endogenous sensitizers. 'Bleaching herbicides', such as norflurazon (**11**) or oxidiazon (**12**), do just that, by inhibiting the enzyme phytoene desaturase.[26]

(**11**)   (**12**)

10.12. ULTRAWEAK BIOLOGICAL PHOTOEMISSION

Most organisms spontaneously emit extremely weak light. This is known as ultraweak photoemission, biological chemoluminescence, or biophoton emission.

Mechanistic studies have been performed on different plant tissues. For example, two-dimensional emission patterns of ultraweak light generated by injured soybean seedlings have been recorded with a photon counting system, accurately determining the wounded regions of the seedling, which exhibit stronger emission. The addition of H_2O_2 strongly enhanced the emission.[27] The light emission of isolated spinach chloroplasts, which had been adapted to the dark, was oxygen-dependent. It was inhibited by superoxide dismutase, suggesting that a transformation of O_2 into O_2^- was an important part of the process. Light emission was also decreased by inhibitors of respiration and suppressed by benzoquinone, but it was enhanced by NADPH. A series of redox reactions in the respiratory chain of chloroplasts was suggested to produce energy transfer to and light emission from chlorophyll molecules.[28]

The measurement of ultraweak biological emissions could not have been made without modern advances in electronics. The principle, however, must have been known at least two centuries earlier, since Jonathan Swift was able to describe in *Gulliver's Travels* the research efforts of one member of the academy of Lagado (on the island of Balnibarbi) who was working on reverse photosynthesis. This scientist was extracting sunbeams out of cucumbers. Although a serious engineering problem still remains to be solved, it now turns out that Swift was right, and cucumber seedlings do actually emit light.[29]

There is an interesting feature in this emission: light piping apparently takes place within the cucumber seedlings, with the implication that ultraweak photon emission could be involved in information transfer in plants.

10.13. GENERATION OF ELECTRONICALLY EXCITED MOLECULES WITHOUT LIGHT: BIOLOGICAL APPLICATIONS

The thermal decomposition of 1,2-dioxetanes has long been known to produce two carbonyl products, both in the singlet state, but one of them electronically excited. The return to the ground state of this excited molecule may be accompanied by luminescence.

Luciferin (**13**) is produced by several organisms, including the firefly, and is converted by the enzyme luciferase into the dioxetane (**14**), which, upon decomposition, yields a thiazole which luminesces. The color observed depends strongly on the reaction conditions, particularly the pH.[30] Thus, yellow-green light is observed at pH 7.9, which allows the formation of a mono- or dianionic species from the ketone generated upon fragmentation of the dioxetanone. Red light is generated at pH 6.

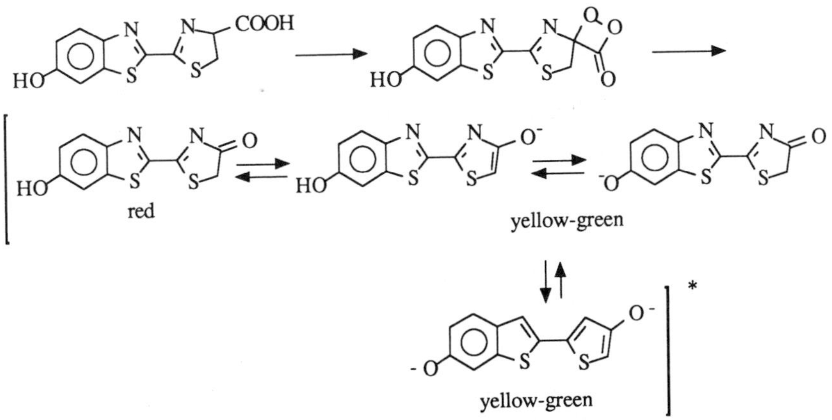

Bioluminescent reactions have been observed in several different kinds of organisms, including bacteria, fungi, insects, fishes, molluscs, medusae, crustaceans, and tunicates. Not all bioluminescence is generated by the organisms in which it observed. For example, fish use symbiotic bacteria in their luminous organs.

Luminol (**15**) is among the best known sources of luminescent chemicals. It can be oxidized, for example by oxygen in alkaline solution, to produce light emission from the 3-aminophthalate ion (16) generated.

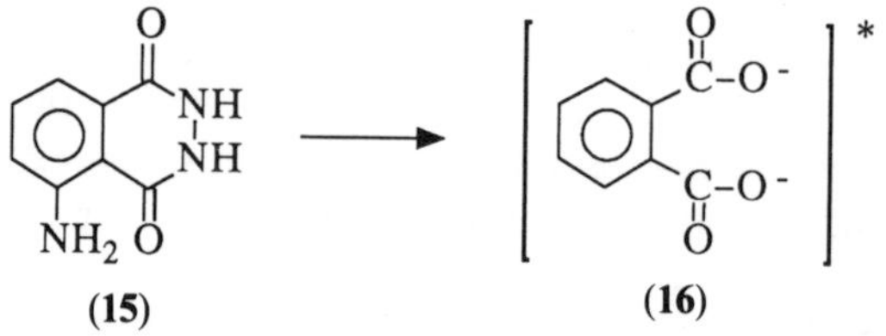

(15)   (16)

The stability of dioxetanes strongly depends on the nature of their substituents. Many water-soluble dioxetanes have been used as reporter molecules in a variety of biological analytical systems, including enzyme-linked immunoassays, nucleic acid probe techniques, and structural determinations. For example, phosphate esters of the type (17) are relatively stable thermally, but the phenols obtained by hydrolysis are not. The detection of fluorescence in the aromatic ester (18) generated upon cleavage of the dioxetane can be a very sensitive method for analyzing phosphatase activity, using (17) as the enzyme substrate. It is even possible to combine the assay with an immunoassay recognition step.[31] Incorporated into a biochemical method using luciferase for detecting DNA, as little as 0.19 pg of plasmid DNA could be measured.[32]

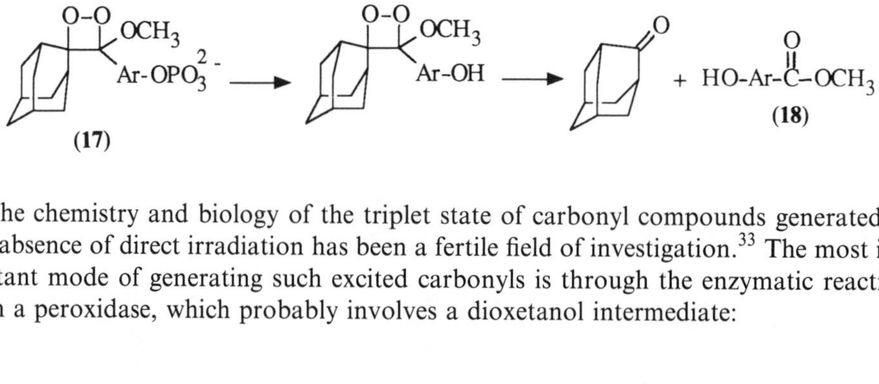

(17)   (18)

The chemistry and biology of the triplet state of carbonyl compounds generated in the absence of direct irradiation has been a fertile field of investigation.[33] The most important mode of generating such excited carbonyls is through the enzymatic reaction with a peroxidase, which probably involves a dioxetanol intermediate:

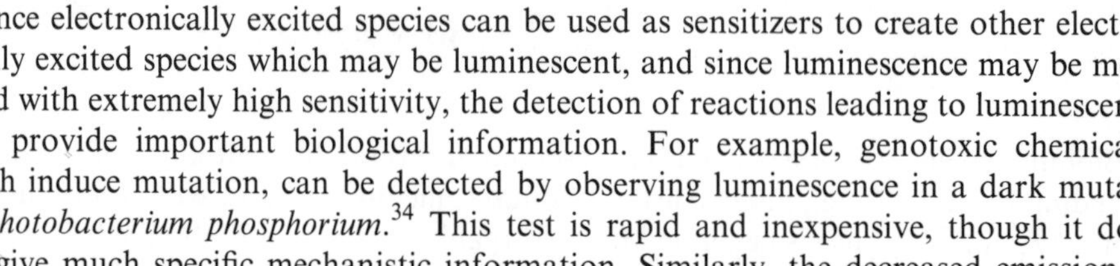

Since electronically excited species can be used as sensitizers to create other electronically excited species which may be luminescent, and since luminescence may be measured with extremely high sensitivity, the detection of reactions leading to luminescence may provide important biological information. For example, genotoxic chemicals, which induce mutation, can be detected by observing luminescence in a dark mutant of *Photobacterium phosphorium*.[34] This test is rapid and inexpensive, though it does not give much specific mechanistic information. Similarly, the decreased emission of light from the *Escherichia coli* containing genes encoding luciferase from the firefly in the presence of the substrate luciferin can be used to detect pollutants.[35] Finally, the generation of an electronically excited triplet state in the dark under natural conditions was observed with the α-oxidase system of young pea leaves. When a long-chain fatty acid was added, both oxygen uptake and chlorophyll luminescence were observed in fractions containing chloroplasts.[36]

10.14. PHOTOINDUCED MUTAGENESIS

Many assays are available for determining mutagenicity, and it is therefore easy to assess whether a photochemical treatment of a sample changes its mutagenicity. Polycylic aromatic hydrocarbons are well known for being mutagenic, a property which requires metabolic activation. However, a typical polycyclic aromatic hydrocarbon such as benzo[a]pyrene is very phototoxic under conditions where it is not obvious that metabolic activation is playing a significant role. For example, it can kill mosquito larvae almost instantaneously in the presence of ultraviolet light at concentrations which do not significantly affect the development of the larvae into adult mosquitoes in the dark,[37] and it rapidly inactivates *E. coli* upon irradiation.[38] Oxygen dependence of the phototoxicity was demonstrated in this organism. Benzo[a]pyrene can sensitize the formation of singlet oxygen as well as the superoxide ion. Singlet oxygen, generated in the gas phase, was shown to increase the mutation frequency induced by benzo[a]pyrene in *Salmonella typhimurium* bacteria.[39] It is interesting that singlet oxygen-induced mutagenesis required that at least one double bond be present in the ring, which is usually transformed into diol epoxides by metabolic activation. Whether this double bond was converted into a dioxetane or epoxide was not established.

10.15. MUTAGENICITY OF DIOXETANES

As mentioned in Chapter 6, the reaction of singlet oxygen with benzofurans produces dioxetanes which are highly mutagenic in *S. typhimurium* strain TA 100. The damage cannot be reversed by irradiation, suggesting that pyrimidine dimers are not involved, but DNA alkylation has been suggested.

The mutagenicity of polycyclic aromatic hydrocarbons such as benzo[a]pyrene is well known to come from metabolic activation which produces diol epoxides. Although this activation requires no light, a parallel has been drawn between this biochemistry and the photochemically generated mutagenesis of the benzofurans. Here, a dioxetane (**19**), formed photochemically is converted into an epoxide, (**20**), by deoxygenation with a sulfide, and this epoxide is mutagenic (the epoxide can also be generated independently by a chemical reaction). This route from an unsaturated substrate to a dioxetane and then to an epoxide has been suggested to have general biological significance.[40] Whether or not it applies to the case of benzo[a]pyrene discussed above is not known.

10.16. PHOTOSENSITIZATION WITH α-TERTHIENYL

In contrast to psoralens, which have important photosensitizing properties, and for which there is an important body of mechanistic work at the molecular level, many powerful photosensitizers have practically unknown targets. The natural product α-terthienyl (**21**) is an example. First isolated from marigold flowers (a time-honored folk remedy against insects), but later found in many plants (particularly in roots) of the family Asteraceae (Compositae), α-terthienyl is toxic to a variety of organisms in the presence of ultraviolet light. These include bacteria, yeasts, nematodes, eggs and immature insects, aquatic organisms including fish, and plants.[41] Individual cell systems and components are also affected, such as human skin, red blood cells, enzymes and other proteins, and DNA. However, there have been almost no data on the molecular basis for the unusually high phototoxicity of this compound, beyond the measurement of a high quantum yield for singlet oxygen formation and a low quantum yield for superoxide ion production. In one exception, α-terthienyl and 2,2′-bithiophene were found to induce peroxidation of squalene and methyl linoleate, which are model systems for lipids.[42] However, not all the photobiochemistry of α-terthienyl depends on the presence of oxygen, as plasmid supercoiled DNA is more readily cleaved *in vitro* anaerobically.[43]

(**21**)

It is in insects that the phototoxic reactions of α-terthienyl have been investigated the most carefully, particularly in larvae and pupae of mosquitoes. The LC_{50} values recorded with several species were almost identical, as were those of strains resistant to other pesticides.[44,45] Localization of the chemical in some organs was observed by fluorescence micrography, and *in vivo* inactivation of specific enzymes such as acetylcholinesterase[46] and superoxide dismutase[47] was also ascertained. However, the mechanism of phototoxicity of α-terthienyl and its derivatives is far from being understood.

It is interesting to note that plants containing α-terthienyl were originally found to protect tulip bulbs from damage by nematodes in soil. When the toxicity of the chemical against nematodes *in vitro* was discovered to depend on a photochemical process, the question of the mechanism for toxic reactions in soil was raised. The most plausible explanation[48] proposed to date involves electronic excitation of α-terthienyl not by direct irradiation, but by energy transfer from a product created by decomposition of a dioxetanone formed by peroxidation of a suitable biological substrate, as described in Section 10.3.

10.17. TARGETING PHOTOTOXIC SENSITIZERS

The lack of target selectivity in the biological activity of all the photosensitizers used to date is very frustrating to those who would like to develop specific applications. Even the sensitizers used in the photodynamic therapy of cancers are far from specific, although they do eventually reach a higher concentration in cancerous cells than in normal cells. All the sensitizers studied specifically in one system have been found to be phototoxic in other systems as well. In many cases, it is because they generate singlet oxygen or other diffusible oxygen species which can damage most classes of biological molecules. The successful design of efficient targeting procedures is a formidable challenge which must be met in order to extend the use of photosensitizers in medicine and biology.

Antigen–antibody interactions are extremely specific. Therefore, delivering to an antigen an antibody carrying a non-toxic moiety which can be photochemically activated after the target is reached promises to be a method for very selective photodynamic therapy. Encouraging results have been described with monoclonal antibody–photosensitizer conjugates of hematoporphyrin against an antigen on the surface of human carcinoma lines.[49]

10.18. SKIN PHOTOSENSITIZATION BY DRUGS

Many important drugs lead to photosensitization when patients are exposed to light. It is often difficult to provide a detailed mechanistic explanation for these phenomena, which might be due to photochemical reactions initiated by the drugs themselves, their photoproducts, their metabolites, or photoproducts derived from these. A detailed study of the photochemistry of the drugs themselves is a necessary first step in order to understand the biological processes. For example, tetracycline antibiotics (22) are well-known photosensitizers which also undergo photochemistry. Lumitetracyclines (23) are formed in high yields from tetracyclins. In the structure shown, R = H or CH_3, R_1 = H, Cl, or $N(CH_3)_2$, and $R_2 = R_3$ = H or OH. The mechanism of this reaction appears to be a nucleophilic displacement of the protonated dimethylamino group by the enol. The conversion occurs only in the absence of oxygen, which makes it of unlikely significance in biological processes, although lumitetracyclines have been reported to be phototoxic.[50] Additional photochemical and photophysical studies have been published.[51]

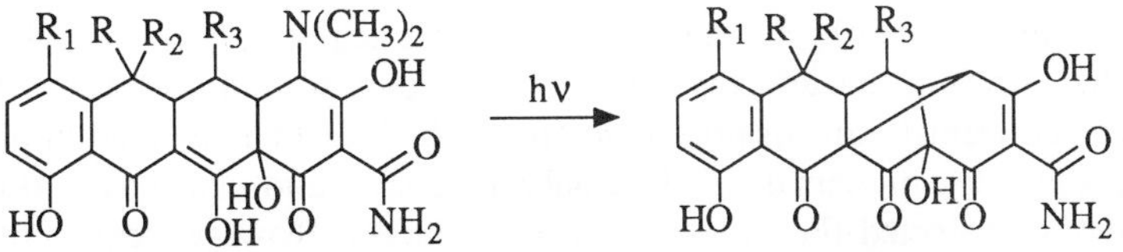

(22)   (23)

The phototoxicity of several tetracyclines has been studied, both *in vitro* and *in vivo*, suggesting the drug's ability to affect membranes, nucleic acids, and mitochondria; the photochemical production of singlet oxygen has been well characterized.[52]

Other known phototoxic drugs include non-steroid anti-inflammatory agents (NSAIDs), thiazide diuretics, and sulfanilamides.

Antimalarial drugs are often photolabile and phototoxic. Chloroquine (24), for example, gives at least seven photoproducts.[53]

Because determining phototoxic mechanisms at the cellular and molecular levels is a very difficult task, simple experiments, such as determining whether or not oxygen is involved, are often performed *in vitro*. Chlorpromazine, mequitazine, and aflaqualone, for example, show high quantum yields of singlet oxygen production: 0.27, 0.28, and 0.14, respectively.[54]

$$CH_3$$
$$NHCH(CH_2)_3N(C_2H_5)_2$$

Cl   N

(24)

10.19. PHOTOCHEMISTRY ON THE SKIN

10.19.1. Non-photosensitized reactions

Exposing the skin to sunlight results in skin roughening which is characteristic of certain outdoor professions, particularly farming. More serious changes are involved too, reflected in the fact that skin cancer is the most prevalent form of cancer in the USA. It has been stated that most of the 500 000 new cases of skin cancer diagnosed each year in the USA result from repeated skin exposure to sunlight, and that most non-melanoma human skin cancers are induced in this manner. Both UVB and UVA have generated skin cancer in animals, but there is very little knowledge of the mechanism(s) of ultraviolet carcinogenesis, which seems to be distinct from that induced by other carcinogens. At the cellular level, oncogenes and tumor suppressor genes have been implicated, but one major difficulty in these mechanistic studies is the long delay between ultraviolet light exposure and cancer appearance.[55]

Although a number of commercially available products offer some protection from solar ultraviolet light, sunscreens are often more effective in filtering UVB than UVA. One of their disadvantages is that they may lead to deficiency in natural formation of vitamin D (cf. Chapters 3 and 11).[56]

Reactive oxygen species formed through photosensitized processes are suspected of being responsible for photo-ageing of the skin, particularly since topical antioxidants and quenchers prevent this effect in mice. Chronic exposure of mice to UVB light leads to increasing concentrations of non-heme iron, which can be involved in photoinduced redox reactions (cf. Chapters 6 and 7). Chelating compounds, such as 1,10-phenanthroline, effectively protected the skin of irradiated mice.[57] However, the photobiology of the skin is extremely complex, and UVA and UVB affect the skin differently, particularly with respect to collagen modifications.[58]

Repair mechanisms play a critical role with respect to photodamage to genetic material, as demonstrated by comparing cells from normal persons with those from individuals suffering from xeroderma pigmentosum, a genetic disorder in which the mechanism for repair of pyrimidine dimers is defective. Skin cancers in these persons are extremely common.

Finally, it should be mentioned that some conditions, such as polymorphic light eruption and solar urticaria, are poorly understood but are clearly associated with exposure to sunlight.

10.19.2. Cutaneous response to light: pigmentation

The response of normal skin to sunlight, particularly to its ultraviolet component, is very complex.[59] It also varies considerably with individuals. the action spectra for erythema and pigmentation are similar, but irradiations with light of specific wavelengths effect each of these processes differently.[60] The criteria used to characterize the types of skin in Caucasian persons are as follows: type I—always burns, never tans; type II—always burns, tans poorly; type III—burns slightly, tans well; type IV—does not burn, always tans. The protecting agent and the dose to be received must be evaluated according to the person's skin type and the expected exposure to sunlight.

10.19.3. Photosensitized reactions

The photosensitization reactions induced by naturally occurring compounds such as psoralens were mentioned earlier. Several classes of chemicals are known to photosensitize either human skin upon simple contact, or animals which graze on plants containing such compounds. For example, photosensitization has been observed in cattle and sheep fed *Ammi majus*, the best known source of 8-methoxypsoralens and related compounds.[61]

Phytophotodermatitis is the term used to describe the skin inflammation produced by contact with plants and subsequent exposure to sunlight. Healing may produce long-lasting hyperpigmentation. The plants involved usually belong to the families Rutaceae or Umbelliferae, less frequently to Moraceae, Compositae, Leguminoseae, or Rosaceae. The clinical symptoms are described either as berloque dermatitis (see Section 10.19.6) or dermatitis bullosa striate pratensis (DBSP). An example of the latter was produced by *Angelica gigas*, which is an important plant used in Korean herb medicine because the root is 'an effective therapeutic agent for anemia, hemorrhoids and diuresis'.[62] Three members of a family tending an angelica garden were found to suffer from DBSP and to be particularly sensitive to the fruit of the plant.

10.19.4. Photoallergy

Topical photoallergy is usually associated with erythematous skin reactions. These are difficult to analyze quantitatively. It is also difficult to predict whether a given molecule will trigger photoallergic reactions. However, a number of chemicals which are known photoallergens have been used to suggest that an ear-swelling test on cyclophosphamide-treated mice could have predictive value.[63]

10.19.5. Skin protection

Wearing proper garments is the simplest method for protecting the skin from excessive exposure to sunlight. Because so many different types of textiles are commonly used, the spectral characteristics of each have been recorded and compared. The study concluded that (1) the protection provided by a given fabric depended more on the nature of the weave than on the particular type of textile, and (2) for maximum protection wool should be avoided, and tightly woven fabrics such as cotton, silk, or high-lustre polyesters were recommended. Clothing generally provides better protection than sunscreens against UVA.[64]

10.19.6. Contact photosensitization with perfumes

A skin condition known as berloque dermatitis is associated with the use on the face and neck of cologne, perfumes, and other cosmetic products which usually contain naturally occurring furanocoumarins. Less commonly, polycyclic aromatic hydrocarbons or other photosensitizers may also be involved. Exposure to sunlight results in phototoxic reactions on the skin that cause pigmentation which may last quite a long time.

10.19.7. Club Med dermatitis

In the absence of more newsworthy events, newspapers have frequently reported on a type of dermatitis prevalent among members of the Club Mediterranée. Vacationers who participate in games in which limes and other citrus fruits have been rubbed against their skin often suffer from the photosensitized effect of the furanocoumarin components. This is a variant of berloque dermatitis.

10.19.8. Bikini dermatitis

Dyes used in the textile industry have been implicated in phototoxic reactions in factory workers and in people who wear certain garments. Bathing suits induce a particularly strong reaction in sensitive persons because they are most often worn in sunlight. A dye such as disperse blue 35 is a mixture of at least 10 components, four of which can generate photocontact dermatitis and produce singlet oxygen *in vitro*. The structures of three of these four are shown ((25)–(27)), the fourth being a monomethylated derivative of one of the last two.[65] Compounds (26) and (27) are also capable of generating the superoxide anion, and it is logical to assume that at least one of these oxygen species is responsible for the skin photosensitization.

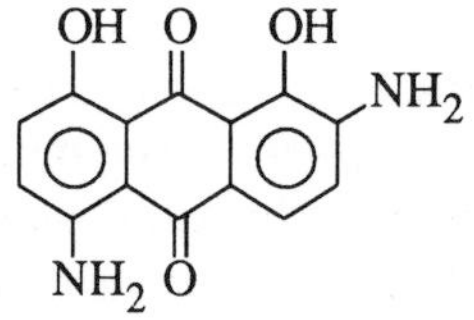

(25) 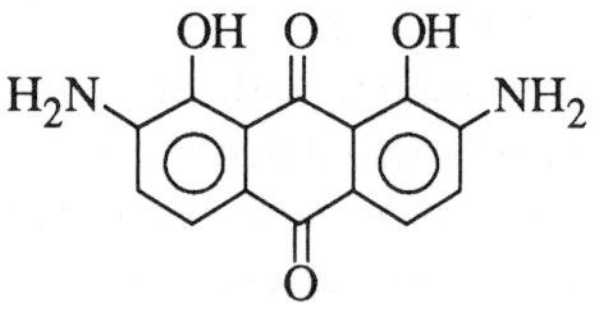   (26) 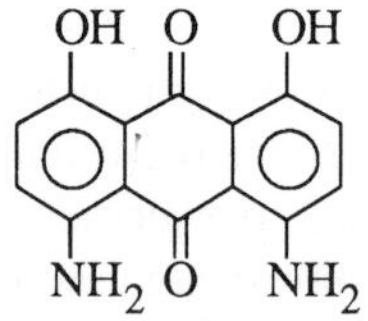   (27)

10.20. ENHANCEMENT OF DRUG CYTOTOXICITY BY PHOTOACTIVATION

Adriamycin and daunomycin and their derivatives (e.g. daunorubicin (28)) are important antitumor agents. Because they are cardiotoxic, their concentration in patients must be carefully limited. It is therefore interesting to note that, at least *in vitro*, the cytotoxicity could be greatly enhanced (by about five orders of magnitude in some cases) by laser irradiation at the absorption maximum of the drugs.[66]

A number of drugs are known to generate photosensitivity in patients, and individual drugs have occasionally been tested *in vitro* for phototoxicity. For example, quinolones such as cinoxacin, nalidixic acid, oxolinic acid, pipemidic acid, rosoxacin, ciprofloxacin, enoxacin, fleroxacin, and ofloxcacin were all found to induce hemolysis of human erythrocytes in the presence of light. At higher concentrations, norfloxacin even induced hemolysis in the absence of light. Avoiding exposure to intense light during treatment with these drugs is therefore recommended.[67] However, the photosensitization does not necessarily occur directly, with the drug itself absorbing light. Drugs such as steroids, antibiotics, hypnotics, and sedatives are known to induce photosensitivity through increased production of uroporphyrin.[68]

10.21. PROTECTION FROM PHOTOTOXIC REACTIONS

Three rational approaches can avoid phototoxic reactions in an organism: (1) prevent light from reaching the cells; (2) quench the electronically excited sensitizers; and (3) quench singlet oxygen or deactivate the superoxide ion produced in photodynamic reactions.

The first approach is followed by nocturnal organisms, or those which can protect themselves from sunlight, such as leaf roller insects. Since most phototoxic agents operate in an oxygen-rich environment in nature, it is quite difficult to distinguish between the quenching of singlet oxygen and that of the excited sensitizer which led to its formation. A number of studies have suggested protection by β-carotene in the diet, although the effect may be very slow and therefore difficult to associate with specific chemical steps. In the case of mice fed β-carotene in their diet, the development of tumors following irradiation with UVB appeared to be much reduced compared to controls; 13-*cis*-retinoic acid provided no such protection.[69] Neither compound, on the other hand, prevented skin ageing resulting from exposure of hairless mice to UVB.[70]

The photoprotective function of structural carotenoids in animals and plants is well known.[71] Another impressive example was provided by *E. coli* genetically engineered to produce carotenoid pigments. Unlike their colorless precursors, the modified carotenoid-producing bacteria were protected from the powerful phototoxic effect of α-terthienyl (Fig. 10.1).[72]

10.22. MISCELLANEOUS PHOTOBIOLOGICAL EFFECTS: PHOTOPHOBIC RESPONSE

The blue-green ciliate *Stentor coeruleus* is sensitive to both the direction and to the intensity of light. When irradiated from one direction, the organism quickly swims away from the light source, even when this requires it to go into an area of greater illumination. The action spectrum correlates with the absorption spectrum of its pigment stentorin.[73]

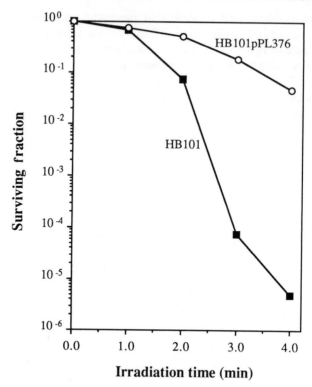

Fig. 10.1. Survival of two strains of HB 101 *E. coli* irradiated in the same conditions, showing the protection by the carotenoid pigments expressed in the strain HB101pPL376.

10.23. CONCLUSION

Many biological processes occur in response to the presence of sunlight and are critically dependent on the presence of certain wavelengths. Unfortunately, the molecular basis for many of the transformations is unknown or very poorly known at best. The small sampling of photochemical processes presented in this chapter has been arbitrarily selected, and it should encourage readers to seek additional information in the literature.

REFERENCES

1. K. D. Olson, C. W. McMahon and R. S. Wolfe, *Proc. Natl Acad. Sci. USA*, **88,** 4099–4103 (1991).
2. M. G. Strakhovskaya, N. S. Belinikina and G. Y. Fraikin, *Mikrobiologiya*, **60,** 292–297 (1991).
3. G. Kelfkens, F. R. de Gruijl and J. C. van der Leun, *Photochem. Photobiol.*, **52,** 819–823 (1990).
4. E. C. De Fabo, F. P. Noonan and J. E. Frederick, *Photochem. Photobiol.*, **52,** 811–817 (1990).
5. M. L. Kripke, *Photochem. Photobiol.*, **52,** 919–924 (1990).
6. C. S. Foote, *Photochem. Photobiol.*, **54,** 659 (1991).
7. T. P. Wang, J. Kagan, S. Lee and T. Keiderling, *Photochem. Photobiol.*, **52,** 753–756 (1990).
8. R. W. Tuveson, G. R. Wang, T. P. Wang, I. A. Kagan, J. Kagan and J. Lam, *Photochem. Photobiol.*, **55,** 63–73 (1992).
9. D. P. Valenzano, J. Trudgen, A. Hutzenbuhler and M. Milne, *Photochem. Photobiol.*, **46,** 985–990 (1987).
10. T. P. Wang and J. Kagan, *Chemosphere*, **19,** 1345–1348 (1989).
11. D. J. Kennaway and C. F. Van Dorp, *Am. J. Physiol.*, **260** (6, pt 2), R1137–R1144 (1991).
12. L. J. Grota, A. J. Lewy, L. A. Godsmith and G. M. Brown, *Ann. NY Acad. Sci.*, **453,** 385–387 (1985).
13. T. Takahashi, K. Yoshihara, M. Watanabe, M. Kubota, R. Johnson, F. Derguini and K. Nakanishi, *Biochem. Biophys. Res. Commun.*, **178,** 1273–1279 (1991).
14. A. Harriman, G. Luengo and K. S. Gulliya, *Photochem. Photobiol.*, **52,** 735–740 (1990).

15. E. W. Palaszynski, F. P. Noonan and E. C. de Fabo, *Photochem. Photobiol.*, **55**, 165–171 (1992).
16. J. Lam, L. P. Christensen and T. Thomasen, *Phytochemistry*, **30**, 1157–1159 (1991).
17. J. W. Dollahite, R. L. Younger and G. O. Hofman, *A. J. Vet. Res.*, **39**, 193–198 (1978).
18. I. A. Kagan, Personal communication.
19. D. Ashkenazy, Y. Kashman, A. Nyska and J. Friedman, *J. Chem. Ecol.*, **11**, 231–239 (1985).
20. M. J. Ashwood-Smith, R. A. Ring, M. Liu, S. Phillips and M. Wilson, *Can. J. Zool.*, **62**, 1971–1976 (1984).
21. J. R. Heitz, *ACS Symp. Ser.*, **339**, 1–21 (1987).
22. C. A. Rebeiz, J. A. Juvik and C. C. Rebeiz, *Pest. Biochem. Phys.*, **30**, 11–27 (1988).
23. C. A. Rebeiz, K. N. Reddy, U. B. Nandihalli and J. Velu, *Photochem. Photobiol.*, **52**, 1099–1117 (1990).
24. S. O. Duke, J. M. Becerril, T. D. Sherman and H. Matsumoto, *ACS Symp. Ser.*, **449**, 371–386 (1991).
25. U. B. Nandihalli and C. Rebeiz, *Pest. Biochem. Physiol.*, **40**, 27–46 (1991).
26. G. Sandmann, I. E. Clarke, P. M. Bramley and P. Boeger, *Z. Naturforsch., C: Biosci.*, **39C**, 443–449 (1984).
27. S. Suzuki, M. Usa, T. Nagoshi, M. Kobayashi, N. Watanabe and H. Inaba, *J. Photochem. Photobiol. B.*, **9**, 211–217 (1991).
28. E. Hideg and H. Inaba, *Photochem. Photobiol.*, **53**, 137–142 (1991).
29. B. Schauf, L. M. Repas and R. Kaufmann, *Photochem. Photobiol.*, **55**, 287–291 (1992).
30. E. H. White and D. F. Roswell, *Photochem. Photobiol.*, **53**, 131–136 (1991).
31. M. Okada, Y. Ashihara, T. Ninomiya and A. Yano (Fujirebio. Inc., Japan), Chemiluminescence EIA using phosphatase as label and dioxetane derivatives as substrates. *Jpn Kokai Tokkyo Koho JP 01,331,272*, 15 December 1989.
32. H. Watanabe and S. Shibata, Detection of DNA probes using luciferase. *Eur. Pat. Appl. EP 386,691*, 12 September 1990.
33. G. Cilento and W. Adam, *Photochem. Photobiol.*, **48**, 361–368 (1988).
34. E. Elmore and M. P. Fitzgerald, *Progr. Clin. Biol. Res.*, **340D**, 379–387 (1990).
35. S. Lee, M. Suzuki, E. Tamiya and I. Karube, *Anal. Chim. Acta*, **244**, 201–206 (1991).
36. M. Salim-Hanna, A. Campa and G. Cilento, *Photochem. Photobiol.*, **45**, 849–854 (1987).
37. J. Kagan and E. D. Kagan, *Chemosphere*, **15**, 243–251 (1986).
38. J. Kagan, R. W. Tuveson and H.-H. Gong, *Mutation Res.*, **216**, 221–242 (1989).
39. J. Seed, L. Seed, K. G. Specht, T. A. Dahl and W. R. Midden, *Photochem. Photobiol.*, **50**, 625–632 (1989).
40. W. Adam, L. Hadjiarapoglou, T. Mosandl, C. R. Saha-Möller and D. Wild, *J. Am. Chem. Soc.*, **113**, 8005–8011 (1991).
41. J. Kagan, *Progr. Chem. Org. Nat. Prod.*, **56**, 87–169 (1991).
42. M. Sasaki, K. Towika, H. Mori, T. Sakata and H. Fujita, *Photomed. Photobiol.*, **12**, 177–178 (1990).
43. T. P. Wang, J. Kagan, R. W. Tuveson and G. R. Wang, *Photochem. Photobiol.*, **53**, 463–466 (1991).
44. B. M. Hasspieler, J. T. Arnason and A. E. R. Downe, *J. Am. Mosq. Contr. Ass.*, **4**, 479–484 (1988).
45. D. Borovsky, J. R. Lindley and J. Kagan, *J. Am. Mosq. Contr. Ass.*, **3**, 246–250 (1987).
46. J. Kagan, M. Hasson and F. Grynspan, *Biochem. Biophys. Acta*, **802**, 442–447 (1984).
47. M. Nivsarkar, G. P. Kumar, M. Laloraya and M. M. Laloraya, *Arch. Insect Biochem. Phys.*, **16**, 249–255 (1991).
48. F. J. Gommers and J. Bakker, *Bioactive Molecules*, **7**, 61–69 (1988).
49. D. He, S. Taniushi, C. H. Sun, M. W. Berns and R. D. Cardiff, *Antibody, Immunoconjugates, Radiopharm.*, **3**, 199–211 (1990).
50. G. Olack and H. Morrison, *J. Org. Chem.*, **56**, 4969–4971 (1991).
51. H. Morrison, G. Olack and C. Xiao, *J. Am. Chem. Soc.*, **113**, 8110–8118 (1991).
52. I. E. Kochevar, *Photochem. Photobiol.*, **45**, 891–895 (1987).
53. K. Nord, J. Karlsen and H. H. Toennesen, *Int. J. Pharm.*, **72**, 11–18 (1991).
54. T. Arai, Y. Nishimura, M. Sasaki, H. Fujita, I. Matsuo, H. Sakuragi and K. Tokumaru, *Bull. Chem. Soc. Japan*, **64**, 2169–2173 (1991).
55. H. N. Ananstaswamy and W. E. Pierceall, *Photochem. Photobiol.*, **52**, 1119–1136 (1990).
56. M. F. Holick and W. Raab, *TW Dermatol.*, **20**, 483–485 (1990).
57. D. L. Bissett, R. Chatterjee and D. P. Hannon, *Photochem. Photobiol.*, **54**, 215–223 (1991).
58. L. H. Kligman and M. Gebre, *Photochem. Photobiol.*, **54**, 233–237 (1991).
59. J. L. M. Hawk and J. A. Parrish, In J. D. Regan and J. A. Parrish (eds.) *The Science of Photomedicine*, pp. 219–260. Plenum Press, New York (1982).
60. R. W. Gange, Y.-K. Park, M. Auletta, N. Kagetsu, A. D. Blackett and J. A. Parrish, In F. Urbach and R. W. Gange (eds.) *Biological Effects of UVA Radiation*, pp. 57–65. Praeger, New York (1986).
61. J. W. Dollahte, R. L. Younger and G. O. Hoffman, *Am. J. Vet. Res.*, **39**, 193–198 (1978).
62. S. K. Hann, Y.-K. Park, S. Im and S. W. Byun, *Photodermatol. Photoimmunol. Photomed.*, **8**, 84–85 (1991).
63. G. F. Gerberick and C. A. Ryan, *Food Chem. Toxicol.*, **28**, 361–368 (1990).
64. J. Robson and B. Diffey, *Photodermatol. Photoimmunol. Photomed.*, **7**, 32–34 (1990).

65. R. Dabestani, K. J. Reszka, D. G. Davis, R. H. Sik and C. F. Chignell, *Photochem. Photobiol.*, **54,** 37–42 (1991).
66. A. Andreoni, A. Colasanti, V. Malatesta, P. Riccio and G. Roberti, *Photochem. Photobiol.*, **53,** 797–805 (1991).
67. B. Przybilla, A. Georgii, T. Bergner and J. Ring, *Dermatologica*, **181,** 98–103 (1990).
68. L. C. Harber, I. E. Kochevar and A. R. Shalita, In J. D. Regan and J. A. Parrish (eds.) *The Science of Photomedicine*, pp. 323–347. Plenum Press, New York (1982).
69. M. M. Mathews-Roth, *Oncology*, **39,** 33–37 (1982).
70. L. H. Klingman and M. M. Mathews-Roth, *Photochem. Photobiol.*, **51,** 733–735 (1990).
71. O. H. Will, III and C. A. Scovel, In N. I. Krinsky, M. M. Mathews-Roth and R. F. Taylor (eds.) *Carotenoids, [Proc. Int. Symp. Carotenoids], 8th 1987*, pp. 229–236. Plenum Press, New York (1989).
72. R. W. Tuveson, R. A. Larson and J. Kagan, *J. Bacteriol.*, **170,** 4675–4680 (1988).
73. D. C. Wood and R. Marinelli, *Photochem. Photobiol.*, **53,** 353–357 (1991).

11. Photochemistry and medicine

11.1. VITAMIN D AND LIGHT

Rickets was a disease brought about by the industrial revolution. As the population became more urbanized, people spent less time outdoors. Children were most visibly affected, suffering from muscle weakness and bone deformity. Eventually, two remedies were found: exposure to sunlight or ingestion of cod liver oil. The first method (without excess) is still excellent. Most readers will not appreciate how fortunate they are that the second method is no longer in vogue. The taste of the oil is still vivid in the author's memory, more than half a century after a single experience with it. Now, milk and other food are frequently enriched with vitamin D, in the USA and many other countries.

The photochemical transformations leading from 7-dehydrocholesterol to vitamin D_3 were discussed in Chapter 3. Wavelengths between 260 and 315 nm can excite the precursor in the epidermis. The thermal isomerization of previtamin D_3 into vitamin D_3, which is slow, is followed by binding to a protein which ensures its circulation. Melanin pigments formed in the skin limit the production of vitamin D in response to excessive exposure to ultraviolet light.[1] Studies with healthy adults totally deprived of sunlight (as would be the case in long submarine voyages, for instance) have proved that inadequate intestinal absorption of calcium and increasingly negative calcium balance are serious after 5–6 weeks. Deficiency in vitamin D (which is really a hormone), on the other hand, can be remedied only by increased exposure to effective ultraviolet radiation and/or dietary intake. Recent data suggest that a significant proportion of the elderly population in the USA is deficient in vitamin D.[2] Note that although sunlight can provide excitation of 7-dehydrocholesterol between 290 and 315 nm, artificial lamps used indoors generally have no emission within this range, which should be of some concern to society.[3] Of equal importance, perhaps, is the observation that short wavelength ultraviolet radiation does not reach the earth with equal intensity throughout the world and throughout the year. Seasonal and geographical differences are very great, and previtamin D_3 photosynthesis may essentially cease in many places for significant periods of time.[4] The activity of the enzyme which produces the vitamin D hormone decreases with advancing age, and inclusion of vitamin D or derivatives has been recommended in the treatment of osteoporosis, a disease in which calcium absorption is insufficient.[5] However, this treatment is still controversial.

11.2. TREATMENT OF HYPERBILIRUBINEMIA IN NEWBORNS

Newborn babies, particularly those born prematurely, are often jaundiced because they have difficulty excreting bilirubin formed through metabolism of heme. Bilirubin is toxic, and high concentrations can induce brain damage or even death. Treatment of neonatal hyperbilirubinemia by visible light has been performed on perhaps millions of infants, making it the most common use of photochemistry in medicine.

Historically, two observations are said to have led to the development of this photo-therapy: bilirubin was found to bleach in a sample left exposed to sunlight, and a nurse noticed that jaundiced babies who were closer to the windows in a nursery recovered more rapidly.[6]

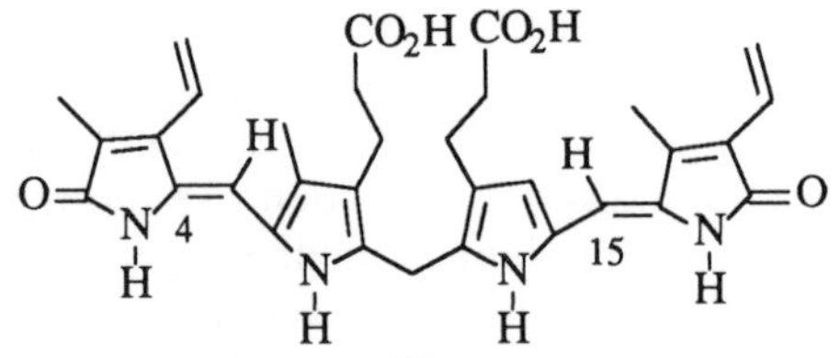

(1)

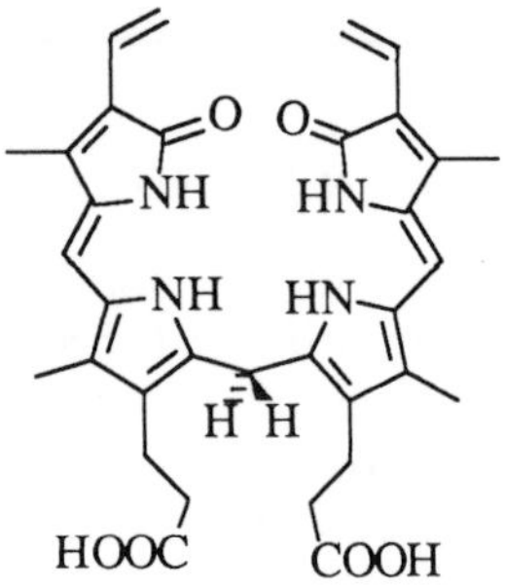

There are four bilirubin diastereomers, $(4Z, 15Z)$-bilirubin (1) being the native one. Each may have many conformations; the conformation shown here is reminiscent of a porphyrin. The naturally occuring bilirubin undergoes (E)–(Z) isomerization at each double bond upon irradiation; *in vitro*, a photoequilibrium is established. It is believed that a reason for the efficacy of the photochemical treatment in babies is that at least one of the photoisomers is more readily excreted; the photoequilibrium is therefore dis-placed to the point where the concentration of bilirubin drops below a critical thresh-old. In time, the infants develop the metabolic capability to cope with the problem and no longer need irradiation.[7] Actually, bilirubin itself is too insoluble and must be conju-gated with glucuronic acid in order to be excreted. Impaired formation or excretion of glucuronides may lead to hyperbilirubinemia. Gunn rats, which are unable to glucuronate bilirubin, are valuable models for studying the chemistry of this phototherapy *in vivo*.

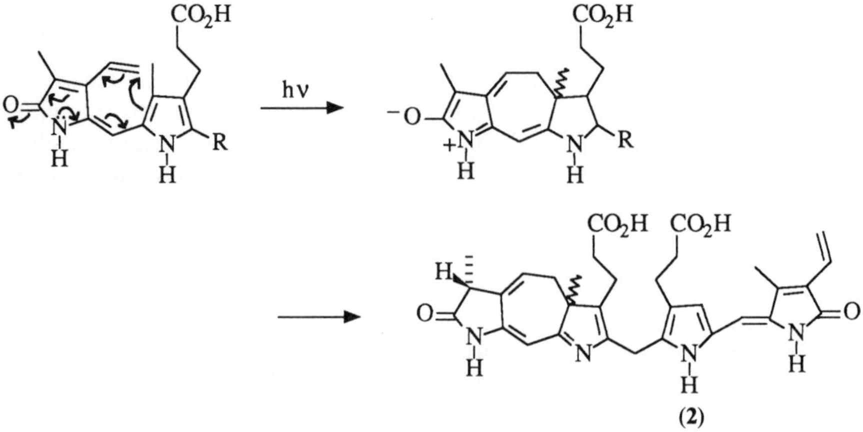

(2)

The photochemistry of bilirubin is complex, particularly its photooxidation, but the relevance of the photoproducts to the medical treatment is not clear. The structure of lumirubin (2), one of the photoproducts, is shown above. An electrocyclic reaction is an obvious route to the fused seven-membered ring moiety.

11.3. LASERS IN MEDICINE

The references in a recent review article on trends in laser medicine were listed under the following headings: medical laser technology and laster–tissue interactions; lasers in dermatology, surgery, and endoscopy; lasers in ophthalmology, vascular applications; laser lithotripsy; dental applications; photodynamic therapy; laser biostimulation; diagnostic applications; and optical fibers.[8] This breadth of topics illustrates the growing role played by lasers in medicine. In contrast to usual light sources used by photochemists, lasers can be strictly monochromatic and very intense. They can be delivered in very short pulses of precise duration, and they can be focused with great accuracy on very small targets from any distance.

Most of the medically beneficial effects of laser treatments come from thermal effects rather than specific photochemical transformations. Much consideration must therefore be given to the dissipation of the heat generated by the laser treatment in a small site, as it spreads to adjacent tissues and may damage them. Damage can be controlled by judiciously selecting the power and duration of each laser pulse.

Lasers are used medically either by precisely aiming at one selected target, or by damaging many targets simultaneously without a need for accurate aim. This is possible when the targets have greater optical absorption than their surroundings. Red blood cells or melanosomes are particularly suitable targets, allowing damage to be restricted to these organelles without affecting other cellular or subcellular structures. In the treatment of melanosomes, the temperature may be so high that explosive vaporization occurs, leading to a whitening of the site.[9] Lasers are therefore particularly useful in dermatology for the removal of birthmarks, lesions, and tatoos. Laser thermal ablation has also been studied in great detail; this process produces superheating of the fluids below the surface, building up pressure until an explosive event removes the tissue.[10]

Lasers are often used in ophthalmology. Argon laser photocoagulation, for example, is the recommended treatment for retinal branch vein occlusion, a frequent retinal vascular abnormality. The protocol recommends a 100 μm spot size at 0.1 second with power to produce a 'medium' white burn at the level of the retinal pigment epithelium, in a random 'grid' pattern, with spacing about one burn width apart.[11] Low-energy exposure to 193 nm laser radiation, which has a very shallow penetration depth, is also used for photokeratitis, or reshaping of the cornea.[12] Finally, a subsequent laser treatment is occasionally necessary in order to treat the clouded membranes that sometimes form in the eye after cataract surgery.

11.4. PHOTOSENSITIZATION WITH PSORALENS: PUVA THERAPY

Many furanocoumarins (also called furocoumarins) are natural plant products which have traditionally been used, some for over 2000 years, in combination with exposure to sunlight, to treat vitiligo, a condition characterized by the absence of skin pigmentation. More recently, the drug has been administered orally rather than topically, and artificial lamps have replaced exposure to sunlight.

Furanocoumarins possessing the psoralen ring system are common; the combined treatment of patients with a psoralen (P) and ultraviolet light (UVA) is called PUVA or PUVA therapy. The psoralens and their synthetic analogs have received extensive mechanistic scrutiny; 8-methoxypsoralen (8-MOP, (3)), 5-methoxypsoralen (5-MOP, (4)), and 4,5,8-trimethylpsoralen (trioxsalen, (5)) are the most common members of this family. The best known effect of irradiating biological systems with psoralens is covalent binding to DNA. Either one or both exposed double bonds in the furan and in the pyrone rings may be involved in [2 + 2] cycloaddition reactions with DNA pyrimidine bases. The latter process is particularly important, as it cross-links DNA strands. The opposite effect, breaking of DNA strands, can also be photosensitized by psoralens, even concomitantly with the formation of adducts.[13] Even when strand breaking is not produced, photolysis of DNA with psoralens in the presence of oxygen may induce DNA lesions which lead to strand cleavage upon subsequent treatment with a base. The use of psoralens has expanded beyond the treatment of vitiligo: these molecules are particularly useful for the treatment of psoriasis, T-cell lymphoma, and other diseases. More recently, PUVA-induced pigmentation has been advocated as a method for protecting skin and blood vessels against damage induced by ultraviolet light.[14]

(3)   (4)   (5)

A lesser known photoreaction of psoralens in biological systems is their interaction with membrane components. Cycloaddition may take place with double bonds of unsaturated fatty acids. Because most psoralens can also sensitize the formation of singlet oxygen, superoxide ion, and derived species (cf. Chapter 6), they can interact with a variety of targets, including unsaturated lipid components. Allylic oxidation of these lipids may thus compete with cycloaddition reactions. Actually, all these reactions are extremely complex; for example, rats' skin photosensitized with a radioactively labeled psoralen led to incorporation of the label into nucleic acids and lipids, as well as into proteins.[15] Interestingly, psoralens do not readily damage the membrane of erythrocytes, which would cause hemolysis. In this case, the photoreciprocity law, which reflects the expectation that, for a well-behaved process, the magnitude of the effect should be proportional to the fluence when the sensitizer concentration is constant, and proportional to the sensitizer concentration when the fluence is kept constant, is not followed. There is a threshold fluence below which hemolysis does not occur, and the hemolysis observed at high fluences has been suggested to result from the participation of photooxidized psoralens (see Chapter 9).[16]

Despite the many reported beneficial effects of PUVA therapy, warnings have been sounded because DNA modification might be accompanied by adverse effects. In mice, for example, ultraviolet light (either UVA or UVB) led to higher carcinogenesis by 8-MOP and significantly decreased their survival time.[17] After 30 years of use in thousands of patients with vitiligo and psoriasis, however, the carcinogenic risk due to PUVA therapy is now considered to be low.[18] However, there is a definite association between cumulative exposure to PUVA and increased risk of squamous cell carcinoma of the skin, particularly on the male genitalia.[19]

11.5. QUANTITATIVE DETERMINATION OF PHOTOTOXIC RESPONSE

In many cases, specific changes occur on the skin in response to irradiation with or without a sensitizer. It has been convenient to measure the *minimal erythemal dose*, also called the minimal phototoxic dose, which is the minimum dose which induces biochemical reactions evidenced by reddening of the skin. It was assumed that there was a direct relationship between a psoralen's ability to elicit erythema and its efficacy as a drug in the PUVA therapy of psoriasis. More recently, however, it has been found that the two activities can be separated, and that there are psoralens which are powerful drugs against psoriasis without producing any erythema. It is therefore much more difficult to determine the doses of drug and light optimal for the PUVA therapy treatment with these drugs.

4,4',6-Trimethylangelicin (TMA, (6)), one of the most active psoralens in terms of its photobinding to DNA, is one of the compounds which are photoactive without generating erythema. Treatment with TMA and UVA modifies dendritic epidermal cells in mice, despite the complete absence of cutaneous phototoxicity.[20] Since hyperpigmentation was observed after 4 weeks of treatment, this combination of properties (no phototoxicity and hyperpigmentation) suggests that TMA might be a valuable drug for the treatment of vitiligo.

(6)

Nitrogen-containing psoralen derivatives, for example with a pyridine ring fused to the pyrone ring,[21] or 4,4',5'-trimethyl-8-azapsoralen,[22] also show no skin phototoxicity but have the light-dependent antiproliferative effect required for treating psoriasis.

The problem of predicting drug photosensitivity extends beyond the photosensitization with psoralens.[23]

11.6. PSORALENS AND IMMUNOLOGY

Epidermal cells can initiate T-cell-dependent immune response, and PUVA therapy is known to alter this significantly. For example, PUVA therapy in man and experimental animals markedly reduces contact hypersensitivity to simple chemical haptens.[24] The epidermal Langerhans cells, which are important for the induction of delayed contact sensitivity, are modified.[25]

The formation of psoriatic plaques involves an autoimmune mechanism.[26] A scale in a patient suffering from psoriasis contains anaphylatoxin C5a, which attracts polymorphonuclear neutrophils. PUVA therapy inhibits this chemotaxis, and a photoproduct from 8-MOP was suggested to be responsible for the inhibition. An epoxide on the furan ring is perhaps an intermediate, since a synthetic product known to be active was formally an addition product of methanol (which had been a solvent in the synthesis) to such an intermediate:[27]

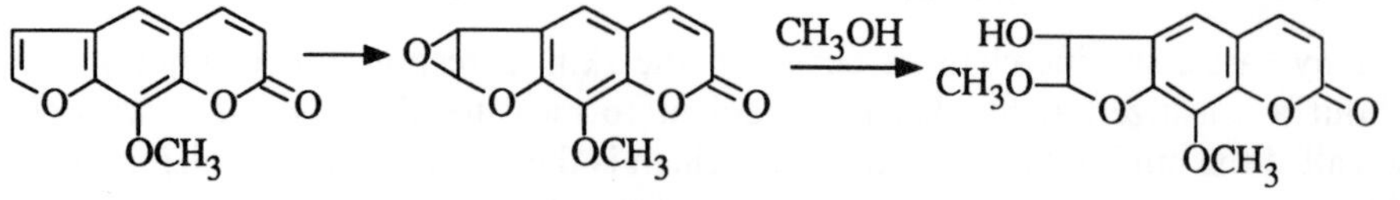

The photosensitized synthesis of a furan epoxide via a dioxetane, which has been recently performed (Chapter 10), is perhaps relevant to this problem. It has also been suggested that biologically active photoproducts from the oxidation of 8-MOP might be useful in treating other diseases involving the immune system, such as lupus erythematosus, rheumatoid arthritis, and acute glomerulonephritis.

The importance of psoralens in the photochemical treatment of psoriasis will perhaps decrease now that the immunodepressant drug cyclosporine has been found to provide excellent relief.[28] Other approaches are also investigated. Amiprilose (7), for example, is a simple glucofuranose derivative which inhibits psoriatic cell proliferation, at least *in vitro*.[29]

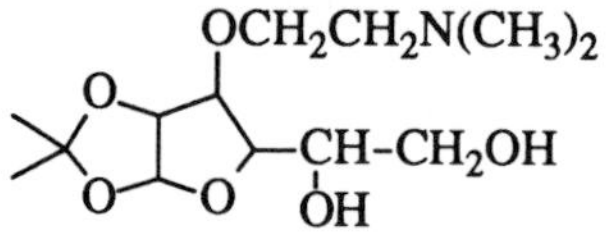

(7)

11.7. PHOTODYNAMIC THERAPY* OF MALIGNANT CELLS

The selective modification of malignant cells is currently one of the most exciting applications of photochemistry in medicine. This procedure is based on the observation that certain organic molecules become concentrated in the malignant cells, because of preferential affinity and/or because of slower metabolism in these cells. A chemical localized in a cell can be lethal to the cell if it creates singlet oxygen or other reactive oxygen species upon exposure to light. Although initially studied in skin diseases where exposure to ultraviolet light could be conveniently performed, the technique has been refined for the treatment of cancers in other tissues, the irradiations being performed by delivering light to the desired locations through fiber optics.

Not all tumors are monoclonal; some consist of several cancer cell populations having different sensitivities to photosensitized treatments.[31] Many variables must therefore be optimized, particularly the light dose, and whether it should be delivered in one or several irradiations.

To be clinically useful in photodynamic therapy, an ideal photosensitizer must be non-toxic and must be activated by light capable of penetrating tissues ($\lambda > 600$ nm). One

---

*PDT is an abbreviation commonly used to describe photodynamic therapy, even though it is not always unambiguous in the medical literature where it competes with meanings such as pancreaticoduodenal transplant or population doubling time.[30]

drawback of most drugs used in photodynamic therapy is that they remain significantly concentrated in normal tissues for a long time, making the patients photosensitive during that period. In some cases, topical application of sensitizers, followed by laser treatment, helps avoid systemic photosensitization, as demonstrated in the successful treatment of human papilloma virus disease with hematoporphyrin.[32]

Photofrin II, currently in advanced clinical trials, is a purified hematoporphyrin derivative which has been widely used, although the chemical structure of its active component(s) is still not firmly established. It is a mixture of monomeric and oligomeric porphyrins. Its activity decreases during storage, while higher-molecular-weight components are formed.[33] However, just as pesticides induce resistance in their intended insect targets, repeated exposure to Photofrin may induce resistance in cells targeted for photodynamic therapy.[34] Extensive mechanistic studies on hematoporphyrin derivatives have been performed,[35] but the relevance of the results obtained *in vitro* to the actual mechanism *in vivo* has been questioned.[36] It is clear, however, that hematoporphyrin can sensitize the formation of singlet oxygen with quantum yields as high as 0.4–0.6. Hydrated electrons are also formed in aqueous solution, particularly in dilute solutions, with either one-photon or two-photon laser excitation.[37]

Many dyes absorbing at long wavelengths have been tested, and some appear to have photosensitizing properties more favorable than Photofrin II.[38] One of the criteria is whether the dye absorbs in the region 600–900 nm, where there is maximum light penetration into tissues. Photofrin II has little absorption in that region, but phthalocyanins absorb strongly at 650–680 nm, and possess tumor-localizing and photosensitizing properties comparable to Photofrin II. for example, the mechanism of tumor necrosis photosensitized by liposome-delivered zinc(*II*) phthalocyanine was studied in mice. The sensitizer is believed to be carried by low-density lipoproteins, and released specifically to malignant cells. Electron microscopy following irradiation showed early photodamage of mitochondria and endoplasmic reticulum in malignant cells. The capillaries supplying the tumor tissue were modified at a much slower rate, most noticeably when the whole tissue became necrotic.[39] Since the mechanism of cell damage depends on oxygen participation, it is not surprising to find that tissue oxygen concentrations were dependent on light dose.[40] The treatment of tumors in areas where the vascular system has been destroyed or is very limited, such as a tumor in the vitreous of the eye, was not very effective because of the limited amount of oxygen diffusing through the tissues.[41]

A comparative study of Photofrin II and Al phthalocyanin tetrasulfonate revealed marked differences in their relative concentrations in different tissues.[42]

It must be noted that knowing the physical properties of a given sensitizer (λ_{max} and ε values) does not suffice to predict whether it will be satisfactory for photodynamic therapy. It is also important to establish that the chemical becomes localized in tumors to be destroyed without producing opacity to such tissues.[43]

One of the most important parameters in the design of superior dyes for photodynamic therapy is the ratio of sensitizer found in cancerous cells compared to normal cells. Potency experiments in cultured carcinoma cells do not provide this information, but recent results suggest that normal tissues may be protected by phosphorylated thiol radioprotective agents during photodynamic therapy, which could lead to improvements in the protocol for photodynamic therapy.[44]

Cancer chemotherapy often uses a combination of drugs. In the same manner, the selectivity and efficacy of photodynamic therapy will perhaps be improved by using a combination of sensitizers. This was suggested from work *in vitro* with L1210 cells which were treated simultaneously with hematoporphyrin derivatives (which damage membranes) and rhodamine 123 (which damages mitochondria), using irradiation at 514 nm.[45]

Treatment by photodynamic therapy of a number of different malignant tumors has been reported, often with excellent results. A clinical study of photodynamic therapy of early stage cancer in head and neck surgery showed that 76% of the patients were histologically free of disease 14 months later.[46] The bladder is a target readily accessible for light delivery through a cystoscope. Treatment by photodynamic therapy in five different centers produced complete remission in 73% of the patients, and partial remission in 19%. However, very large tumors could not be treated by this method.[47] Other types of tumor that have been treated include brain,[48,49] colon,[50] lung, and genital tract tumors.[51]

11.8. PHOTODYNAMIC THERAPY IN DERMATOLOGY

PUVA therapy is a most important photochemical treatment in dermatology, particularly in view of the large number of persons suffering from psoriasis and vitiligo. Sensitizers other than psoralens also show activity in dermatology. For example, photoreactive compounds such as hematoporphyrin derivatives or Photofrin II have been claimed to be useful for the photodynamic therapy of hypervascular dermal lesions, such as port-wine stains or spider veins.[52]

11.9. TOO MUCH PORPHYRIN: TREATMENT WITH CAROTENOIDS

Some diseases, such as erythropoietic protoporphyria or congenital porphyria, are associated with deficient metabolism of porphyrin pigments. Their accumulation leads to severe photosensitization of the patients' skin upon exposure to visible light. In addition to avoiding exposure to light, these patients have also been treated with high doses of β-carotene, a powerful quencher of singlet oxygen.[53]

11.10. PHOTOSENSITIZED INACTIVATION OF VIRUSES

11.10.1. Inactivation of viruses in blood components

Many viruses have been inactivated by photochemical treatment in the presence of a sensitizer. A potential application of such treatments is in sterilizing blood and blood products used in medicine. This is a very important goal, in view of the millions of units of blood processed every year. Unfortunately, serious diseases have often been transmitted by transfusion of contaminated blood products. HVB (hepatitis virus type B), HSV-1 (*Herpes simplex* virus 1), HTLV-1 (a leukemia virus) and HIV-1 (human immunodeficiency virus type 1—the AIDS virus) are among the viruses enclosed within a lipid envelope. The goal of inactivating the virus in the presence of red blood cells by photosensitizing treatments constitutes a challenge, in view of the ease with which the latter hemolyze in the presence of singlet oxygen generated photochemically. Fortunately, it turns out that the concentration of Photofrin II required for virus inactivation is at least one order of magnitude smaller than that necessary for producing hemolysis. Other photosensitizers which require visible light for activity have also been found effective in the sterilization of banked blood products.[54,55]

The photosensitized treatment of blood plasma avoids the risks of erythrocyte hemolysis. Treatment with the common phenothiazine dyes methylene blue or toluidine blue completely removed the infectivity of HIV-1.[56] The activity of clotting factors and other plasma proteins was only slightly decreased. Actually, the use of methylene blue or of the xanthene dyes rose bengal and eosin Y has been patented for the selective inactivation of viruses, particularly HIV.[57]

Bacteriophages have been proposed as laboratory models for animal viruses, and their inactivation sensitized with merocyanin 540 was investigated. One, $\phi6$, gave results qualitatively similar to those determined with human HSV.[58]

11.10.2. Other examples of virus inactivation

Many photosensitizers affect herpes viruses. For example, sapphyrin derivatives (**8**) are efficient for *in vitro* inactivation of HSV-1.[59]

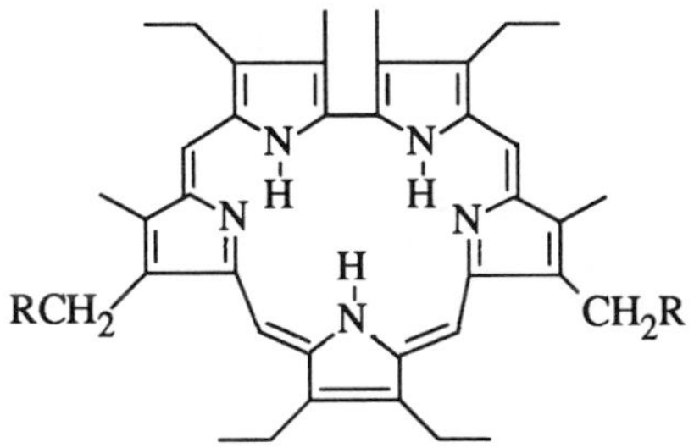

$R = Me$

$R = CO_2H$

(**8**)

A well-known natural product responsible for the photosensitization of cattle grazing on plants from the genus *Hypericium*, particularly *H. hirstutum*, is hypericin (**9**).[60] It also has bactericidal activity and has been used as an antidepressant. Hypericin was shown to be active against retroviruses, inhibiting the production of infectious virus from chronically infected cells as well as reverse transcriptase activity of mature virions. The antiviral activity is greatly enhanced by irradiation with fluorescent light in cases of murine cytomegalovirus (MCMV), Sindbis virus, HIV-1,[61] and equine infectious anemia virus.[62] Earlier reports of such activity had failed to recognize the absolute requirement for irradiation, as no controls had been carried out without exposure to room lights.

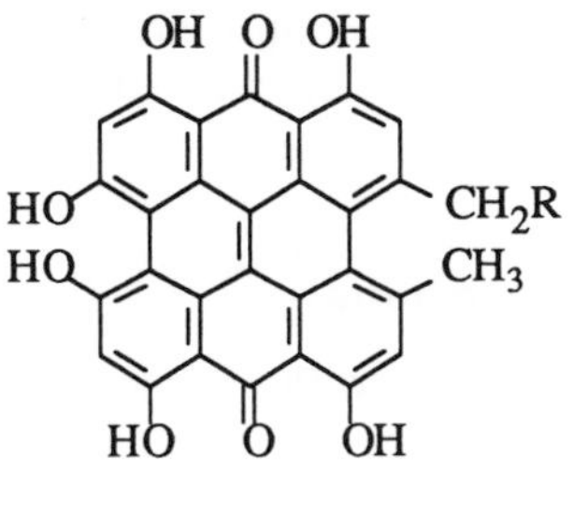

$R = H$ (**9**)

$R = OH$ (**10**)

In addition to the direct effect against the virus itself, hypericin inhibited cells infected with viable MCMV, and the effect was enhanced by exposure to visible light. However, there were no adverse effects on cultured cells at the concentrations where the compound had effective antiviral activity. One should also note that pseudohypericin (**10**) and cercosporin, molecules which are closely related structurally, are also powerful photodynamic compounds.

11.10.3. Furanocoumarins and furanochromones as antiviral agents

Psoralens may have antiviral activity, as suggested earlier. Actually, both DNA and RNA viruses are susceptible to inactivation by many different psoralens, as are virus-infected cells.[63] When a virus such as MCMV was inactivated by 8-MOP, its viral genome could still penetrate into the cell nucleus, but without inducing viral DNA, RNA, or protein synthesis. The mechanism(s) of inactivation by psoralens are not clear. 8-MOP is known to cross-link with DNA in the presence of ultraviolet light while angelicin (**11**), which is based on an isomeric furanocoumarin structure, generally does not. Even though angelicin is not photoactive against some DNA and RNA viruses, in contrast to 8-MOP, both sensitizers are equally active against the single-stranded DNA phage M13. The newly discovered ability for psoralens to cleave DNA may be particularly important in this case.

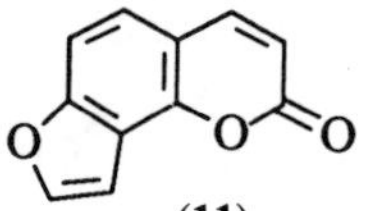

(**11**)

Furanochromones have a backbone isomeric with that of the psoralens, a 4-pyrone replacing the 2-pyrone ring. Two naturally occurring furanochromones are particularly well known, khellin (**12**) and visnagin (**13**), the former for its significant activity in protecting against atherosclerosis.

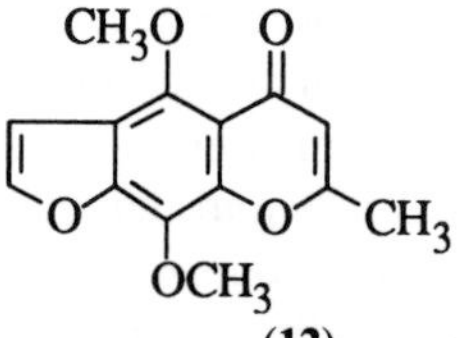

(**12**) 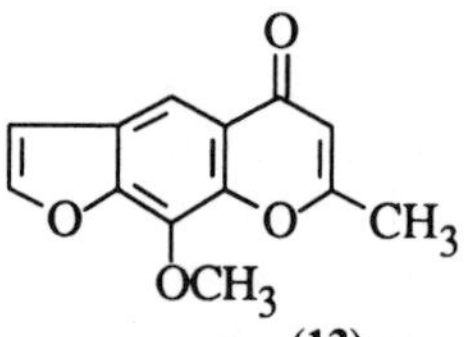   (**13**)

Visnagin is almost as phototoxic as 8-MOP against viruses, khellin much less so. A virus photosensitized with either compound retained its structural integrity and penetrated cultured mouse cells, but it did not replicate.[64]

11.10.4. Antiviral activity of gilvocarcins

The number of biological reactions discovered with psoralens reflects the large amount of research which has been devoted to this class of compounds. One should not conclude that other classes of molecules have less diverse photochemistry and photobiology. For example, gilvocarcins are bacterial products somewhat related to the above molecules, in the sense that they contain a coumarin system fused to aromatic rings, but with no furan rings. Gilvocarcin V (**14**) is a powerful photosensitizer which inhibits the capacity of cells to support plaque formation by HSV-1 at a concentration about five orders of magnitude lower than 8-MOP.[65] In contrast to psoralens, its photochemistry has barely been investigated.

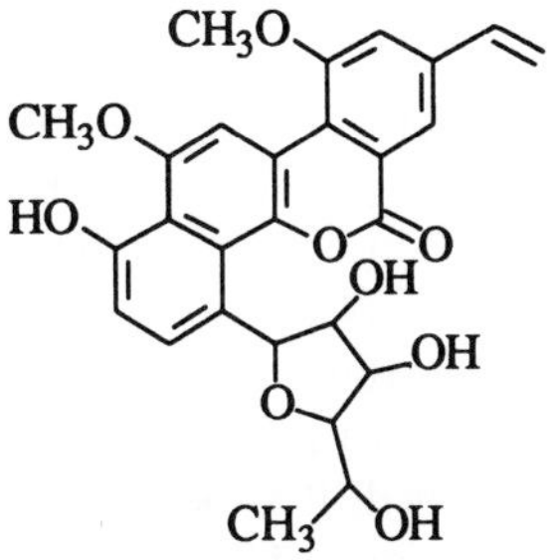

(14)

11.11. ELIMINATION OF RESIDUAL TUMOR CELLS FROM AUTOLOGOUS BONE MARROW GRAFTS

Most circulating human blood cells have a short lifespan and must be replenished from stem cells located in the bone marrow. Marrow damage is a major problem in cancer therapy when ionizing radiation and cytotoxic drugs are used, but bone marrow may be transplanted into leukemia patients whose stem cells have been destroyed. Transplantation of a patient's own marrow avoids the immunological problems usually associated with tissue transplants from donors. For the treatment to be successful, however, the marrow to be transplanted must be free from tumor cells. Thus, it should be collected when healthy cells are still significantly numerous. After destruction of all his or her remaining marrow cells, for example with radiation, the patient receives a transplant of his or her own healthy marrow. A critical requirement in this procedure is the removal of most, if not all, neoplastic cells from the graft. This can be done using photodynamic therapy, by treating marrow cell samples with merocyanin 540 (15) and light prior to transplanting them. An appreciable difference in sensitivity between normal and neoplastic cells leads to predominant killing of the latter.[66] Clinical trials of this procedure are in progress.

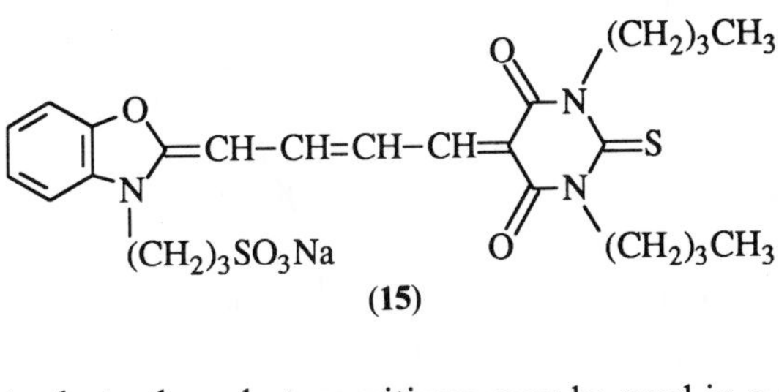

(15)

One must also note that other photosensitizers may be used in a similar manner for purging cancerous cells from marrow to be used in autologous bone marrow transplantation. A derivative of pyrene, a non-carcinogenic polycyclic aromatic hydrocarbon, was recently shown to have properties similar to merocycanin 540. In that work, 12-(1-pyrene)dodecanoic acid was found to be highly phototoxic to leukemic cells in the presence of normal cells.[67]

11.12. OTHER PHOTOSENSITIZED TREATMENTS WITH MEROCYANIN 540

Experimental results on photosensitization with merocyanin 540 in mice were most impressive. These groups were treated with a lethal dose of radiation. The group without further treatment died within 20 days. The group that received a mixture of normal and L1210 (leukemia) cells died within 33 days. However, the mice in the group that had received the same mixture of cells previously photosensitized with merocyanin 540 were all alive and free of disease 160 days later.[68]

Although the detailed mechanisms of the light-dependent cytotoxic activity of merocyanin 540 are not yet fully established, it is believed that reactive oxygen species are responsible. Singlet oxygen can be formed[69] and has been held responsible for the photosensitized damage to cell membranes demonstrated with human erythrocytes.[70] Electron spin resonance (ESR) spectroscopic studies proved that merocyanin 540 was also capable of inducing Type I photodynamic reactions.[71] In aqueous solution, merocyanin 540 occurs as monomeric and aggregated molecules. Although the triplet state of merocyanin 540 could be obtained either by energy transfer from anthracene or by direct irradiation in the ultraviolet region, it was not detected upon irradiation with visible light.[72] In methanol solution, merocyanin 540 was found to produce a triplet inefficiently and to undergo cis–trans isomerization from the singlet state. It underwent redox actions in the presence of electron acceptors.

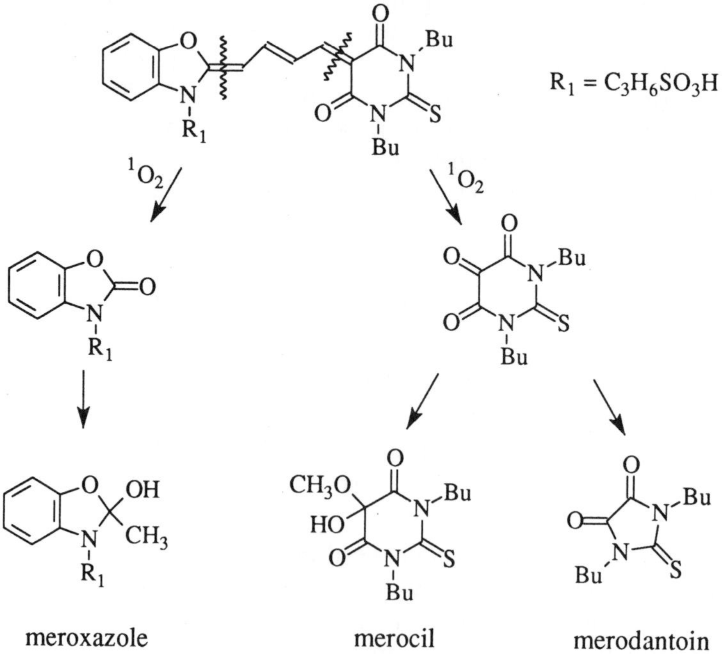

$R_1 = C_3H_6SO_3H$

meroxazole   merocil   merodantoin

The most probable mechanism for phototoxicity was thought to involve either local thermal disruption of cell membranes, or generation of toxins derived from the breakdown of the dye.[73] The latter hypothesis was strengthened by observations that irradiation of merocyanin 540 produced material which retained its activity for at least 30 days and which killed tumor cells in the dark with little cytotoxicity toward normal human cells.[74] The structure of the active component(s) is still unknown. However, three photooxidation products of merocyanin 540 have now been identified, resulting from oxidative cleavage of the hexatriene system near each of the heterocyclic moieties. The structures are shown in the scheme on p. 196. As expected from Chapter 5, additional decarboxylation of the molecule possessing three adjacent carbonyl groups was also encountered. The structure of the isolated products reflect additional chemistry due to the presence of the solvent methanol.[75]

Although the sequence of events leading to cytotoxicity is not fully known, the beginning may be more firmly established because the action spectrum of the antileukemic (in L1210 cells) and antiviral (in *Herpes simplex*) activities of the compound matched the absorption spectrum of the membrane-bound monomeric sensitizer.[76]

11.13. MEROCYANIN 540 AS AN ANTIMALARIAL AGENTS

Like practically all phototoxic molecules, merocyanin 540 has little specificity. As related earlier, this sensitizer has been advocated for the photosensitized inactivation of viruses in blood samples. The same approach was investigated for the removal of malaria pathogens from red blood cells. Malarially infected erythrocytes bound more merocyanin 540 and were more sensitive to irradiation than non-infected cells. Extracorporeal exposure of infected erythrocytes to light could preven the transmission of the disease.[77]

11.14. OTHER PHOTOSENSITIZERS

Many different tetrapyrrole derivatives have satisfactory properties for potential use in photodynamic therapy. Bacteriochlorophyll *a* derivatives, for example, accumulate in tumor cells when administered systemically and are phototoxic. They can also be used *in vitro*, for example to destroy infectious cells or viruses in stored blood.[78]

The method of delivery of a sensitizer to the cells to be destroyed is also quite important. Chlorin e_6, covalently bound to polystyrene microspheres, was efficiently taken up by human bladder carcinoma cells *in vitro* and became localized in the cytoplasm, in contrast to the unconjugated sensitizer, which was found within most cellular membranes. The intracellular concentration of the dye, using microsphere conjugation, was thus increased 20-fold. The chlorin e_6–microsphere conjugates were highly phototoxic while the unconjugated dye was not, as measured by inhibition of colony formation.[79] The photobiological activity did not directly correlate with the quantum yield of singlet oxygen formation, such this quantum yield decreases upon covalent bonding of the dye to the microspheres.

Many other chemicals which produce photosensitization *in vivo* have not been listed in this survey. It is safe to predict that a long time will elapse before many of them are used clinically.

11.15. PHOTOPHERESIS WITH PSORALENS

Extracorporeal photosensitized blood treatment has been advocated for many conditions, particularly leucocyte-mediated diseases such as scleroderma, systemic lupus erythematosus, cutaneous T-cell lymphoma, AIDS, and psoriasis. Psoralen and 8-MOP have been recommended for such treatments.[80] The patients ingest the drug and, 60–90 minutes later, some of their blood is withdrawn and treated to separate the red blood cells (which are returned to the bloodstream) from the white cells. These are diluted and irradiated with ultraviolet light (UVA) and then returned to the blood stream.[81] The remarkable aspect of this treatment when used against T-cell lymphoma is that the cancerous cells are not in the blood, and yet it is a photochemical treatment of the blood which triggers clinical improvement, perhaps through reactions which affect the immune system.[82]

11.16. SUNLIGHT AND HERPES LABIALIS

One should remember that N. R. Finsen, a Danish physician who lived between 1860 and 1904, received a Nobel prize in 1903 in recognition of his discovery that exposure to ultraviolet light was useful in the treatment of skin lesions associated with tuberculosis. Following this example and so many other examples where sunlight, in combination with some sensitizer, has a biological activity which suppresses the development of viruses, cells, or organisms, it may come as a surprise to find that herpes-dependent lip affections known as herpes labialis are actually *enhanced* by exposure to ultraviolet light. This is not an insignificant problem, as it has been reported that about 150 million Americans have latent HSV-1 infection. The condition is particularly unpleasant to practitioners of skiing and other winter sports, who are exposed to higher than average levels of UVB radiation. A very high correlation was established experimentally between exposure to UVB and recurrence of herpes labialis. Fortunately, sun-blocking agents can provide complete protection, and only topical protection is needed.[83] In other words, skiers who protect their lips with sun-blocking agents can avoid suffering from the herpes problem while still being able to tan their faces, should they be foolish enough to desire it. A cream containing 1% coumarin has been claimed to provide longer lasting protection than sun-blocking agents, but the mechanism is not clear.[84]

REFERENCES

1. M. F. Hollick, *Ann. NY Acad. Sci.*, **453**, 1–14 (1985).
2. R. M. Neer, *Ann. NY Acad. Sci.*, **453**, 14–20 (1985).
3. L. Thorington, *Ann. NY Acad. Sci.*, **453**, 28–54 (1985).
4. Z. Lu, T. C. Chen, L. Kline, T. Markestad, J. Prettifor, C. Mautalen and M. F. Holick, *Photoderm. Photoimmun. Photomed.*, **8**, 30 (1991).
5. H. F. DeLuca, *Metab., Clin. Exp.*, **39**, 3–9 (1990).
6. R. H. Dobbs and R. J. Cremer, *Arch. Dis. Child.*, **50**, 833–836 (1975).
7. D. A. Lightner and A. F. McDonagh, *Acc. Chem. Res.*, **17**, 417–424 (1984). J. F. Ennever, *Photochem. Photobiol.*, **47**, 871–876 (1988).
8. J. A. Parrish and B. C. Wilson, *Photochem. Photobiol.*, **53**, 731–738 (1991).
9. S. L. Jacques and D. J. McAuliffe, *Photochem. Photobiol.*, **53**, 769–775 (1991).
10. A. J. Welch, M. Motamedi, S. Rastegar, G. L. LeCarpenter and D. Jansen, *Photochem. Photobiol.*, **53**, 815–823 (1991).
11. The Branch Vein Occlusion Study Group, *Am. J. Ophthalmol.*, **99**, 218–219 (1985).
12. H. D. Sliney, R. K. Krueger, S. L. Trokel and K. D. Rappoport, *Photochem. Photobiol.*, **53**, 739–744 (1991).
13. J. Kagan, X. Chen, T. P. Wang and P. Forlot, *Photochem. Photobiol.*, **56**, 185–193 (1992).
14. T. B. Fitzpatrick. In T. B. Fitzpatrick, P. Forlot, M. A. Pathak and F. Urbach (eds.), *Psoralens: Past, Present and Future of Photochemoprotection and Other Biological Activities*, pp. 5–10. John Libbey Eurotext, Paris (1989).

15. F. Dall'Acqua and P. Martelli, *J. Photochem. Photobiol. B.*, **8**, 235–254 (1991).
16. A. Y. Potapenko, M. A. Agamalieva, A. I. Nagiev, E. P. Lysenko, L. N. Bezdetnaya and V. L. Sukhorukov, *Photochem. Photobiol.*, **54**, 375–379 (1991).
17. P. D. Forbes, R. E. Davies, F. Urbach and J. K. Dunnick, *J. Toxicol., Cutaneous Ocul. Toxicol.*, **9**, 237–250 (1990).
18. T. B. Fitzpatrick. In T. B. Fitzpatrick, P. Forlot, M. A. Pathak and F. Urbach (eds.) *Psoralens: Past, Present and Future of Photochemoprotection and Other Biological Activities*, pp. 5–10. John Libbey Eurotext, Paris (1989).
19. R. S. Stern, *Photochem. Photobiol.*, **53S**, 43S (1991).
20. J. Alcalay, F. Dall'Acqua and M. L. Kripke, *Br. J. Dermatol.*, **122**, 9–14 (1990).
21. L. Dubertret, D. Averbeck, E. Bisagni, J. Moron, E. Moustacch, C. Billardon, D. Papadopoulo, S. Nocentini, P. Vigny, J. Blais, R. V. Bensasson, J. C. Ronfard-Haret, E. J. Land, F. Zajdela and R. Latarjet, *Biochimie*, **67**, 417–422 (1985).
22. D. Vedaldi, F. Dall'Acqua, S. Caffieri, F. Baccichetti, F. Carlassare, F. Bordin, A. Chilin and A. Guiotto, *Photochem. Photobiol.*, **53**, 143–148 (1991).
23. L. C. Harber, *J. Invest. Derm.*, **77**, 65–70 (1981).
24. W. Aberer, N. Romani, A. Elbe and G. Stingl, *J. Immunol.*, **136**, 1210–1216 (1986).
25. H. Okamoto and T. Horio, *J. Invest. Dermatol.*, **77**, 345–346 (1981).
26. E. H. Beutner, M. Jarzabek-Chorzelska, S. Jablonska, T. P. Chorzelski and G. Rzesa, *Arch. Derm. Res.*, **261**, 123–134 (1978).
27. N. Mizuno, K. Esaki, J. Sakakibara, N. Murakami and S. Nagai, *Photochem. Photobiol.*, **54**, 697–701 (1991).
28. C. N. Ellis, *et al.*, *N. Engle, J. Med.*, *324*, 333–334 (1991).
29. M. L. Chapman, S. D. Dimitrijevich, J. C. Hevelone, D. Goetz, J. Cohen, G. E. Wise and R. W. Gracy, *In Vitro Cell. Dev. Biol.*, **26**, 991–996 (1990).
30. J. D. Spikes, *J. Photochem. Photobiol. B.*, **10**, 371–371 (1991).
31. T. Patrice, *J. Photochem. Photobiol. B.*, **9**, 372–374 (1991).
32. M. J. Shikowitz, Topical hematoporphyrin composition for photodynamic papilloma treatment, *Eur. Pat. Appl. EP 350,036*, 10 January 1990. *Chem. Abstr.*, *113*: 138517.
33. Y.-K. Ho, J. R. Missert and T. J. Dougherty, *Photochem. Photobiol.*, **54**, 83–87 (1991).
34. G. Singh, B. C. Wilson, S. M. Sharkey, G. P. Browman and P. Deschamps, *Photochem. Photobiol.*, **54**, 307–312 (1991).
35. B. W. Henderson, *Photodermatology*, **6**, 200–211 (1989).
36. T. J. Dougherty, *Photochem. Photobiol.*, **45**, 879–889 (1987).
37. G. Grabner, N. Getoff, T. Gantchev, D. Angelov and M. Shopova, *Photochem. Photobiol.*, **54**, 673–681 (1991).
38. R. K. Pandey, D. A. Bellnier, K. M. Smith and T. J. Dougherty, *Photochem. Photobiol.*, **53**, 65–72 (1991).
39. C. Milanesi, C. Zhou, R. Biolo and G. Jori, *Br. J. Cancer*, **51**, 846–850 (1990).
40. B. J. Tromberg, A. Orenstein, S. Kimel, S. J. Barker, J. Hyatt, J. S. Nelson and M. W. Berns, *Photochem. Photobiol.*, **52**, 375–385 (1990).
41. R. W. Lingua and J.-M. Parel, *J. Photochem. Photobiol. B.*, **9**, 119–122 (1991).
42. Q. Peng, J. Moan, M. Kongshaug, J. F. Evenson, H. Anholt and C. Rimington, *Int. J. Cancer*, **48**, 258–264 (1991).
43. T. J. Dougherty and W. R. Potter, *J. Photochem. Photobiol. B.*, **8**, 223–225 (1991).
44. J. Bedwell, P. T. Chatlani, A. J. MacRobert, J. E. Roberts, H. Barr, J. Dillon and S. G. Brown, *Photochem. Photobiol.*, **53**, 753–756 (1991).
45. M.-T. Foultier, T. Patric, C. Tanielian, C. Wolff, S. Yactayo, A. Berrada and A. Combre, *J. Photochem. Photobiol. B.*, **10**, 119–132 (1991).
46. J. Feyh, A. Goetz, W. Müller, R. Königsberger and E. Kastenbauer, *J. Photochem. Photobiol., B.*, **7**, 353–358 (1990).
47. C.-W. Lin, *SPIE (New Directions in Photodynamic Therapy)*, **847**, 51–56 (1987).
48. M. O. Dereski, M. Chopp, J. H. Garcia and F. W. Hetzel, *Photochem. Photobiol.*, **54**, 109–112 (1991), and references cited therein.
49. P. Muller and B. Wilson, *J. Photochem. Photobiol. B.*, **9**, 117–119 (1991).
50. H. Barr, P. Chatlani, C. J. Tralau, A. J. McRobert, P. B. Boulos and S. G. Bown, *Gut*, **32**, 517–523 (1991).
51. R. V. Lobraico, *Proc. SPIE—Int. Soc. Opt. Eng.*, **1353**, 125–130 (1990).
52. M. W. Berns, Hematoporphyrin derivatives for photodynamic therapy of hypervascular dermal lesions. *Can. Pat. Appl. CA 2,012,175*, 30 September 1990.
53. M. M. Mathews-Roth, *Fed. Proc.*, **46**, 1890–1893 (1987).
54. M. M. Judy, J. L. Mathews, J. T. Newman, H. L. Skiles, R. L. Boriack, J. L. Sessler, M. Cyr, B. G. Maiya and S. T. Nichol, *Photochem. Photobiol.*, **53**, 101–107 (1991).
55. H. C. Neyndorff, D. L. Bartel, F. Tufaro and J. G. Levy, *Transfusion (Philadelphia)*, **30**, 485–490 (1990).

56. B. Lambrecht, H. Mohr, J. Knuever-Hopf and H. Schmidt, *Vox Sang.*, **60**, 207–213 (1991).
57. R. F. Shinazi and R. A. Floyd, Antiviral therapy using thiazine and xanthene dyes. *PCT Int. Appl. WO 90 13,269*, 15 November 1990.
58. C. D. Lytle, A. P. Budacz, E. Keville, S. A. Miller and K. N. Prodouz, *Photochem. Photobiol.*, **54**, 489–493 (1991).
59. M. M. Judy, J. L. Matthews, J. T. Newman, H. L. Skiles, R. L. Boriack, J. L. Sessler, M. Cyr, B. G. Maiya and S. T. Nichol, *Photochem. Photobiol.*, **53**, 101–107 (1991).
60. N. Pace and G. Mackinney, *J. Chem. Soc.*, **63**, 2570–2574 (1941).
61. I. Lopez-Bazzocchi, J. B. Hudson and G. H. N. Towers, *Photochem. Photobiol.*, **54**, 95–98 (1991).
62. S. Carpenter and G. A. Kraus, *Photochem. Photobiol.*, **53**, 169–174 (1991).
63. J. B. Hudson and G. H. N. Towers, *Pharmac. Ther.*, **49**, 181–222 (1991), and references cited therein.
64. J. B. Hudson, E. A. Graham, L. L. Hudson and G. H. N. Towers, *Planta Med.*, **54**, 131–135 (1988).
65. L. E. Bockstahler, R. K. Elesperu, V. M. Hitchins, P. E. Carney, K. M. Olvey and C. D. Lytle, *Photochem. Photobiol.*, **51**, 477–479 (1990).
66. F. Siebert, *Photochem. Photobiol.*, **46**, 71–76 (1987).
67. E. Fibach, S. Gatt and E. A. Rachmilewitz, *Exp. Hematol.*, **18**, 89–93 (1990).
68. F. Siebert, *Photochem. Photobiol.*, **46**, 71–76 (1987).
69. B. Kalyanaraman, J. B. Feix, F. Sieber, J. P. Thomas and A. W. Girotti, *Proc. Natl. Acad. Sci. USA*, **84**, 2999–3003 (1987).
70. D. P. Valenzano, J. Trudgen, A. Hutzenbuhler and M. Milne, *Photochem. Photobiol.*, **46**, 985–990 (1987).
71. J. B. Feix and B. Kalyanaraman, *Photochem. Photobiol.*, **53**, 39–45 (1991).
72. A. Seret, M. Hoebeke and A. Van de Vorst, *Photochem. Photobiol.*, **52**, 601–604 (1990).
73. J. Davila, A. Harriman and K. S. Gulliya, *Photochem. Photobiol.*, **53**, 1–11 (1991).
74. K. S. Gulliya, S. Pervaiz, R. M. Dowben and J. L. Matthews, *Photochem. Photobiol.*, **52**, 831–838 (1990).
75. B. Franck and U. Schneider, *Photochem. Photobiol.*, **56**, 271–276 (1992).
76. J. M. O'Brien, R. J. Singh, J. B. Feix, B. Kalyanaraman and F. Sieber, *Photochem. Photobiol.*, **54**, 851–854 (1991).
77. O. M. Smith, D. L. Traul, L. McOlash and F. Sieber, *J. Infect. Dis.*, **163**, 1312–1317 (1991).
78. B. W. Henderson, A. B. Sumlin, B. L. Owczarczak and T. J. Dougherty, *J. Photochem. Photobiol., B.*, **10**, 303–313 (1991).
79. R. Bachor, C. R. Shea, R. Gillies and T. Hasan, *Proc. Natl. Acad. Sci. USA*, **88**, 1580–1584 (1991).
80. M. O. Bachynsky, M. H. Infield, R. J. Margolis and D. A. Perla (Hoffman-La Roch, F., und Co. A.-G., Switz.). Psoralen formulation for UV-A photophoresis. *Eur. Pat. Appl. EP 392429 A1*, 17 October 1990.
81. P. Heald, M. Perez, G. McKiernan, I. Christiensen and R. Edelson, *Photodermatology*, **6**, 171–176 (1989).
82. R. L. Edelson, *J. Photochem. Photobiol. B.*, **10**, 165–171 (1991).
83. J. F. Rooney, Y. Bryson, M. L. Mannix, M. Dillon, C. R. Wohlenberg, S. Banks, C. J. Wallington, A. L. Notkins and S. E. Straus, *Lancet*, **338**, 1419–1422 (1991).
84. R. D. Thornes, *Lancet*, **339**, 133 (1992).

12. Applied photochemistry

A small selection of practical applications of photochemistry relevant to our daily lives is presented here. This area is certainly destined to grow in the future.

12.1. LIGHT AND DYES; ONE STORY

Photochemistry is responsible for many changes noticeable in our daily experience. Colored fabrics and materials usually fade much more where they have been exposed to sunlight. People looking at old color photographs of relatives cannot help noting the color changes. The importance of this problem should not be underestimated, as much artwork which we hope to see preserved forever is doomed by the very fact that it is observed and that light used for observation induces photochemical reactions, for example through formation of the very reactive singlet oxygen. The story of the Apocalypse pictured on the series of tapestries kept in Angers, France, provides a striking example. Completed around 1380, the 84 tapestries were kept through the years in dark castles and churches. Even after their acquisition by the French government at the end of the 19th century, their large size and the insufficient space available allowed the exhibition of only a portion of the remaining 72 tapestries, and 'only under very poor lighting'. While distinctly faded, the colors were still quite impressive when the tapestries were first exposed a few years ago in a gallery built especially for them in the beautiful medieval castle of Angers. The windows were wall-to-wall and floor-to-ceiling, and the tapestries were thus exhibited with 'the lighting that did justice to them' (again quoting from the official history by R. Planchenault). A little later, it was noticed that the back sides of the exposed tapestries, which had been covered with fabric lining, had kept vivid colors and had as fine a design as the front sides. The lining was then removed and the reversed tapestries exposed to the public. Unfortunately, it did not take long for a dramatic fading of the colors in the bright light to be noticed. The exhibition gallery had to be closed and remodeled, with all the windows eliminated, and the tapestries, now protected in nitrogen-filled cases, are viewed under very dim artificial lighting.

Of course the bleaching power of sunlight has been well known through the generations; even now, many shun electric driers because laundry dried outdoors is perceived as being distinctly whiter (an attitude cleverly recognized by the manufactor of the detergent Sunlight, which for good measure is guaranteed to be 100% phosphate-free).

12.2. ROYAL PURPLE AND OTHER NATURAL FABRIC DYES

Before synthetic dyes became available, a few natural ones were used. Royal purple is perhaps the oldest known, going back to at least the 13th century BC. The precursors (colorless), (1), are found in the hypobranchial secretions of certain marine mollusks. It has been written that dyeing textiles with molluscan purple was the first photochemical industrial process in the history of mankind. Here, photooxidation reactions are of major importance. The biological precursors are enol sulfate esters (with R = Br), which are hydrolyzed enzymatically. The initial product can condense with its oxidation product, yielding the immediate precursor (2) to the indigo. This product is not colored, but forms the insoluble purple colored dye (3) upon oxidation, particularly in the presence of light. Thus, one method of dyeing consisted of the direct application of the molluscan secretions to a textile, followed by exposure to sunlight.[1] Any reaction

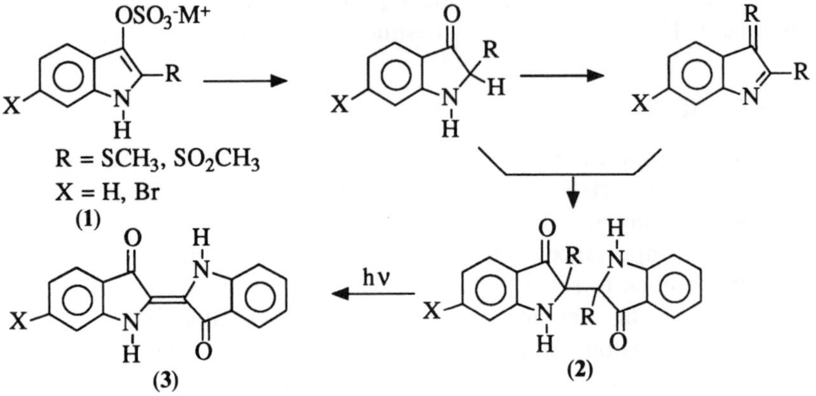

R = SCH$_3$, SO$_2$CH$_3$
X = H, Br
(1)

modifying the central double bond should affect the color. For example, photoinduced *cis–trans* isomerization leads to a change in the spectral characteristics, but it is rapidly reversed thermally. One would predict $^1O_2$ to react with indigo, cleaving the double bond through a dioxetane intermediate, leading to fading of dyed fabrics. Indeed, photofading is accelerated in the presence of $^1O_2$ sensitizers, and retarded by $^1O_2$ quenchers.[2] Of course, any photoinduced electron transfer reaction generating the leuco form would lead to fading, but reoxidation would cancel this effect.

It is interesting that plants (*Indigofera tinctoria*) have also been the source of indigo (R = H) used for dying fabrics blue. Woad (*Isatis tinctoria*) cultivated in England and Scotland provided most of the blue dye used in England. The plant material was allowed to ferment both before being dried in the sun and afterwards before use.

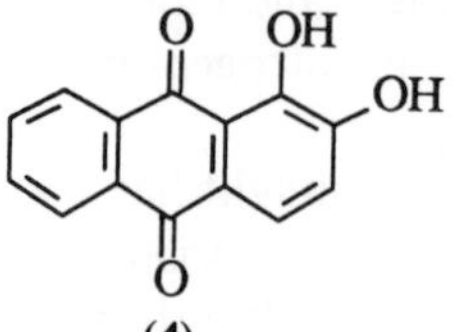

(4)

Other organic dyes were used long before synthetic ones became available. Alizarin (4), a diglucoside of a dihydroxyquinone used to produce red fabrics, was extracted from the roots of *Rubia tinctorum*. Although the photochemical degradation of the molecule is not known in detail, one may speculate that oxidation of the *o*-dihydroxyphenyl moiety, producing an *o*-quinone, would make the molecule susceptible to base-catalyzed cleavage with an attendant spectral shift. Carmine, also a red dye, was extracted from the scale insect *Coccus cacti*. Carminic acid (**5**), its main component, is somewhat more complex, being a *C*-glucoside, but it contains the same structural

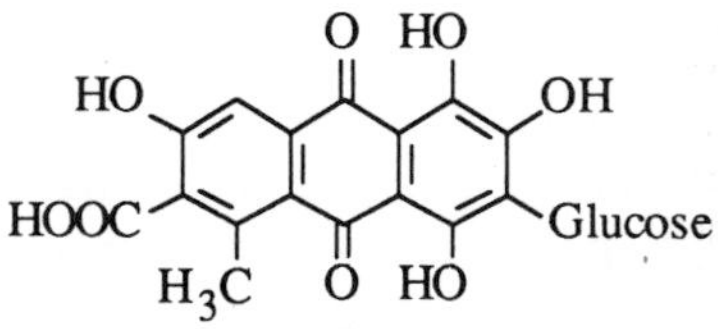

(**5**)

feature. Hematoxylin (**6**), from the heartwood of the logwood *Haematoxylon campechianum*, turns red in the presence of light and produces black fabrics when treated with

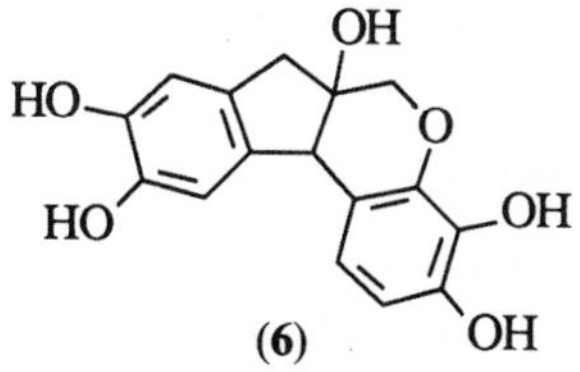

(**6**)

iron salts. Luteolin (**7**) and quercetin (**8**) are among the most common yellow flavonoid pigments in plants.

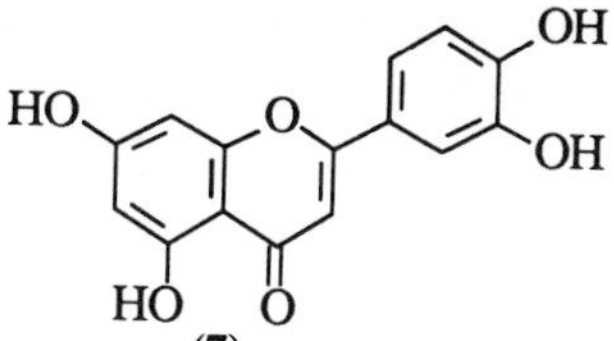

(**7**) 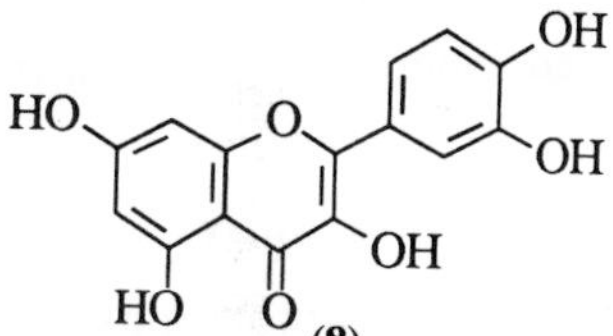   (**8**)

All the above dyes are capable of strongly chelating metal salts, and it is the photochemistry of these salts which is probably most relevant to the light-induced fading of dyed fabrics.

It is important to note that all dyes are potentially capable of sensitizing the formation of $^1O_2$. This is because the longest wavelength in the visible spectrum, 800 nm, corresponds to 36 kcal/mol (see Fig. 1.8), whereas the energy difference between the triplet ground state and singlet excited states of oxygen is about 23 kcal/mol. The $^1O_2$ generated may be damaging in two different manners: (1) it could bleach the dye that helped produce it, and (2) it could degrade the fabric that supports it.

12.3. PHOTOGRAPHY, PHOTOLITHOGRAPHY, AND REPROGRAPHY

Photography, particularly color photography, has touched the lives of almost everyone. In a strange way, it was a by-product of the presence of petroleum in the Middle East which led to the initial photographic processes. Nicephore Niepce who, ironically, had to give up a career in the French military because of his poor eyesight, discovered that when a layer of an asphalt component called bitumen of Judea was spread over a metal plate it turned from black to white where it had been exposed to light, and that it was no longer soluble in lavender oil in those areas. Removing with lavender oil the bitumen which had not polymerized exposed the metal, which could then be etched with acid. The areas protected by the polymerized bitumen resisted the effect of the acid. This observation provided the name 'resist' which is now commonly used.*

The photoengraved plate could be used for printing. Later on, starch with chromium salts, albumen, or gelatin was used as resist material. The technique of halftone reproduction, which broke down an image into dots of varying size separated by white areas, allowed the printing of images possessing widely ranging shades of darkness. Eventually, the printing of colored materials became possible by photographing the material through filters for the three primary colors, producing three printing plates. Overlaying the printing of the three images using appropriately colored inks produced the final color reproduction.

The preparation of printing masters is equally important in large-scale reproductions and in microfabrication of electronic devices.[4]

12.3.1. Photography and xerography

Organic photochemistry is not involved in the primary steps of these image reproduction systems. In photography, it is the photoreduction of silver ions into silver metal which is the main phenomenon. Color photography is based on the same chemistry, where silver halide is in superimposed gelatin layers, each sensitive to only one of the primary colors.

Xerography is based on a physical phenomenon, which creates changes in electrical charges on the surface of a suitable material. The electrostatic latent image is treated with a charged toner, which is then transferred onto paper, where it is baked.

12.3.2. Photolithography

The principle of photolithography was introduced earlier (Section 5.9.2). Resists may be either negative or positive, depending upon the type of chemistry induced by light in the medium irradiated through a suitable mask. A negative resist is obtained when the substrate in the irradiated areas becomes less soluble and therefore remains untouched in the subsequent development step; with a positive resist the irradiated substrate becomes more soluble and is removed in the development.

12.3.2.1. Negative resists

The dimerization of cinnamic acid moieties attached to polymers has been used extensively for making negative resists. Cross-linking in a polymer, as shown below, reduces its solubility and allows the formation of a negative resist:

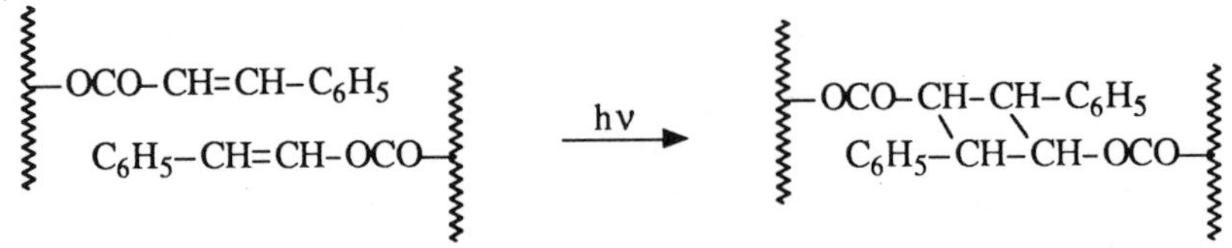

$$\text{—OCO–CH=CH–C}_6\text{H}_5 \qquad \text{C}_6\text{H}_5\text{–CH=CH–OCO—} \xrightarrow{h\nu} \text{—OCO–CH–CH–C}_6\text{H}_5 \qquad \text{C}_6\text{H}_5\text{–CH–CH–OCO—}$$

---

*The frequently repeated assertion that bitumen of Judea was used by the Egyptians in mummification has not been supported by chemical analyses.[3]

Variations on this theme, for example by increasing the conjugation of the double bond, allow the absorption characteristics of the polymer to be matched with the emission characteristics of the light source. The cyclobutane ring formation shown above may also be induced by a photosensitized reaction rather than by direct excitation.

In another approach, cross-linking of a hydrocarbon backbone is achieved through the photolysis of an azide. The triplet nitrene created can abstract a hydrogen atom from the substrate; coupling the radicals formed produces a polymer of reduced solubility. The photolysis must be performed in the absence of oxygen, which would react with the nitrene. This approach is very similar to the photoaffinity labeling process described in Section 9.12.

Another photochemical transformation, also mentioned in Chapter 8, has been used for making negative resists:

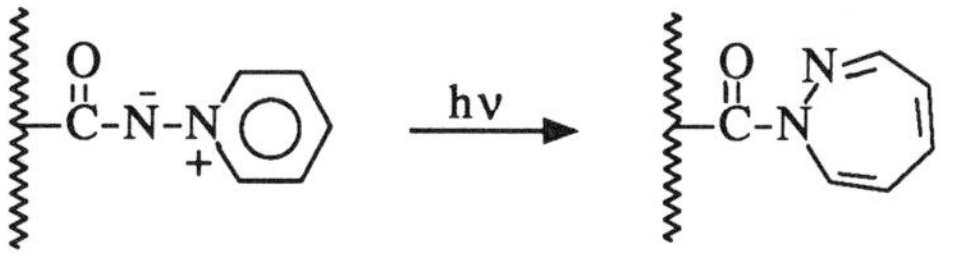

This isomerization converts a material which is water-soluble into one which is not.

12.3.2.2. Positive resists

Diazoketones (**9**) in general, and a diazoquinone in particular, have been used to produce positive resists. The chemical reaction which increases the solubility in the exposed area is a Wolff rearrangement (Section 5.9.2). The dissolution in a base of the carboxylic acid (**10**) produced allows its removal.

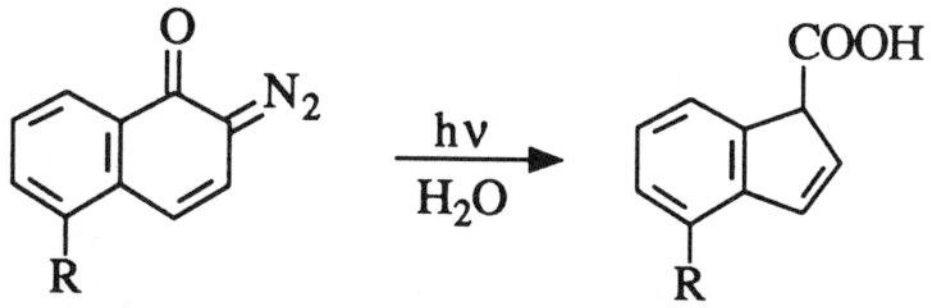

Another approach is also based on a reaction encountered earlier (see Chapter 8), in which a nitrobenzyl ester (e.g. (**11**)) is cleaved photochemically.

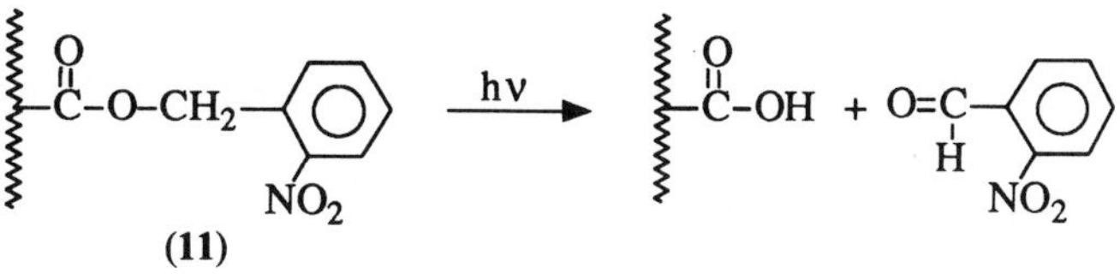

When a resist is formed on a silicon support, it becomes possible to etch the exposed surface and treat it before applying a new layer of resist material. One can irradiate again through a different mask, etch the surface, and repeat the sequence several times in order to create an integrated circuit chip.

12.4. PHOTOIMAGING USING ORGANOMETALLIC POLYMERS

The rich photochemistry of organometallic molecules has been deliberately left out of this book. An exception must be noted, however, because it provides an alternative to the photolithography process described above and in Chapter 5. As outlined by Gould et al.,[5] electropolymerization over conductive substrates produces thin films of polypyridyl complexes of ruthenium(II) which retain many of the properties of the individual building blocks. For example they undergo ligand loss upon irradiation, as illustrated by the following reaction (bpy is 2,2'-bipyridine):

$$[Ru(bpy)_3]^{2+} + 2Cl^- \xrightarrow{h\nu} Ru(bpy)_2Cl_2 + bpy$$

Ruthenium complexes containing 4-vinylpyridine (vpy) units can be polymerized. In the case of poly$[Ru(bpy)_2(vpy)_2]^{2+}$ (**12**), irradiation produces a loss of metal complex from the films. This disruption of the polymeric network allows the underlying substrate to become exposed. Irradiation through a mask can therefore create an image on the supporting medium.

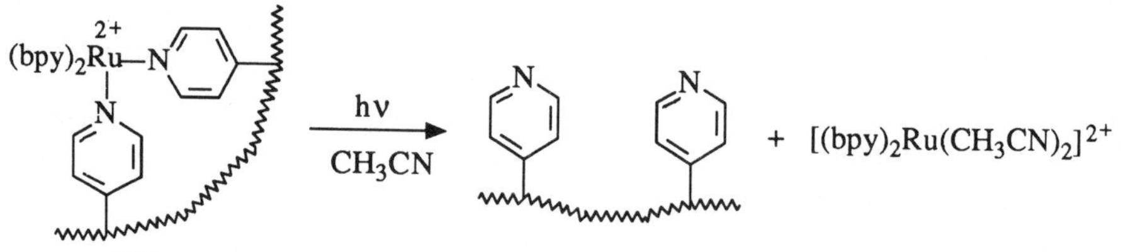

(12)

12.5. MICROMANIPULATION OF PARTICLES

Particles about 1 μm in size can be optically manipulated along any direction by a laser beam. When a laser beam is refracted through a particle, a radiative force is exerted in the opposite direction of the light momentum change. If its refractive index is higher than that of the surrounding medium, the particle can be optically trapped in the vicinity of the focal point of the laser beam. Moving the laser beam can therefore lead to motion of the particle. This phenomenon can be coupled with chemical reactions induced in a single microcapsule containing a sensitizer. This was demonstrated with an encapsulated toluene solution of pyrene. A 355 nm laser beam led to deformation and ejection of pyrene/toluene droplets. Since many reagents can be encapsulated and since individual microcapsules can be manipulated and decomposed with lasers, the authors point out that site-specific chemical modification of surfaces will be possible, for example on microelectrodes or integrated circuits.[6]

12.6 PHOTOCHEMICAL REACTIONS OF ENVIRONMENTAL INTEREST

Many environmental modifications may be ascribed to photochemical reactions. Some of these reactions take place in the upper atmosphere, for example chain reactions with fluorinated hydrocarbons in the ozone layer.

12.6.1. Polycyclic aromatic hydrocarbons

Many pollutants transported in the atmosphere, such as polycyclic aromatic hydrocarbons (PAHs) produced by incomplete combustion of organic matter in power plants, motors and other sources, also generate singlet oxygen in the gas phase[7] as well as in solution. Anthracene, phenanthrene, fluoranthene, pyrene, benzo[a]pyrene, chrysene, and benz[a]anthracene are typical examples of such photodynamic sensitizers, some of more than 200 members of the family of PAHs found in the environment. It has been estimated that more than 200 000 tonnes of PAHs enter the world's oceans and surface waters every year. Most of these molecules are powerful photosensitizers, phototoxic to aquatic organisms such as daphnia, brine shrimps, tadpoles, mosquito larvae and fish as well as microorganisms. The extent to which our environment is contaminated by PAHs is truly astonishing. In Great Britain, for example, the concentration of PAHs in soil is usually between 0.1 and 54 mg/kg, and even greater in urban areas or near sources of emission. Some of this contamination is picked up by plants; fresh carrots, for example, have been found to contain 13 μg/kg (of which 2.2 μg was benzo[a]pyrene).[8]

'Acid rain' is the common term describing atmospheric deposition of acidic products, sulfuric and nitric acids, which are the end-products from burning organic substances containing nitrogen and sulfur impurities. PAHs are generated simultaneously, transported into the atmosphere and deposited in the same manner. The death of forests in and around highly industrialized countries and the death of aquatic species have been frequently ascribed to acid depositions. However, based on the results in laboratory experiments, it is conceivable that some of the ecological damage usually associated with acid rain might instead be due to photodynamic reactions.[9]

Because PAHs have very low solubility in water, they eventually settle along with other sediments in natural aquatic environments. There is renewed danger to aquatic organisms when dredging or other activities disturb these sediments, resuspend the PAHs and allow them to exert their phototoxic effects.[10] The danger is not indefinite, as it has been suggested that aquatic organic matter is degraded through photochemical reactions induced by sunlight, yielding volatile organic compounds and carbon monoxide. Since this photodegradation is the rate-limiting step for the removal of a large fraction of dissolved organic carbon compounds in oceans, this rate is expected to increase if there is an increase in solar ultraviolet radiation due to damage to the ozone layer.[11] The oceanic residence time of photoreactive dissolved organic compounds has been estimated to be between 500 and 2100 years.

The opposite question is occasionally asked, namely, why is a molecule that was expected to be found in the environment actually missing? One example is that of the fullerene hydrocarbon C_{60}. Although readily generated in a variety of high-temperature reactions, it has not been observed on the earth. This absence has been postulated to result from ultraviolet light degradation in the presence of oxygen, shown separately to yield many products, including aldehydes and ketones (however, the absence of C_{60} in chimney soot, which is unlikely to have undergone much photochemistry, may weaken this theory).[12]

12.6.2. Halogenated hydrocarbons

Photochemical reactions involving chlorinated hydrocarbons, another class of widely distributed environmental pollutants, produce some of the symptoms associated with acid rain in trees, and might be relevant to the death of forests in industrialized countries.[13]

The photochemistry of halogenated molecules may be quite diverse. The photodegradation of toxic members of this family is one topic of environmental relevance. Polychlorinated dioxins and dibenzofurans, their environmental fate and methods of decontamination have received much attention. 2,3,7,8-Tetrachlorodibenzo-p-dioxin (13), for example, has a very low water solubility and volatility, and it is extremely stable thermally. It is also extremely stable when irradiated as a solid or in pure water. However, it undergoes photoreduction with ultraviolet light in the presence of organic solvents acting as hydrogen donors, or in the presence of surfactants.[14] Many photoproducts are formed, with competing hydroxylation and removal of halogen atoms. 2-Hydroxybenzoic acid (14) has also been reported to be a final product in

the sequence. The nature of solid components present at the time of photolysis may significantly affect the outcome. The degradation of polychlorinated dioxins and dibenzofurans adsorbed over dry fly ash is very much slower than in the presence of solvent, and there are variations depending upon the composition of the ash.[15]

As expected from a system capable of undergoing photo-redox reactions, the simple model of chlorobenzene (15) photolyzed as an aqueous suspension in the presence of TiO_2 revealed the formation of hydroxylation and dechlorination products.[16]

The variety of photochemical reactions known in the laboratory, and the marked effects of inorganic components (semiconductors, clays, etc.) which was reported in earlier chapters suggest that many organic molecules in the environment should be susceptible to phototransformations (therefore photodegradation) in the presence of sunlight. One obstacle to a rapid photodegradation of chemicals in soil may be the lack of mechanical mixing, which would have exposed layers normally found below the surface. The detoxification of hazardous chemicals which pollute groundwater can be approached more easily, for example by pumping the water through glass tubes exposed to sunlight (or artificial ultraviolet light). Photosensitizers may be added as needed. The trade literature has reported that this approach successfully removed trichloroethylene from groundwater.[17]

12.6.3. Photodehalogenation of a fungicide

Captan (16), an effective fungicide used for the protection of fruits, vegetables, grapes, and citrus crops, is rapidly decomposed by sunlight. Here again, the presence of a solvent with good hydrogen donor properties, such as 2-propanol, accelerates the photodecomposition.[18]

(16)

12.6.4. Photodegradation of other agrochemicals

The photodegradation of agrochemicals is always a topic of concern, as it is one of the factors determining their persistence. It is important to note that experiments performed on pure samples may not be representative of the behavior of the commercial preparations which have been formulated for specific applications.[19]

12.6.5. Natural defense mechanisms

Organisms have evolved means of coping with the effect of naturally occurring photosensitizers. Some of the best-studied examples concern furanocoumarins and sulfur-containing natural products, and how insects and other creatures, such as monkeys, avoid damaging exposure to sunlight or manufacture inactivating enzyme systems. However, the mechanisms of biological protection against natural photosensitizers may be quite selective, as revealed in a study of the parsnip webworm *Depressaria pastina-cella*, which feeds exclusively on plants containing furanocoumarins but which cannot tolerate natural photosensitizers found in other plants, notably α-terthienyl.[20]

12.6.6. Photoherbicides

One goal in the design of photoactive biocides is to create phototoxic metabolites against which organisms have no protection. One spectacular example concerns the conversion of an aminolevulinic acid into phototoxic tetrapyrroles, in order to control weeds, as described in Chapter 10, Section 10.11.

Laboratory experiments which successfully demonstrate phototoxicity have been scaled up for practical use. For example, treatment with methylene blue was used to affect lake restoration. Green algae and diatom levels decreased significantly within 1 day of treatment, but blue-green algae did not decline.[21]

12.7. PHOTOCHEMISTRY AND ALCOHOLIC BEVERAGES

The hop (*Humulus lupulus*) is indispensable in the manufacture of beer. It contributes many natural products, some of which undergo facile photochemical transformations, most notably humulone (**17**), which is converted into the so-called iso-α-acid (**18**). The first step may be viewed as a di-π-methane rearrangement (with a C=O bond replacing one C=C bond); it is followed by a cyclopropanol ring opening and by ketonization of one of the ring enols. The photoproduct is as bitter as quinine sulfate.

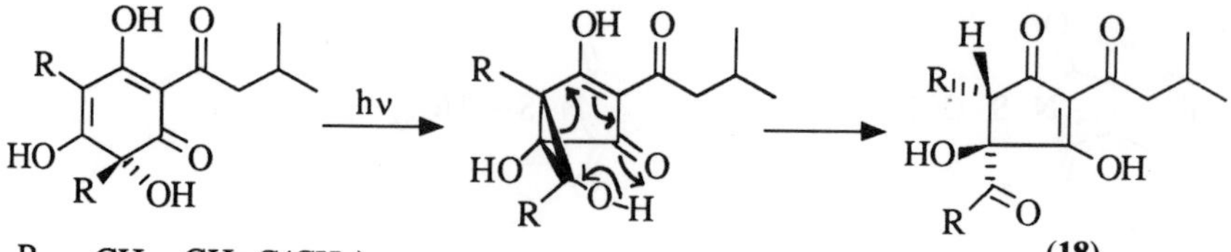

R = CH₂–CH=C(CH₃)₂

$R = CH_2-CH=C(CH_3)_2$

(17)

(18)

Actually, humulone can be converted into both (18) and one of its isomers by thermal means. Since this conversion occurs in the brewing process, bitter components are always present, but their taste is masked by the other ingredients. Additional formation of α-acids through exposure to light, however, can distinctly affect the characteristics of beer by producing what is known as the 'light-struck flavor'. This results from the photochemistry of humulone described above, followed by a Norrish type I cleavage of the RCO side-chain, decarbonylation and reaction with protein components to produce the offending thiol $(CH_3)_2C=CHCH_2SH$.[22]

This photochemistry explains why beer is usually stored in metal cans or colored glass. Brown bottles protect better than green ones. When drinking beer outdoors, do not keep your glass in the sun for very long if you want to avoid rapid degradation of the taste and flavor!

A similar kind of photodegradation has been noted with champagne. French experts observed that the quality was distinctly inferior when the bottles were sold in supermarkets as opposed to traditional liquor stores. Eventually, it was discovered that the intense fluorescent lighting traditionally present in large retail stores produced a light-struck flavor ('goût de lumière'), triggered by photochemical transformations involving sulfur components, such as methionine and cysteine, which produce H_2S, CH_3SH, and $(CH_3)_2S$.[23] These compounds have also been implicated in light-struck flavors in milk and beer. The phototransformations leading to the offending products in champagne can be prevented by the addition of a trace amount of ascorbic acid.[24]

Those who love to drink gin and tonic outdoors should realize that quinine (19), the key ingredient in tonic water, can also undergo photochemical decomposition. In aqueous solutions containing citric acid many primary and secondary photolysis products have been identified, comprising products of photoreduction, as well as addition products and fragmentation products.[25,26] The main photoproducts are those where

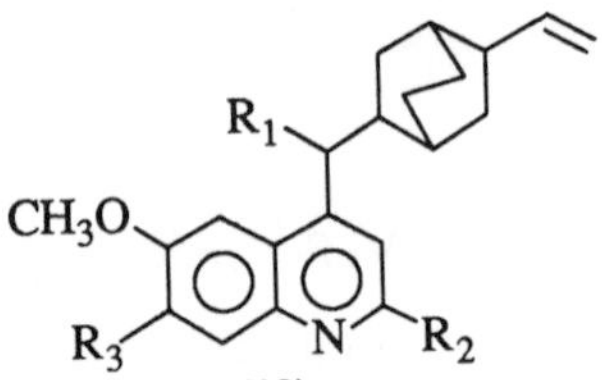

$R_1 = R_2 = R_3 = H$

(19)

$R_1 = R_2 = R_3 = H$, or where $C(OH)(CH_2CO_2H)_2$ is at R_1, R_2, or R_3, the other two groups being hydrogens, R_1 being H or OH. Participation by 2 mol of citric acid could also be demonstrated by the isolation of the product (**20**).

(**20**)

12.8. OTHER EXAMPLES OF PHOTODEGRADATION IN FOODS

The quality of the processed liver of geese or ducks, known as 'foie gras', is claimed to degrade upon storage in glass containers. In order to prevent photodegradation, experts in the producing region of Perigord in France recommend foie gras be kept in metal cans.

12.9. CYALUME CHEMICAL LIGHTS

The chemical generation of dioxetanes was described earlier. The transfer of energy from electronically excited carbonyl fragmentation products to dyes allows the observation of the phosphorescence of these dyes. The chemicals can be confined inside a plastic shroud, emitting visible light which can be quite bright. The main virtue of such systems is that the risks of sparks usually associated with electrical light sources are eliminated. Commercial sticks are available in different lengths and different colors (orange, red, blue, yellow, white), different intensities, and different durations (from 5 minutes to 12 hours). They are sold by Aldrich as Cyalume chemical lights.

12.10. USE OF LUMINESCENCE FOR DETECTION

The examples selected here illustrate the types of application recorded in the literature.

   One of the most generally used techniques in biochemistry and biology involves the detection of radioactive compounds by scintillation counting. The fluorescence of diphenyloxazole and/or analogs is stimulated by radiation (usually β) emitted by the radioactive molecules.

   Traditionally, immunoassays are performed with radioactive markers. However, chemiluminescence techniques may be as sensitive. For example, cyclosporine may be measured in whole blood by incubating samples with cyclosporin C-hemisuccinate-aminobutyl-N-ethylisoluminol, antibody, and magnetic particles coated with a second antibody. After magnetic separation and washing, the samples are incubated with NaOH, and the chemiluminescence generated by treatment with a peroxidase and H_2O_2 measured.[27] Likewise, a detection technique for ATP using a firefly luciferase reagent is so sensitive that 80 femtomole/l can be reliably detected, and that it is contamination of deionized water by ATP which may limit the sensitivity of the technique.[28]

   One interesting use of fluorescence detection techniques is in forensic science, for fingerprint analysis.[29] Ninhydrin reacts with α-aminoacids to prdouce a purple-blue product, a reaction which allows the development of latent fingerprints on porous surfaces, but the sensitivity of the detection technique is much enhanced when the ninhydrin-treated prints are treated with zinc chloride: a coordination compound is formed, which is highly fluorescent when excited with light from an argon laser.

Because luminescence detection is extremely sensitive, it is used extensively in analytical chemistry. For example, nitropyrenes and nitrosopyrenes eluted from high-performance liquid chromatography (HPLC) column could be detected after electrochemical reduction to aminopyrenes. They reacted with the electronically excited carbonyl produced from bis(2,4,6-trichlorophenyl)oxalate reacting with H_2O_2. The detection limits were in the femtomole range.[30]

Finally, there are many less sophisticated applications of luminescence detection techniques in our daily lives, ranging from watch dials which glow in the dark to the detection of leaks in batteries which contain a fluorescing pyrene derivative in the electrolyte mixture.[31]

12.11. FLUORESCENT WHITENERS

Colorless molecules which absorb light in the long-wavelength ultraviolet region are likely to emit light in the visible range when they luminesce. Their presence in white paper or fabrics tends to make them appear brighter in natural or fluorescence light. Such molecules are called fluorescent brighteners or optical brighteners. In the case of garments, they may be introduced either during the manufacture of the fabrics or during the laundry process.

Many different kinds of molecules have been used, usually rigid and polycyclic. For example, 4,4′-bistriazinylaminostilbenedisulfonic acid derivatives, pyrazolyl- and triazolylstilbene derivatives, benzoxazoles, thiazoles, imidazoles, 1,3,4-oxadiazoles, coumarins, pyrazolines, and naphthalimides were among the earlier compounds used.[32,33] As a testimony to the economic importance of the fluorescent whiteners, 644 references were listed in a review covering the period 1973–1985.[34]

12.12. MEROCYANIN DYES, PROTON TRANSPORT, AND PHOTOVOLTAGE GENERATION

Irradiated merocyanine dyes (e.g. (**21**)) can be used as proton pumps when incorporated in a planar membrane of oxidized cholesterol. Starting from *trans*-merocyanine, a com-

(**21**)

plete reaction cycle was obtained by protolytic, photochemical and/or thermal reactions. First, a membrane was produced on a glass filter. One side was in contact with $0.1\,M$ HCl, the other with saturated $0.1\,M$ I_2/I^- solution, and platinum electrodes were placed symmetrically across the membrane. Illumination produced a voltage which decreased slowly when the light was turned off. The sequence accounting for the proton transport is as follows:

$$M_{trans} \leftrightarrows MH^+_{trans} \leftrightarrows MH^+_{cis} \overset{-H^+}{\leftrightarrows} M_{cis} \rightarrow M_{trans}$$

In this sequence, protons are accepted on one side of the membrane (in contact with HCl), bound to the phenoxide end of the dye, transferred across the membrane by photochemical *cis–trans* isomerization, and accepted by I⁻ on the other side. The working direction of the proton pump is determined by the irreversibility of the last step (*cis* to *trans* isomerization).[35] A photovoltage of about 320 mV with sustained storage capacity was obtained, suggesting the possibility of using the system for the design of low-cost solar cells.

12.13. PHOTOINDUCED TRANSPORT OF AMINOACIDS ACROSS LIPID MEMBRANES

Charge separation is often associated with photochromism, as shown in the isomerization of the spiropyran (22) into the purple merocyanin dye (23). Both (22) and (23) dissolve much more easily in organic solvents, such as hexane and octanol, than in

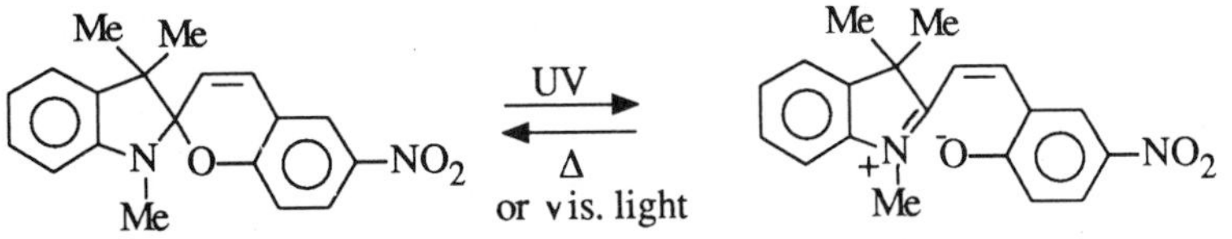

water. The fact that the zwitterionic species (23) can form a complex with another zwitterionic molecule such as an aminoacid has been used to transport an aminoacid across a liposome bilayer membrane of egg phosphatidylcholine containing the spiropiran. Phenylalanine encapsulated in the interior of the liposome could be transported to the exterior upon successive irradiations with ultraviolet and visible light. This transport was explained by the formation of a complex between the photogenerated form (23) and the aminoacid at the interface between the bilayer and the interior of the liposome. The complex, having no net charge, could diffuse within the bilayer, and was dissociated when visible light treatment isomerized (23) back into (22). The aminoacid released at the interface between the bilayer and the exterior of the liposome was then released into the medium.[36]

12.14. PHOTOINDUCED TRANSPORT OF METAL IONS

The reversible ring opening of spiropyrans (24) has also been used to transport metal ions (Zn^{2+}, Cu^{2+}, and Cd^{2+}) in a U-tube from one aqueous phase, first through a chloroform phase, and then into another aqueous phase (this may be viewed as a crude model for transport across a biological membrane). In this case, the metal cation was tightly complexed in the product of pyran ring opening by the phenolate ion and the *ortho* nitrogen substituents. the cation uptake occurred upon irradiation with ultraviolet light, its release with visible light.[37]

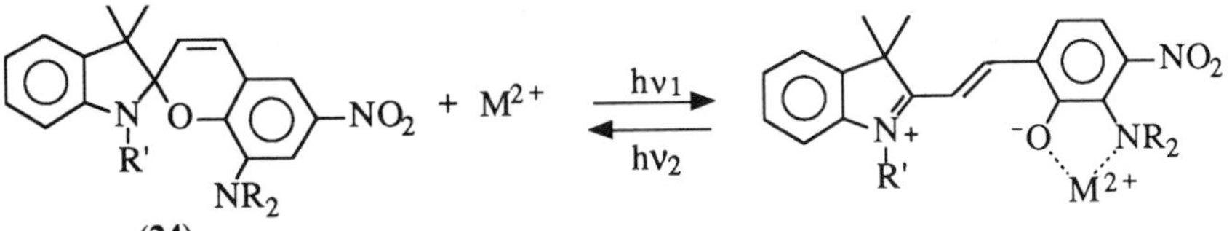

12.15. *CIS–TRANS* ISOMERIZATION AND ION TRANSPORT

As illustrated in the synthesis of helicenes (see Chapter 8), a *cis–trans* isomerization reaction can bring together groups which are originally quite far from each other. One elegant application of this structural isomerization allowed a metal ion to be extracted from an aqueous phase and transported into another aqueous phase across an organic phase (CH$_2$Cl$_2$) in a U-tube. Here, the *syn–anti* isomerization of an azo compound occurs with a butterfly-like motion. The *anti* isomer (**25**) was soluble only in the organic phase, while the *syn* isomer (**26**) was partially soluble in the aqueous phase as well. In the *syn* form, the crown ether substituents came close enough to form a cooperative complex with an alkali metal cation. The complexed ion was picked up in the aqueous phase, as with tweezers, transported across the organic phase, and released into the aqueous phase on the other side of the U-tube upon thermal isomerization of the *syn* into the *anti* form. Irradiation with ultraviolet light regenerated the *syn* form, which could be partitioned again into the organic phase, from which it could return into the first aqueous phase, complex with another cation, and repeat the transport cycle.[38]

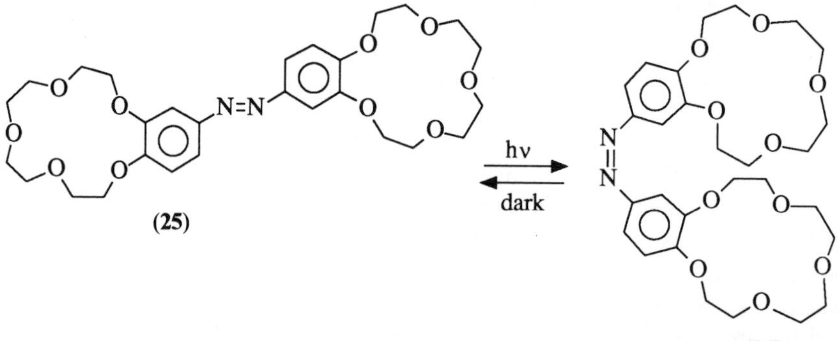

The large difference in geometry introduced by the *syn–anti* photoisomerization of azo compounds has also been used as a switch for cation complexation in crown ethers.[39] Here, the azo linkage was attached to the periphery in such a manner that, in the *anti* form (**27**), a substituent was located in the cavity of the crown ether. The substituent is moved away from the crown ether in the *syn* form (**28**), allowing the complexation

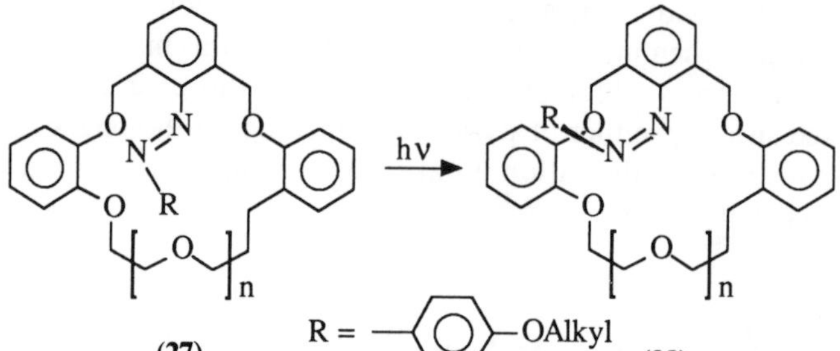

of metal cations. In this photoproduct, one of the nitrogen atoms in the azo group was also found to coordinate when a metal cation was complexed to the oxygen atoms.

Note that several different approaches have achieved light-dependent transport of metal ions. One that uses photochromic spiranoindolines was described earlier (Section 12.12).

12.16. PHOTOCHROMIC COMPOUNDS

Photochromic molecules have potential applications for optical storage and handling of information. Many classes of photochromic molecules are known,[40] the most famous being the spiropyrans, (29), which undergo reversible heterolytic cleavage (examples of application were discussed earlier in 12.13 and 12.14). The negative charge of the enolate ion in the open form is always stabilized by conjugation, while an adjacent heteroatom is usually present to stabilize the positive charge.

(29)

Electrocyclic reactions cause other classes of molecules to change color. The fulgide described as an actinometer in Chapter 1, Section 1.17 is another. Compounds which undergo *cis–trans* isomerization form yet another class (the azo compound described earlier in this chapter for transferring ions across an organic phase is an example). The osazone (30) was recently described and provided another example of photochromism due to tautomerism. Yellow in the solid state, it turns red upon exposure to sunlight.[41]

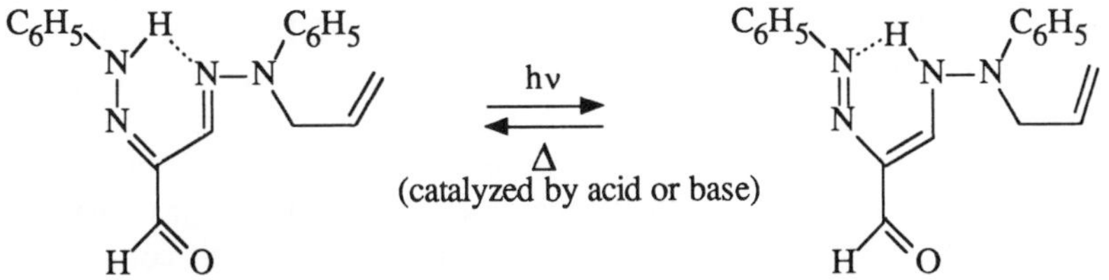

(30)

Photochromic compounds have gained many applications. One of the more unusual concerns a photochromic flesh-colored pigment created by heating a mixture of 90% anatase and 10% FeO(OH) at 750–850°C. Applied to the skin, the color of the formulated pigment was reported to be an elegant pale indoors and a tan-simulating brown outdoors. Non-cosmetic uses suggested for these products include military camouflage, photography, and paints for vehicles.[42]

12.17. ERASABLE OPTICAL MEDIA

Attached to polymeric materials, photochromic compounds allow for the design of erasable optical media. On disks or tapes, for example, they provide high storage capacity because lasers can easily inscribe 1 μm data tracks separated by 1 μm spaces.

Useful erasable optical media will require two essential properties: thermal stability of the photochromic system and high fatigue resistance. Further investigation of the photochromic fulgide systems already mentioned in Section 1.16 uncovered one example where both forms were stable up to at least 300°C:[43] The photochromic system could be

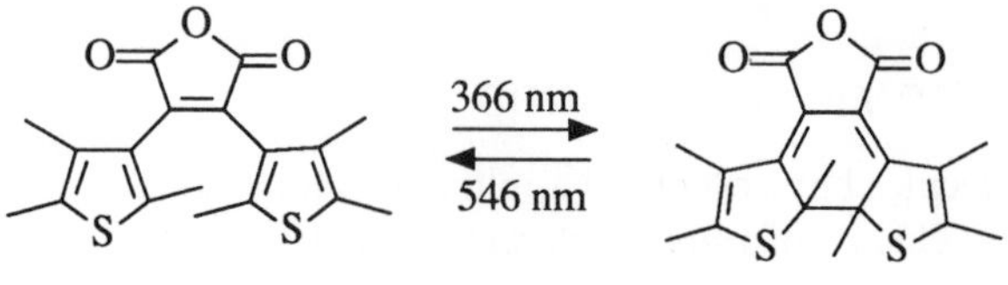

bound to a polystyrene resin (via imide formation with an amino group placed at the *para* position on the phenyl rings).

Another approach utilizes spiropyran compounds attached to a resin, (**31**), which turn reversibly from yellow to blue upon irradiation.[44] The reverse reaction occurs thermally too, but the rate is controlled by the viscosity of the medium.

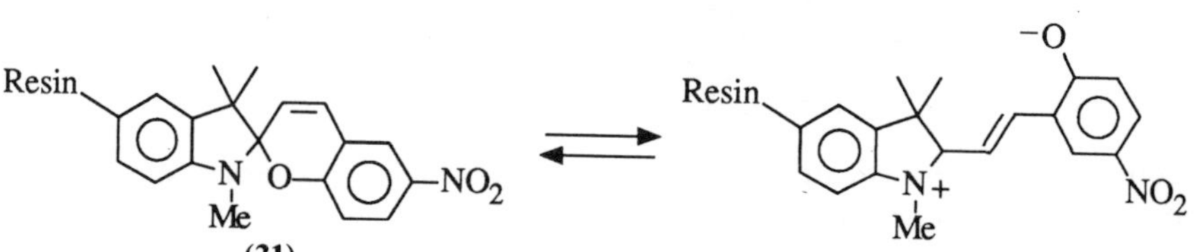

(**31**)

12.18. SYNTHETIC PHOTORECEPTORS

Rhodopsin is the biological photoreceptor involved in vision. It includes a protein component and a retinal chromophore undergoing $(E)-(Z)$ (*trans–cis*) isomerizaton. It is possible to obtain light-harvesting rhodopsin from bacterial sources, and this bacteriorhodopsin has been used for building devices capable of sensing changes in light intensity. In one approach, a photocell is created by placing an electrolyte gel between a gold-coated glass plate and a film of bacteriorhodopsin-rich bacterial membrane coated onto a transparent tin oxide electrode supported by a glass plate. Exposing the device to light causes a current to flow in an external circuit connecting the two electrodes.[45] It is believed that the current is created because light-induced $(E)-(Z)$ isomerization of the retinal produces a change in dipole moment, producing a charge displacement. The action spectrum of the photocurrent matches the absorption spectrum of bacteriorhodopsin. The quantum efficiency of the system is about 0.01.

When the tin oxide electrode was replaced by a 64-pixel lithium–tin oxide array, a photoreceptor capable of detecting changing images was obtained (these devices responded only to changes in light intensity producing sterochemical changes, but did not produce a sustained current during constant irradiation).

Rhodopsin-based detectors might be precursors to devices for detecting motion, for artificial vision, or for constructing optical computers.

12.19. PHOTOCHEMISTRY IN THE PAINT INDUSTRY

Paints cover a large portion of our surroundings. In 1990 alone, about 4×10^9 liters were produced in the USA, with a value of $11.6 billion. The pressure to reduce the emission of volatile organic compounds has accelerated the development of water-based paints which are cured by exposure to ultraviolet light.[46] The production growth for this type of paint is estimated to be 10–15% per year. Thus, concerns about photochemistry enter when painting takes place, as well as throughout the lifetime of the painted objects, as stability to continuous exposure to natural light is an important property of the paint.

In parallel, the growth of this industry has generated research into the design of suitable light sources, which cost up to $200 000 (for comparison, a system for electron beam curing of paints may cost up to $2 million), but which make the energy efficiency of ultraviolet curing up to 100 times greater than that of oven curing.

The key photochemical reaction in the curing of paint is the polymerization of one or more unsaturated molecules, usually acrylic derivatives. A photoinitiator must be used, such as the ketone (32), which easily generates radical fragments upon Norrish type I fission induced by ultraviolet light. As described below, both cationic and anionic initiators may also be valuable.

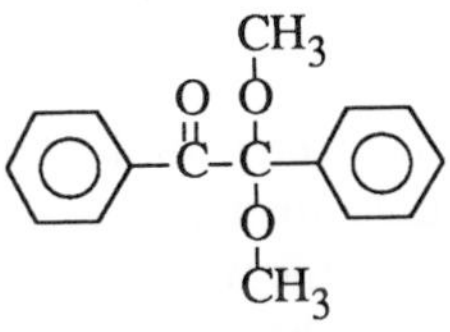

(32)

12.20. PHOTOGENERATION OF ACIDS; USE IN POLYMER CHEMISTRY

12.20.1. Photoionization of hydrocarbons

The photolysis of a hydrocarbon in which heterolytic fragmentation produces a proton and a carbanion has been described. In the example of suberene (33), facile H^+ exchange was observed in D_2O, and the singlet excited molecule was found to have a pK_a value close to -1.[47]

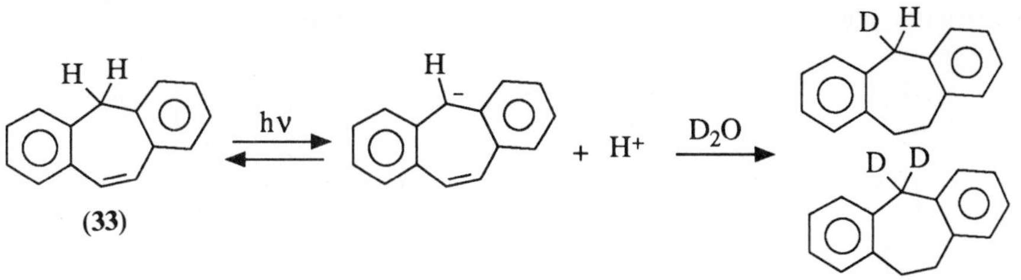

(33)

12.20.2. Photogeneration of acids in polymer chemistry

In some situations it is desirable to activate at an appropriate time a catalyst already present in a reaction mixture. For the generation of an acid, the idea is to introduce the catalyst in the form of a salt, which is not active, and to decompose this salt photochemically. This approach can be used for both Lewis acids and protic acids.

The photochemistry of sulfonium, ammonium, phosphonium, arsonium and iodonium salts has been investigated. As illustrated with the example salt (34), the acid is generated quantitatively, but the mixture of organic photoproducts is complex. In the rearrangement shown, the benzylic substituent becomes attached to the anthracene moiety at all possible positions. Fragmentation products are also created, namely a sulfide and a product of reaction between the benzylic moiety and the solvent.[48]

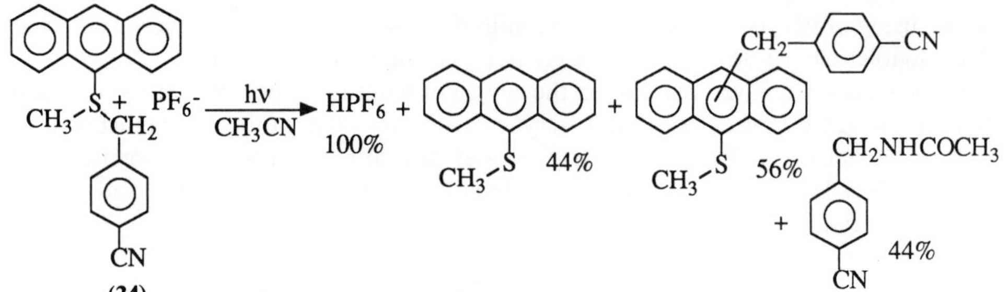

The generation of acids from anthracene-containing sulfonium salts was found to be quite sensitive to the wavelength used for irradiation. In one case, this quantum yield increased by a factor of 28 in changing the wavelength from 400 to 280 nm.[49] The mechanism is believed to involve competing heterolytic and homolytic cleavage reactions in the initial step, followed by reaction of the generated species with the solvent:

$$\overset{+}{Ar\text{-}S}\text{-}R_2 \ X^- \overset{h\nu}{\longrightarrow} Ar\text{-}S\text{-}R_1 + R_2^+ + X^-$$
$$\underset{R_1}{|}$$

$$\overset{+}{Ar\text{-}S}\text{-}R_2 \ X^- \overset{h\nu}{\longrightarrow} \overset{+\cdot}{Ar\text{-}S}\text{-}R_1 + \cdot R_2 + X^-$$
$$\underset{R_1}{|}$$

In addition to the acid, many organic products are obtained by photolysis of diphenyliodonium salts (35) in acetonitrile or aqueous acetonitrile. The distribution of the organic compounds depends on the choice of anion.[50]

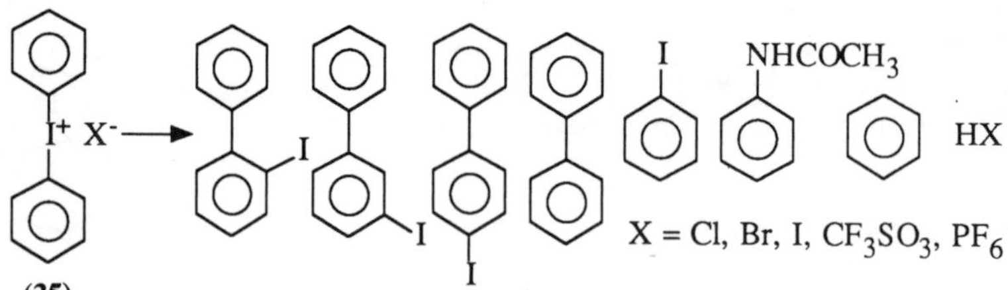

A third example shows the photogeneration of a Lewis acid. The decomposition of an aryl diazonium fluoroborate (Chapter 7) produces BF_3, N_2, and an aryl fluoride:

$$[Ar\text{-}N{\equiv}N^+BF_4\text{-} \overset{h\nu}{\longrightarrow} BF_3 + N_2 + ArF$$

When this reaction is used in the presence of an epoxide monomer, it leads to the polymerization of the monomer by the Lewis acid photogenerated. An appropriate choice of diazonium fluoroborate allows control over the wavelengths at which the irradiation proceeds. The epoxide photopolymers have been used in the manufacture of printing plates, the recording of images, and the fabrication of printed and integrated circuits.

12.21. PHOTOGENERATION OF BASES; USE IN POLYMER CHEMISTRY

The photogeneration of bases has been little studied. One exception is the production of NH_3 from metal–ammine complexes, which suffers from the low solubility of these molecules in organic media. More recently, the photodecomposition of carbamates (**36**)

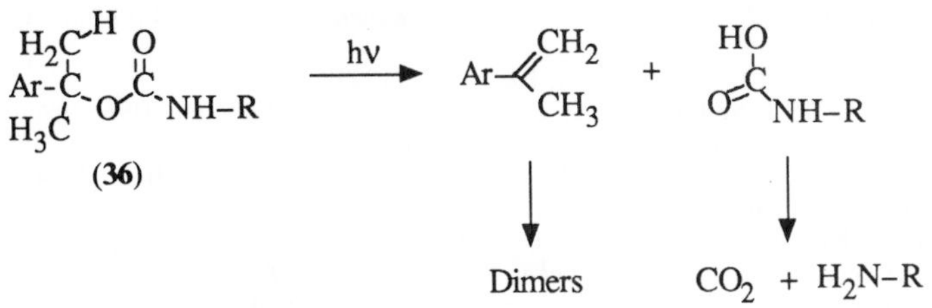

was reported.[51] Presumably, Norrish type II fragmentation was followed by decarboxylation. A poly(methylmethacrylate) film was prepared by irradiation of the monomer in the presence of the sensitizer in which Ar = 3, 5-dimethoxyphenyl and R = cyclohexyl. Cyclobutane dimers from the styrene were by-products. Symmetrical dicarbamates underwent a similar photodecomposition.

o-Nitrobenzyl carbamates (**37**) were used in a slightly different approach, taking advantage of the photochemical intramolecular abstraction of a benzyl hydrogen by the nitro group, eventually leading to *o*-nitrosobenzaldehyde, CO_2, and the amine either in solution or in the solid state.[52]

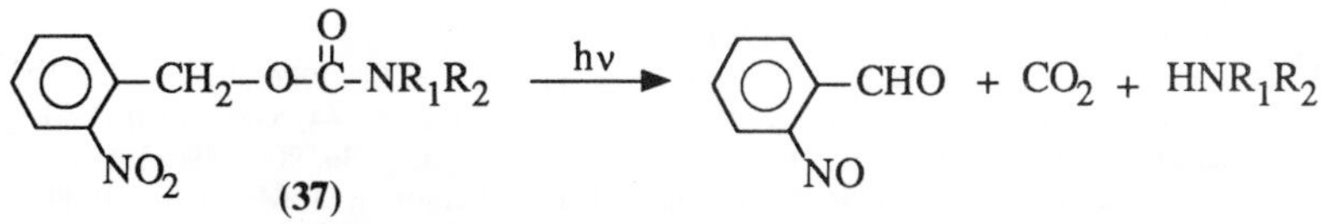

It may be noted that the photochemical reactions described here were discussed earlier in the section on photoremovable protecting groups (see Chapter 8) since, in practice, the starting materials are themselves obtained from amines.

12.22. PHOTOCHEMISTRY OF NUCLEIC ACIDS APPLIED TO MOLECULAR BIOLOGY TECHNIQUES

Traditionally, the immobilization of DNA on a solid support has been accomplished by baking nitrocellulose membranes in a vacuum at 80°C for 2 hours, a treatment believed to create a strong hydrophobic bond. Alternatively, a covalent bond can be formed upon ultraviolet irradiation between the amino groups of a nylon membrane and the thymine groups of a nucleic acid.[53] The chemistry involved was described in Section 9.7.2. The ultraviolet cross-linking technique can also be applied to nitrocellulose membranes.[54]

12.23. AN ENDING WITH A GOOD TASTE

One of the more unusual benefits of photochemistry may be found in the report claiming that the nutritional quality of fresh cherry juice was significantly improved after treatment with pulsed laser irradiation in a constant magnetic field. The acidity was reduced, the concentration of ascorbic acid was increased, and the storability and taste were improved.[55] What a nice project for further investigation!

REFERENCES

1. P. E. McGovern and R. H. Michel, *Acc. Chem. Res.*, **23**, 152–158 (1990).
2. N. Kuramoto and T. Kitao, *J. Soc. Dyers Colour.*, **95**, 257–261 (1979).
3. A. Lucas, *Preservative Materials used by the Ancient Egyptians*, National Printing Department, Cairo (1911).
4. A. Reiser, *Photoreactive Polymers*, Wiley, New York (1989).
5. S. Gould, T. R. O'Toole and T. J. Meyer, *J. Am. Chem. Soc.*, **112**, 9490–9496 (1990).
6. H. Misawa, N. Kitamura and H. Masuhara, *J. Am. Chem. Soc.*, **113**, 7859–7863 (1991).
7. W. C. Eisenberg, K. Taylor and R. W. Murray, *Carcinogenesis*, **5**, 1095–1096 (1984).
8. S. R. Wild and K. C. Jones, *Chemosphere*, **23**, 243–251 (1991).
9. J. Kagan, E. D. Kagan, I. A. Kagan and P. A. Kagan, *ACS Symp. Ser.*, **327**, 191–204 (1987).
10. R. Davenport and A. Spacie, *J. Great Lakes Res.*, **17**, 51–56 (1991).
11. K. Mopper, X. Zhou, R. J. Kieber, D. J. Kieber, R. J. Sikorski and R. D. Jones, *Nature (London)*, **353**, 60–62 (1991).
12. R. Taylor, J. P. Parsons, A. G. Avent, S. P. Rannard, T. J. Dennis, J. P. Hare, H. W. Kroto and D. R. M. Walton, *Nature*, **351**, 277 (1991).
13. H. Frank and W. Frank, *Naturwissenschaften*, **72**, 139–141 (1985).
14. E. J. Dougherty, A. L. McPeters and M. R. Overcash, *Chemosphere*, **23**, 589–600 (1991).
15. M. Tysklind and C. Rappe, *Chemosphere*, **23**, 1365–1375 (1991).
16. H. Kawagushi and M. Furuya, *Chemosphere*, **21**, 1435–1440 (1990).
17. *Chem. Engin. News*, p. 20, 26 August 1991.
18. W. Schwack and H. Flösser-Müller, *Chemosphere*, **21**, 905–912 (1990).
19. P. Meallier, A. Namouni and M. Mansour, *Chemosphere*, **21**, 913–917 (1990).
20. M. R. Berenbaum and K. Lee, *Chemoecology*, **1**, 81–85 (1990).
21. R. J. Watts and R. H. French, *Chemosphere*, **20**, 663–667 (1990).
22. M. Verzele and D. De Keukeleire, *Chemistry and Analysis of Hop Bitter Acids*. Elsevier, Amsterdam (1991).
23. A. Maujean and N. Seguin, *Sci. Aliments*, **3**, 589–601 (1983).
24. N. Seguin, Thesis, University of Reims (1964).
25. W. A. Laurie, D. McHale and K. Saag, *Tetrahedron*, **42**, 3711–3714 (1986).
26. W. A. Laurie, D. McHale, K. Saag and J. B. Sheridan, *Tetrahedron*, **44**, 5905–5910 (1988).
27. T. V. Stabler and A. L. Siegel, *Clin. Chem. (Winston-Salem, N.C.)*, **36**, 906–908 (1990).
28. W. J. Simpson, J. L. Fernandez, J. R. M. Hammond, P. S. Senior, B. J. McCarthy, P. H. Jago, S. Sidorowicz, S. A. A. Jassim and S. P. Denyer, *Lett. Appl. Microbiol.*, **11**, 208–210 (1990).
29. E. R. Menzel, *ASTM Spec. Tech. Publ. (Laser Tech. Lumin. Spectros.)*, **1066**, 36–49 (1990).
30. N. Imaizumi, K. Hayakawa, Y. Suzuki and M. Miyazaki, *Biomed. Chromatogr.*, **4**, 108–112 (1990).
31. R. B. Dopp, Alkaline batteries containing fluorescent dyes. *U.S. 4,999,265*, 12 March 1991.
32. E. Hemingway, *Rep. Progr. Appl. Chem.*, **54**, 150–158 (1969).
33. A. Wagner, *Naturwissenschaften*, **55**, 533–538 (1968).
34. A. E. Siegrist, H. Hefti, H. R. Meyer and E. Schmidt, *Rev. Prog. Color. Relat. Top.*, **17**, 39–55 (1987).
35. R. Datta and P. Nandy, *Photochem. Photobiol.*, **52**, 907–909 (1990).
36. J. Sunamoto, K. Iwamoto, Y. Mohri and T. Kominato, *J. Am. Chem. Soc.*, **104**, 5502–5504 (1982).
37. J. D. Winkler, K. Deshayes and B. Shao, *J. Am. Chem. Soc.*, **111**, 769–770 (1990).
38. S. Shinkai, T. Nakaji, T. Ogawa, K. Shigematsu and O. Manabe, *J. Am. Chem. Soc.*, **103**, 111–115 (1981).
39. S. Shinkai, K. Miyazaki and O. Manabe, *J. Chem. Soc. Perkin Trans. I*, 449–456 (1987).
40. G. H. Brown, *Photochromism*. Wiley, New York (1971).
41. K. Hatano, T. Uno, T. Kato, T. Takeda, T. Chiba and S. Tejima, *J. Am. Chem. Soc.*, **113**, 3069–3071 (1991).
42. K. Ohno, S. Kumagai, F. Suzuki and N. Tsujita (Shiseido Co., Ltd), Anatase-containing photochromic pigment for cosmetics. *Ger. Offen. DE 4,038,258*, 6 June 1991.
43. M. Irie and M. Mohri, *J. Org. Chem.*, **53**, 803–808 (1988).

44. S. Yitzchaik, I. Cabrera, F. Buchholtz and V. Krongauz, *Macromolecules*, **23,** 707–713 (1990).
45. T. Miyasaka, K. Koyama and I. Itoh, *Science,* **255,** 342 (1992).
46. M. S. Reisch, *Chem. Engin. News*, 28–58, 14 October 1991.
47. D. Budac and P. Wan, *J. Chem. Soc.*, **57,** 887–894 (1992).
48. F. D. Seava and D. T. Breslin, *J. Org. Chem.*, **54,** 712–714 (1989), and references cited therein.
49. X. He, W.-Y. Huang and A. Reiser, *J. Org. Chem.*, **57,** 759–761 (1992).
50. N. P. Hacker, D. V. Leff and J. L. Dektar, *J. Org. Chem.*, **56,** 2280–2282 (1991).
51. J. F. Cameron and J. M. J. Frechet, *J. Org. Chem.*, **55,** 5919–5922 (1990).
52. J. F. Cameron and J. M. J. Frechet, *J. Am. Chem. Soc.*, **113,** 4303–4313 (1991).
53. G. M. Church and W. Gilber, *Proc. Natl. Acad. Sci. USA*, **81,** 1991–1995 (1984).
54. E. W. Khandjian, *Bio/Technology*, **5,** 165–167 (1987).
55. A. I. Belskii, *Elektron. Obrab. Mater.*, 65–67 (1991); *Chem. Abstr.*, **115,** 7328q (1991).

13. Conclusion

13.1. GENERAL COMMENTS

The multitude of light-dependent biological phenomena make it impossible to discuss them all. Those who believe that biological phenomena will be fully explained by unraveling the chemistry of the systems must feel gratified that much fundamental understanding of photochemistry has been gained in the laboratory, and that the fundamental study of light-dependent biological phenomena is attracting attention from an increasingly greater fraction of the scientific community. As the medical applications of photochemistry become more widely publicized and directly touch a greater number of families, the field of photochemistry should attract even more practitioners and become more commonly taught in colleges and universities. Because photochemistry can be so intricately associated with a visual component, it is also uniquely suited for inclusion into elementary and high school programs.

13.2. FURTHER READING

13.2.1. Textbooks

N. J. Turro, *Molecular Photochemistry*, Benjamin, New York (1965).

R. O. Kan, *Organic Photochemistry*, McGraw-Hill, New York (1966).

D. C. Neckers, *Mechanistic Organic Photochemistry*, Reinhold, New York (1967).

J. G. Calvert and J. N. Pitts, Jr., *Photochemistry*, Wiley, New York (1967).

A. Schönberg, G. O. Schenck and O.-A. Neumüller, *Preparative Organic Photochemistry*, Springer-Verlag, New York (1968).

A. Cox and T. J. Kemp, *Introductory Photochemistry*, McGraw-Hill, New York (1971).

D. R. Arnold, N. C. Baird, J. R. Bolton, J. C. D. Brand, P. W. M. Jacobs, P. de Mayo and W. R. Ware, *Photochemistry, An Introduction*. Academic Press, London (1974).

D. O. Cowan and R. L. Drisko, *Elements of Organic Photochemistry*, Plenum Press, New York (1976).

J. A. Barltrop and J. D. Coyle, *Principles of Photochemistry*, Wiley, Chichester (1978).

K. K. Rohatgi-Mukherjee, *Fundamentals of Photochemistry*, Halsted Press (1978).

N. J. Turro, *Modern Molecular Photochemistry*, Benjamin/Cummings, New York (1978).

J. D. Coyle, R. R. Hill and D. R. Roberts (eds), *Light, Chemical Change and Life: A Source Book in Photochemistry*, The Open University Press, Milton Keynes (1982).

W. M. Horspool (ed.), *Synthetic Organic Photochemistry*, Plenum Press, New York (1984).

J. M. Coxon and B. Halton, *Organic Photochemistry*, Cambridge University Press, Cambridge (1987).

J. D. Coyle, *Introduction to Organic Photochemistry*, Wiley, Chichester (1986).
R. P. Wayne, *Principles and Applications of Photochemistry*, Oxford, Oxford University Press (1988).
K. C. Smith (ed.), *The Science of Photobiology*, 2nd edn, Plenum Press, New York (1991).
A. Bilbert and J. E. Baggott, *Essentials of Molecular Photochemistry*, CRC Press, Boca Raton (1991).

13.2.2. Book series

Many book series present reviews on specific aspects of organic photochemistry. The best known are *Advances in Photochemistry*, *Organic Photochemistry*, *Bioorganic Photochemistry*, and the *Specialist Periodical Reports on Photochemistry* from the Royal Society of Chemistry. Photobiologists interested in the topic should also consult *Singlet O_2*, vols 1–4, edited by A. A. Frimer (CRC Press, Boca Raton, 1985).

13.2.3. Primary journals

The original literature dealing with organic photochemistry may be found in all the journals which have a general audience, such as *Journal of the American Chemical Society* and other organs of national chemical societies, or more specialized journals addressing topics in organic chemistry, such as *Journal of Organic Chemistry*, *Tetrahedron* and *Tetrahedron Letters*, *Synthesis*, *Heterocycles*, *Journal of Heterocyclic Chemistry*, *Nucleic Acid Research*, *Chemosphere*, and all the major journals in chemistry, biochemistry, plant physiology, and radiation biology. In addition there are journals specifically concerned with photochemical investigations, such as *Journal of Photochemistry and Photobiology*, parts A and B, and its predecessor, *Photobiochemistry and Photobiophysics*, *Photochemistry and Photobiology*, *Photodermatology Photoimmunology Photomedicine* (which succeeded *Photodermatology*), *Photomedicine and Photobiology*, *Vision Research* and *Polymer Photochemistry*. Finally, a delightful quarterly publication, *The Spectrum*, is published by the Center for Photochemical Sciences, Bowling Green State University, Bowling Green, OH 43403.

Index

(Chapter titles are in capital letters)

Aberchrome actinometers 14
Absorbance, definition 10
Absorptivity, definition 10
Acenaphthyne 72
Acetone photochemistry 55
Acetylcholinesterase 142, 146
Acetylenes
  cyclic 72
  cyclization 44, 118, 120
  cycloaddition 133
Acetylenic ketones
  cyclization 127
Acids, photogeneration 217
Acridine orange 84
9-Acridone-2-sulfonic acid 94
Actinometers 14
Actinometry
  in mechanistic studies 27
  in photomedicine 15
Action spectrum
  definition 2
  hypericin 142
  stentorin 181
Adamantylideneadamantane
  in singlet oxygen quenching 22
  singlet oxygen detection 88
Adenine
  in DNA cleavage 155
  structure 147
Adenosine
  structure 147
Adriamycin 180
Aflatoxin B_1, mutagenicity 91
Affinity labeling (see photoaffinity labeling)
Agroclavine 117
Alcohols
  from carbonyls 60
  as hydrogen donors 63
Alkali-labile sites
  in DNA photochemistry 109
Alkenes
  addition of amines 101
  addition to aromatics 52
  cis-trans isomerization 34, 169
  cycloaddition 36
  sensitized reactions with alcohols 99
Alkyl halides, fragmentation 67
Amines
  addition to alkenes in SET reactions 101
  in electron transfer reactions 103
  photochemistry with carbonyls 105
  reactions with thymidine 151

Aminoacids
  photochemistry 145
  transport 213
Aminoacridines
  DNA cleavage 153, 155
δ-Aminolevulinic acid
  insecticide 171
  herbicide 172
Amiprilose 190
Annulenes 117
Antarafacial, definition 37
Anti-inflammatory agents 108
Anti-Markovnikov addition 102
Antibody, regulation of activity 142
Antimalarial agents 127, 177, 197
Antiviral agents 192–195, 197, 198
Apocalypse 201
Argon matrix 23
Aromatic compounds
  addition to alkenes 52
  boron halides 67
  [4 + 4] cycloaddition 118
  dimerization 52
  halogenation 65
  isomerization 51
  photoreduction 109
Aromatic substitution
  $S_{RN}1$ mechanism 107
Artemisinin 127
Ascaridole 88
Asymmetric syntheses 16, 135
Asymmetric transformations
  and polarized light 24
Azepines 78
Azides
  lithography 205
  photolysis 77
Aziridylimines 74
Azirines, from aryl nitrenes 78
Azo derivatives 75
Azobenzene-4-carboxylic acid 142
2,2′-Azo-bis-isobutyronitrile 75
Azocines synthesis 121

Bacteria inactivation 85, 193
Bacteriochlorophyll-a 197
Bacteriophage inactivation 154
Bacteriorhodopsin photoreceptors 216
Barellene 50
Barton reaction 69
Base labile lesions in DNA 153
Bases, photogeneration 219
Beer–Lambert's law 10
Benzene

cycloaddition to pyrimidine 132
dimer 88
1,4-endoperoxide 88
photochemistry 51
Benzo[a]pyrene phototoxicity 175, 207
Benzo[c]acridine synthesis 117
Benzocyclobutadiene 68
Benzofurans
dioxetanes 175
mutagenesis 175
photooxidation 129
Benzpinacol 61
p-Benzoquinone, in quenching experiments 22
Benzvalene 51
Benzyl groups, deprotection 137
Benzylic radicals 56
Benzylsulfonamides, deprotection 138
Benzyne 24, 79
Beverages 209
[3.1.0]-Bicyclohexanes 116
[1.1.1]-Bicyclopentanes 38
Bilirubin 36, 186
Bioluminescence 173
2,2′-Bipyridine 206
2,2′-Bipyridyl herbicides 172
Biradicals 68
BLB lamps 26
Bleomycin 158
Blood
see erythrocytes
viruses inactivation 192
Blueprints 71
Bond energies 8
Bone marrow
photosensitized cancer treatment 195
Boron halides photolysis 67
β-Bourbonene 123
Brighteners 129
Bullvalene 50
1,3-Butadienes
cycloaddition reactions 41
dimerization 119
electrocyclic reactions 39
from cyclobutenes 45

Caffeic acid 169
Cancer
photodynamic therapy 190, 192, 195, 197
Captan 208
Carbamates 219
Carbanions
in aromatic substitutions 107
in SET reactions 101
Carbapenem 74
Carbenes
from acetylenic compounds 120
from alkenes 47
from azo precursors 75
from diazo precursors 71

from tosylhydrazones 74
fulvene synthesis 51
isomerization 74
nitrene interconversion 78
photoaffinity labeling 159
Carbon disulfide cycloaddition 76
CARBON–CARBON DOUBLE BOND 34
CARBONYL GROUP 55
Carbonyls
cycloaddition to alkenes 60, 128
dioxetane decomposition 23, 174
isotopic labeling 130
photochemistry with amines 105
Carbonyls, α, β-unsaturated
cycloaddition to alkynes 44
deconjugation 16, 136
Fenestrane 44
reactions 122
Carbonyls, β, γ-unsaturated 125, 126
Carbonyls, γ, δ-unsaturated 126
Carbonyls, δ, ε-unsaturated 127
Carotenes
protection 181, 192
quenching 22, 82
singlet oxygen trapping 83
Carvone 43
Catalase 93
Cataracts 145–147
Cation-radicals 24, 99
Cedrene 52
Cembratrienone 125
Cephalosporin 74
Chain reactions 13
Champagne 210
Charge-transfer complex 100
Chemical evolution 112
Chiral sensitizers 136
Chloramines 70
Chloranil 101
Chlorin e$_6$ 197
Chlorophyll
luminescence 175
biosynthesis 141
synthesis 89
Chloroquine 177
Chlorpromazine
bacteriophage inactivation 154
skin sensitization 177
Cholesterol
hydroperoxides 144
simglet oxygen detection 144
Chromophores
definition 2
Chymotrypsin photoregulation 142
Cinnamic acid 170
dimerization 43
Circadian rhythms 168
Circularly polarized light 24, 136
Cis-trans isomerization

alkenes, in biology 169
bilirubin 35
biosynthesis of coumarins 169
carbon-carbon double bond 34
enantioselective 136
ion transport 214
tumor treatment 169
Citral 43
Clays and chemical evolution 112
Co complexes
DNA cleavage 153
Codeinone rearrangement 160
Conrotatory 40
Cordopatine 170
Coriolin 120
Corrin 61
Coumarins
biosynthesis 169
cyclobutadiene synthesis 131
DNA addition 162
Cross-linking
DNA-DNA 60, 188
DNA-proteins 109, 148, 151
Crown ether
ions transport 214
synthesis 119
Crystallins 146
Cubanes
substitution 67
synthesis 44
Cucumbers, luminescence 173
Curtius rearrangement 77
Cyalume 211
Cyanoaromatic sensitizers 99–102, 104, 129, 163
Cycloaddition reactions 36
[2 + 3], acetylenic compounds 120, 133
aromatics 118
carbonyl and alkene 60, 128
cubane synthesis 44
cyclophane syntheses 44
intramolecular 43–44
nitriles 121
in plants 170
2-pyridones 119
pyrimidines 132, 148
sensitized 44
singlet oxygen *see entry*
thiocarbonyls 96
ylides 133
Cyclobutadiene 24
photochemistry 131
synthesis in host molecule 131
Cyclobutanes
DNA dimers 149
from alkenes 37
Cyclobutanols 59
Cyclobutanones
to tetrahydrofurans 59

Cyclobutene 39
ring opening 45
Cyclodextrin 56, 64
1,3-Cyclohexadienes ring opening 115
Cyclohexadienone 49
Cyclohexane, NOCl reaction 66
Cyclohexanone oxime 66
Cyclohexene, cis–trans isomerization 35
Cyclooctatetraene 50
Cyclophanes
synthesis 44, 63
synthesis of annulenes 117
Cyclopropanation of alkenes 68
Cyclopropanes
from alkenes 46
oxidation to dioxolanes 129
ring opening 160
Cyclopropenes
and oxygen atoms 80
synthesis 121
and vinyl carbenes 73
Cyclopropenone 72
Cycloreversion reactions 45
Cyclosporine 211
Cytidine structure 147
Cytosine
deamination 149
hydration 149
photohydration 112
structure 147

DABCO
singlet oxygen trapping 83
Daunomycin 180
DNA cleavage 153
Deamination
cytosine 149
Decarbonylation 56, 59, 72, 79, 80
Decarboxylation 79, 131
Decay 7
7-Dehydrocholesterol 48
Deoxygenation reactions 107
Deoxyribose 147
Dermatitis 179, 180
Dermatology
photodynamic therapy 192
Deuterated water
DNA cleavage 154
singlet oxygen detection 83
Dewar benzene 51, 133
isomerization 22
synthesis 24
Dewar furan and thiophene 23
Dextranesucrase
affinity labeling 160
Di-π-methane rearrangement 49
Diazapyrenium cations
DNA cleavage 153
Diazirines 75

Diazo compounds
  photolysis 71
Diazoketones
  photolysis 71
Diazomethane 71
Diazonium ions 71
Dibenzofurans 208
Dicyanoanthracene sensitizers 101, 129
Dicyanobenzene sensitizers 102
Diels-Alder reactions 41, 101
Dihydroconessine 70
Dihydrophenanthrenes from stilbenes 116
Diiodomethane
  cyclopropane synthesis 68
1,25-Dihydroxyvitamin D_3 49
1,25-Dihydroxyvitamin D_3 receptor
  affinity labeling 161
7,12-dimethylbenz[a]anthracene
  DNA cleavage 153
Dimethyldiazaperopyrenium dication
  DNA cleavage 154
Dioxaziridine 78
Dioxetanes
  decomposition to carbonyls 23
  in singlet oxygen quenching 22
  in vitamin D synthesis 49
  isotopic labeling of carbonyls 130
  from benzofurans 129
  from hindered alkenes 88
  from luciferin 173
  from hydroperoxides 128
  from psoralens 91
  mutagenicity 91, 175
Dioxins 208
Dioxolanes 129
Dipeptides 146
Diphenylcyclopropane 47
Disrotatory 40
1,2-Dithietane 96
Divinyl protoporphyrin IX
  insecticide 171
DNA
  cross-linking to proteins 109, 148, 151
  cycloaddition 60
  detection 174
  immobilization 219
  luminescence of Ru complex 12
  structure 147
  vaporization 152
DNA adducts
  mapping 159
DNA cleavage 152, 163
  9-aminoacridine 153, 155
  bleomycin 158
  complexes 153
  daunomycin 153
  diazapyrenium cations 153
  7,12-dimethylbenz[a]anthracene 153

dimethyldiazaperopyrenium dication 154
ellipticine 155
esperamicin 155
hydroxyl radical 157
Λ-tris(4,7-diphenyl-1,10-phenanthro-
    line)cobalt(III) 156
mechanisms 154
Methylene blue 153–154, 156
NADH 153
neocarzinostatin 155
1-nitropyrene 153
nitro compounds 158
nitrobenzamide 153
phenothiazine 155
porphyrin 153–154
proflavine 153
psoralens 153
rose bengal 153
$Ru(bpy)_3^{2+}$ 155
sodium azide 154
superoxide 157
α-terthienyl 153, 176
topology 156
uranyl salts 153
DNA damage
  bacteriophages 154
  repair 141, 166
DNA dimers
  deamination 149
DNA photoadducts
  ring cleavage 103
DNA photochemistry 147
  effect of 5-bromouracil 109
  strand breaks 109
  psoralens 161
DNA repair 153
DNA sequencing 158
  fluorescent labels 159
Dyes 201

E–Z isomerization (see cis–trans isomerization)
Electrocyclic reactions 39
Electron transfer
  back reaction 102
  in biological systems 112
  cyanoaromatic sensitizers 99
  electron rich sensitizers 102
  from strained single bonds 101
  intramolecular 109
  and nucleic acids 112
  in proteins 146
  with amines 103
ELECTRON TRANSFERS 99
Electrosynthesis 139
Ellipticine, DNA cleavage 155
Enamides cyclization 117
Enantioselective syntheses 16, 24, 101, 135
Endoperoxides 88

singlet oxygen production 22, 86
Enolate ions
  in aromatic substitutions 107
  enantioselective syntheses 136
Environmental photochemistry 206
Enzyme activity
  light dependence 141, 166
Enzyme inactivation
  α-terthienyl 142
  succinoxidase 142
Eosin, virus inactivation 193
Epimerization
  estrone 56
  penem 65
Episulfides 96
Epoxides
  from ketocarbenes 73
  from singlet oxygen reaction 88
Epoxychalcones 62
Erythrocytes
  affinity labeling 160
  malaria 197
  peroxidation 143
  sensitization 167, 188
Erythrosin B
  singlet oxygen sensitizer 84
  insecticide 171
Esperamicin
  DNA cleavage 155
Estrone 56
Excimer 11, 99
Exciplex 11, 99
Excitation, vertical 6
Excited molecules from dark reactions 22
Excited states
  fate 19
  lifetime 18
EXPERIMENTAL TECHNIQUES 26
Extinction coefficients 10
Eye
  cataracts 145, 146
  protection 26
  laser treatments 187

Fading (colors) 201
Fenestrane 44
Fenton reaction 93
Ferricytochrome c 94, 112
Filters 30
Fingerprint analysis 211
Firefly luciferase detection 211
α-Fission (see Norrish type I)
β-Fission (see Norrish type II)
Fluoranthene 85
Fluorescence
  definition 6
  DNA sequencing 159
  nucleic acids analysis 150
  see also luminescence 18

Fluorescent whiteners 212
Fly ash 208
Footprinting 159
Formamide deprotection 138
N-Formylkynurenine 145
FRAGMENTATION REACTIONS 65
Fries rearrangement 61
Fulgide
  actinometry 14-15
  photochromism 215
Fullerene 207
Fulvene 51
Fungicide (Captan) 208
Furanochromones 194
Furanocoumarins (see also psoralens)
  cycloaddition 161
  and melatonin production 168
  defense mechanisms 209
  virus inactivation 194
Furans
  low temperature photochemistry 23
  cycloaddition 161
  oxidation 129, 162, 189
  singlet oxygen trapping 83
FUV, definition 8

Garryine 77
Germicidal lamps 26
Gilvocarcins
  DNA cleavage 153
  DNA protein cross-link 153
  phototoxicity 95
  virus inactivation 194
Gorgonian organisms 126
Guanine 147, 156
  in DNA cleavage 154
  oxidation 152
Guanosine 147
Gulliver's Travels 173

Hückel transition state 40
Haber–Weiss reaction 93
Halogenation 65
Hapten 142
Heavy atom effect 8, 16
Helicenes 25, 116
Hematoporphyrin 191
Heme group
  electron transfer 146
Herbicides
  bleaching 172
  α-terthienyl 176
  methylene blue 209
  porphyrins 172
Herpes labialis 198
Hexachlorocyclohexane 65
Hexahelicene
  asymmetric synthesis 25
Hexamethylphosphoric triamide
  solvent in photoreductions 127

1,3,5-Hexatriene cyclization 115, 116
Hirsutene 127, 130
Histidine
  in cataracts 146
  in singlet oxygen quenching 83
Histones
  cross-linking to DNA 151
HIV inactivation 192, 193
Hückel transition state 39
Humulone 209
Hydration, pyrimidines 112, 149
Hydrogen abstraction 57, 59, 60, 62, 63, 64, 122, 134,
    137,143, 157, 209, 219
Hydrogen peroxide
  chemiluminescence 211
  epoxidation 163
  from Fenton and Haber-Weiss reactions 93
  from photolytic ozonation 95
  from superoxide 82, 93, 144
  inactivation of bacteria 152
  reaction with catalase 93
  singlet oxygen synthesis 86
  TiO_2 degradation of chlorinated compounds 111
  ultraweak photoemission 172
Hydroperoxides
  allylic, from singlet oxygen reaction 87
  decomposition 127, 144
  from lipids 106
  rearrangement 145
  singlet oxygen production 87
  synthesis 127
  tocopherol 142
Hydroperoxyl radicals
  and malignant cells 112
2-Hydroxybenzophenone
  polymer protection 64
8-Hydroxyguanine 154
3-Hydroxykynurenine 145
Hydroxyl radical 93, 144
  DNA cleavage 155, 157
  and malignant cells 112
  in pulsed radiolysis 24
  in water treatment 95
Hyperbilirubinemia 36, 185
Hypericin 141
  virus inactivation 193
Hyrax 171

Immobilization of DNA 219
Immunoassays 174
  luminescence 211
Immunology
  psoralens 189
  urocanic acid 167, 170
Indolines 104
Industrial applications 115
Ingenanes 124
Insecticides 141, 171, 176, 209
Integrated circuit 72, 205

Internal conversion 6, 169
Intersystem crossing 5, 7
Isocyanates from acyl azides 77
Isomerization (see also rearrangements)
  alkene to alkene 45
  alkene to cyclopropane 46
  alkenes to carbenes 47
  amines 104
  aromatic compounds 51
  benzyne 79
  bilirubin 36
  carbenes and nitrenes 78
  carbonyl compounds 63, 123
  carvone 43
  cis–trans or E–Z 4, 6, 34, 122, 126, 167, 169
  cis–trans, asymmetric 16, 136
  citral 43
  cyclization 38, 115–116, 118–120,
  diene to cyclobutane 44
  eneyne to cyclobutene 44
  heterocyclic compounds 52, 150
  hydrogen abstraction 64
  indene 50
  N-iminopyridinium ylides to 1,2-diazepines 132
  merocyanins 212
  parthenin 57
  penem 65
  phenyl nitrene 77
  photochromic compounds 213, 215
  2-pyrrolones 131
  steroid 56
  syn-anti, azobenzenes 142, 214
  thiophenes 52
Isotopic labeling
  photoaffinity labeling 159
  carbonyls 130

Jablonski diagram 6

Ketenes
  in Norrish reactions 55
  in Wolff rearrangement 71
Khellin 194
Kinetic studies 18
Kynurenic acid 145
Kynurenine 145

L-tris(4,7-diphenyl-1,10-phenanthroline)cobalt(III)
  DNA cleavage 156
β-Lactams 132
Lactonization
  in electron transfer reactions 102
Ladenberg benzene 51
Lasers
  carbene formation 75
  cherry juice 220
  DNA cleavage 152
  DNA sequencing 158
  dyes 15

endoperoxide decomposition 87
drug activation 180
flash photolysis 8, 16, 19, 23
forensic analysis 211
micromanipulation 206
in medicine 187, 191
optical media 215
singlet oxygen 92
Light
circularly polarized 24
piping 173
sources 26
Light-struck flavor 210
Lipid membranes, transport 213
Lipid peroxidation 106, 143
Liposomes, peroxidation 143
Lithography 72, 203
Luciferase
analysis of pollution 174
Luciferin 173
Lumi-estrone 56
Luminescence
definition 6–7
detection 211
genotoxity 174
in kinetic studies 18
quantum yield 19
quenching 20
spectra 11
Luminol 173

Malaria
artemisinin (quinghaosu) 127
chloroquine 177
merocyanin-540 197
Markovnikov
addition of alcohols to alkenes 102
Markovnikov (anti) addition 100
Melatonin 168
Melittin 146
Membrane
cholesterol 144
crossing 141
damage 167, 177, 192, 196, 197
DNA immobilization 219
modifications 188
transport 160, 212
treatment 187
photoreceptor 216
peroxidation 143
TiO$_2$ localization 112
Mercury sensitization 16
Merocyanin 540
erythrocytes 168
malaria 197
marrow treatment 195
photochemistry 196
virus inactivation 193

Merocyanins
aminoacids and metal ions transport 213
proton pump 212
Mesitylene isomerization 52
Metal ions, transport 213
8-Methoxypsoralen
carcinogenesis 188
inactivation by insects 171
circadian rythms 168
oxidation 162
photopheresis 198
photoproduct 162
photosensitization 179
PUVA therapy 187
Methyl linoleate hydroperoxides 144
Methylene blue
DNA cleavage 153–154, 156
herbicide 209
singlet oxygen 84
virus inactivation 193
Methylindenes 50
2-(Methylthio)ethanesulfonic acid reductase 166
Micelles 16, 32
Micromanipulation 206
Mitomycins 134
Modhephene 126
Monochromators 30-31
Mosquito
aromatic hydrocarbons phototoxicity 175, 186
α-terthienyl phototoxicity 176
enzyme inactivation 146
pheromone 57
Möbius transition state 40
Multiplicity
experimental determination 17
Mutagenicity
aflatoxin 91
benzofuran derivatives 129, 175
polycylic aromatic hydrocarbons 175
psoralens 91
UV light 60

N-Bromosuccinimide 69
N-Iminopyridinium ylides 132
NADH, DNA cleavage 153
Nematodes 176
Neocarzinostatin
DNA cleavage 155
Nitrenes
azides 77
carbenes interconversion 78
photoaffinity labeling 159
resists 205
Nitriles
cycloaddition 121
N-sulfides 76
sensitizers 99–102, 104
Nitrite esters 69
Nitroaromatics 70

intramolecular reactions 134
protecting groups 137
resists 205
Nitrobenzamide
DNA cleavage 153
Nitro blue tetrazolium 94
Nitro compounds
DNA cleavage 158
1-Nitropyrene
DNA cleavage 153
Nitroso compounds
cyclization 134
Nitrosyl chloride 66
Nonsteroid anti-inflammatory agents 177
Norbornadiene from quadricyclene 101
Norbornane substitution 67
Norrish type I reactions 55
Norrish type II reactions 58
Nucleic acids (see also DNA)
and electron transfers 112
cleavage 106
cycloaddition 60,132
fluorescence analysis 150
photochemistry 147
protection and deprotection 138
Nucleosides, structures 147
Nucleotides, structures 147
NUV definition 8

Optical density, definition 10
Orbitals
atomic and molecular 2
Woodward–Hoffman rules 39
Ortho-xylylene 68
Osazone photochromism 215
Oxa-di-π-methane rearrangements 126
Oxacarbenes 121
Oxazoles, with singlet oxygen 90
Oxetane 150
Oximes
from cyclohexane 67
deprotection 138
Oxirene 73
Oxygen (see also singlet oxygen)
aromatization reactions 116
atom 80
control 32
electron transfer 24
electronically excited 82
hydrogen abstraction 58, 63, 137
multiplicity 17
nitrene addition 78
photodynamic reactions 82
superoxide formation 92, 105
transfer with pyridine oxides 138
1,2,4-trioxane synthesis 128
OXYGEN PHOTOCHEMISTRY 82
Ozone
in synthesis of singlet oxygen 86

layer 147
water treatment 95

Paints 75, 216
Papain photoregulation 142
Parthenin 57
Paterno Büchi reaction 60, 128
pBR322 cleavage 152, 156
PDT (see photodynamic therapy)
Penem
epimerization 65
isomerization 65
1,4-Pentadienes isomerization 38
Pentalenolactone P 126
Peptide bond photochemistry 146
Peptide synthesis
solid support 72
Perhydrohistrionicotoxin 69
Peroxidase 49, 174, 211
Peroxidation
lipids 143, 176
membranes 143
Peroxides
analysis 145
cyclic 88
decomposition 63, 144
formation 64, 82, 106
rearrangement to diepoxides 88
sulfides 89
Phage
inactivation 154, 193, 194
mutation 149
1,10-Phenanthroline
herbicide 172
skin protection 178
Phenothiazine
DNA cleavage 155
Phenyl azide
in photoaffinity labeling 77
Phenylheptatriyne 95
Pheromone, photocatalytic synthesis 139
Phosphatase 174
Phosphorescence
definition 6-7
see also luminescence 18
Photoaffinity labeling 75, 77, 159
Photoallergy 179
Photocatalytic electrosynthesis 139
Photochemistry
EXPERIMENTAL TECHNIQUES 26
FUNDAMENTALS 1
Photochromic compounds 14, 142, 213, 215
Photodynamic reactions (see also singlet oxygen and
superoxide)
definitions
Photodynamic therapy
blood marrow 195
cancers 190
dermatology 192

Photoelectrochemical cells 113
Photofrin 191
Photogalvanic cells 113
Photography 203
Photolyase 103, 141, 149
PHOTOMEDICINE 185
  actinometry 15
Photon energy 8
Photopheresis 197
Photoreceptors, synthetic 216
Phototoxicity 95, 112, 120, 166, 171, 172, 175, 176,
    177,180, 181, 189, 194, 196, 197, 207, 209
Phthalimide fluorescence 12
Phthalocyanine 191
Phytoene desaturase 172
Pigmentation 188
Pinacols 60, 105
Piperylene 17
Polarized light 25, 136
Polyamide 66
Polycyclic aromatic hydrocarbons
  endoperoxides 86, 88
  photosensitization 91, 153, 175, 180, 195, 207
  syntheses 118
Polymers
  acid catalysis 217
  base catalysis 219
  protection 64
Porphyrins 84
  DNA cleavage 153–154
  insecticides 171
  herbicides 172
  metabolites 180, 192
  photodynamic therapy 176, 190
  sensitizers 84, 148, 181
Previtamin D3 48
Prismane 51
Proflavine, DNA cleavage 153
Protecting groups, photoremoval 137
Proteins
  cross-link to DNA 109, 148, 151
  damage 146
  photochemistry 112
Proteus mirabilis, DNA cleavage 153
Protochlorophyllide reductase 141
Psoralens (see also furanocoumarins)
  and immunology 189
  cycloaddition 161
  cytotoxicity 91
  DNA cleavage 153, 163
  DNA reactions 161
  photodynamic reactions 162
  photopheresis 197
  phototoxicity 171
  PUVA therapy 187
  reaction with singlet oxygen 91, 162
  skin photosensitization 162
  superoxide formation 162
Psoriasis 162, 188-189

Pulse radiolysis 24, 87
Purines 147
PUVA therapy 91, 161, 163, 187, 189, 192
Pyrazoline-3,5-dione 78
12-(1-Pyrene)dodecanoic acid 196
Pyridine oxides, photoreduction 138
2-Pyridones, [4 + 4] cycloaddition 119
Pyridopsoralen 162, 163, 189
Pyrimidines
  cycloaddition 132
  cyclobutane formation 148
  dimers 178
  fragmentation 133
  photohydration 112
  structures 147
Pyrimidone Dewar isomer 151
2-Pyrrolones, isomerization 131

Quadricyclene isomerization 101
Quantum yields 12, 13, 19, 101
Quenching 17, 18, 20
Quinghaosu 127
Quinine 210
Quinolones, photosensitization 181

Radical-anions 24, 102, 107, 108, 127, 129
Radical-cations 24, 102, 105, 109, 129, 130, 152
Radical-radical recombination 23
Rayonet reactors 26
Rearrangement
  alkene oxides 87
  di-π-methane 49, 135, 209
  enamide 61
  Fries 62
  oxa-di-π-methane 126
  oxime to amide 67
  endoperoxides 86
  previtamin D3 48
  vinyl cations 68
Red blood cells
  see erythrocytes 167
Reduction reactions
  aromatics 109
  carbonyls 60, 130
Regioselectivity
  in amine addition to alkenes 101
Remote functionalization 62
  Barton reaction 69
  steroids 134
  with N-chloramines 70
Reprography 204
Resists 72, 204–205
Rhodamines 123, 158, 192
Rhodopsin
  affinity labeling 75
  isomerization 35
Ribose 147
Rose bengal 84, 156
  DNA cleavage 153

on solid support 106
singlet oxygen sensitizer 85
virus inactivation 193
Royal purple 201
Ru complexes
  DNA cleavage 153, 155
  luminescence 12
  singlet oxygen synthesis 85
Rydberg states 25

Sapphyrin 193
Semi-bullvalene 50
Sensitizers
  biological activity 92
  decomposition 16
  electron poor 99
  electron rich 99, 102
  in kinetic studies 18
  on solid supports 106
  targeting 176
  triplet 15
  triplet energies 33
Sensitox I 85
Single electron transfer (SET) reactions 99
Singlet oxygen 145, 202
  carbonyl labeling 130
  cytotoxicity 91
  from dark reactions 49
  defense mechanism 85
  definition 17, 82
  detection 82, 84
  diagnostic tests 144
  DNA cleavage 154
  endoperoxide decomposition 22
  erythrocyte damage 168
  fading 202
  furan reactions 129
  from psoralens 162
  gas phase production 85
  herbicides 172
  in chemical laser 92
  in protein photochemistry 146
  lifetime 83
  merocyanin 540 196
  mutagenicity 175
  photodynamic therapy 190
  production 144, 145
  quenching 20
  radical trapping to 1,2-dioxolanes 129
  reactions, 87, 115, 124, 145
  reaction with proteins 146
  reactions with sulfides 89
  syntheses 84
  trapping 83, 142
Singlet state, definition 4
Skin
  photochemistry 178
  photosensitization 162, 177, 179

protection 179
  types 179
Sodium azide
  DNA cleavage 154
  singlet oxygen trapping 83
Sodium hypochlorite 86
Sodium molybdate 86
Solvatochromic dyes 10
Solvent effect
  on electron transfer reactions 102
  on singlet oxygen lifetime 83
Solvolysis 109
Spatanes 124
Spinach chloroplasts 172
Spiropyrans photochromism 215
$S_{RN}1$ aromatic substitution 107
Stentor 181
Stern-Volmer 20, 83
Steroids (see also vitamin D and previtamin D_3)
  epimerization 56
  remote functionalization 62
  Wolff rearrangement 71
Stilbenes
  cyclization 116
  dimerization 43
  electron transfer reactions 104
  helicene synthesis 25
Strand break
  in DNA photochemistry 109
Succinoxidase 142
Sulfanilamides 177
Sulfides, reaction with singlet oxygen 89
Sulfochlorination 66
Sulfur dioxide
  elimination 66, 76
  synthesis of sulfonyl chlorides 66
  trapping 128
Sulfur monoxide 96
Sulfur, diatomic 95, 96
Sunscreens 111, 178, 198
Superoxide 106
  definition 82
  detection 93
  diagnostic tests 144
  disproportionation 144
  DNA cleavage 157
  from psoralens 162
  generation 94
  membrane damage 143
  photodynamic reactions 92
  singlet oxygen production 87
Superoxide dismutase 82, 93-94, 146
  in mechanistic studies 18
  inactivation 142
Suprafacial, definition 37
Suprasterols 116
Switching, enzymatic activity 142
Syn-anti (see isomerization)

T-cell lymphoma 188
Tachysterol 48
Taxinine 122
α-Terthienyl
  DNA cleavage 153
  enzyme inactivation 142, 146
  insecticide 176, 209
  phototoxicity 95
  sensitization 176
Tetracyclines 177
Tetrahydrofurans
  from cyclobutanones 59
Tetrahydroquinolines 104
Tetraphenylporphines 85
2,3-Thiadiazine 1,1-dioxide 76
Thiatriazoles 76
Thiazide diuretics 177
Thietane 76
Thietane S,S-dioxide 76
Thiobarbituric acid, peroxide analysis 145
Thiocarbonyls, S_2 synthesis 96
Thioketal deprotection 138
Thiophenes
  isomerization 52
  low temperature photochemistry 23
  singlet oxygen 90
Thymidine
  reaction with amines 151
Thymine
  dimers 163
  in DNA cleavage 155
  structure 147
Time scale 8
Titanium oxide
  chemical evolution 112
  chlorobenzene degradation 208
  malignant cells 112
  photochemistry 111
  sunscreens 111
  water treatment 111
Titanium salts
  photolysis and enzyme activity 166
Tocopherol 82, 142
Toluidine blue 193
Tosylates deprotection 137
Tosylhydrazone salts 74
Transannular reaction
  cyclodecanone 62
Transition states, aromaticity 39
Transport
  aminoacids 213
  metal ions 213
Triaziridine 78
Trichloroethylene degradation 208
4,4',6-Trimethylangelicin 189
4,4',5'-Trimethyl-8-azapsoralen 189
1,2,4-Trioxane 128
1,2,4-Trioxolanes 90
Triplet

definition 4
  ground state molecules 17
  sensitizers, definition 15
Triquinanes
  synthesis 126
Trithiolanes 96
Truxillic acid 170
Truxinic acid 170
Trypsin inhibitor 146
Tryptophan 146
  oxidation 145
  photochemistry 124
Type I (see superoxide)
  photodynamic reactions 24, 92, 167
Type II (see singlet oxygen)
  photodynamic reaction 92, 167
Tyrosine 146

Ultraweak biological photoemission 172
Units 9
Uracil
  cycloaddition to benzene 132
  dimers 150
  structure 147
Uranyl salts 153
Uridine 147
Urocanic acid
  immunology 167, 170
UVA, definition 8
UVB, definition 8
UVC, definition 8

Vessels 29
Vindorosin 124
Vinyl carbenes and cyclopropenes 73
Vinyl cations, formation and rearrangement 68
Virus inactivation 192–195, 197, 198
Visnagin 194
Vitamin D 47, 49, 116, 185
Vitamin D binding protein
  affinity labeling 160
Vitamin E 82, 142

Water titration 10, 12
Water treatment
  photolytic ozonation 95
  with singlet oxygen 85
  with TiO_2 111
Wavelength dependence (see also action spectra)
  of chemical reactions 12, 125
Wolff rearrangement 71, 74, 205
Woodward–Hoffman rules 36, 39

Xanthine oxidase
  in generation of superoxide 94
Xeroderma pigmentosum 178

Ylides 1322, 133

Zeolites 32, 35, 63, 85
Zinc oxide in sunscreens 111